Immigrant Geographies of North American Cities

Edited by
Carlos Teixeira
Wei Li
Audrey Kobayashi

OXFORD
UNIVERSITY PRESS

OXFORD
UNIVERSITY PRESS

Oxford University Press is a department of the University of Oxford.
It furthers the University's objective of excellence in research, scholarship, and education
by publishing worldwide. Oxford is a registered trade mark of Oxford University Press
in the UK and in certain other countries.

Published in Canada by
Oxford University Press
8 Sampson Mews, Suite 204,
Don Mills, Ontario M3C 0H5 Canada

www.oupcanada.com

Copyright © Oxford University Press Canada 2012

The moral rights of the author have been asserted

Database right Oxford University Press (maker)

First Edition published in 2012

Library and Archives Canada Cataloguing in Publication

Immigrant geographies of North American cities / edited by Carlos Teixeira,
Wei Li, Audrey Kobayashi.

Includes bibliographical references and index.
ISBN 978-0-19-543782-9

1. Immigrants—Canada—History. 2. Immigrants—United States—History.
3. Canada—Emigration and immigration—Government policy. 4. United States—Emigration
and immigration—Government policy. 5. Immigrants—Canada—Social conditions.
6. Immigrants—United States—Social conditions.
7. Immigrants—Canada--Economic conditions.
8. Immigrants—United States—Economic conditions.
I. Teixeira, José Carlos, 1959– II. Li, Wei, 1957– III. Kobayashi, Audrey, 1951–

JV6351.I46 2011   304.8'7003   C2011-902433-0

Cover image: ©iStockPhoto.com/ hatman12

Oxford University Press is committed to our environment. This book is printed on Forest
Stewardship Council® certified paper, harvested from responsible sources.

Printed and bound in Canada.

1 2 3 4 — 15 14 13 12

CONTENTS

Preface   viii
Acknowledgements   xi
Contributors   xiii

Introduction

      Immigrant Geographies: Issues and Debates   xiv
      Audrey Kobayashi, Wei Li, and Carlos Teixeira

Part I   **The Internationalization of North American Cities and Suburbs   1**

    1   Going Local: Canadian and American Immigration Policy in the New Century   2
        Helga Leitner and Valerie Preston

    2   Immigration Trends in the United States and Canada: A Historical Perspective   22
        Dirk Hoerder and Scott Walker

Part II   **The Imprint of Immigration in North American Cities and Suburbs   47**

    3   Immigration and Urban and Suburban Settlements   48
        Robert A. Murdie and Emily Skop

    4   The Spatial Segregation and Socio-economic Inequality of Immigrant Groups   69
        Joe Darden and Eric Fong

    5   Immigrants, Refugees, and Housing   91
        Thomas Carter and Domenic Vitiello

    6   Economic Experiences of Immigrants   112
        Lucia Lo and Wei Li

    7   How Gender Matters to Immigration and Settlement in Canadian and US Cities   138
        Brian Ray and Damaris Rose

    8   Immigration, Health, and Health Care   158
        Lu Wang, Elizabeth Chacko, and Lindsay Withers

9   Immigrant Political Incorporation in American and
     Canadian Cities   179
     Els de Graauw and Caroline Andrew

**Part III   Immigrant Groups in North American Cities
and Suburbs   207**

     10   Contemporary Asian Immigrants in the United States
          and Canada   208
          Shuguang Wang and Qingfang Wang

     11   Contemporary Profiles of Black Immigrants in the United States
          and Canada   231
          Thomas Boswell and Brian Ray

     12   Latin American Immigrants: Parallel and Diverging
          Geographies   256
          Luisa Veronis and Heather Smith

     13   Crossing the 49th Parallel: American Immigrants in Canada
          and Canadians in the US   288
          Susan Hardwick and Heather Smith

Conclusion

     A Review and Some Significant Findings   312
     James Allen and Carlos Teixeira

Glossary   332
Index   343

TABLES

I.1   Population and International Migrants in the World, Canada, and the US, 1990–2010   xv

I.2   Number and Percentage of International Migrants, Selected OECD Countries   xvii

I.3   Immigration Admission Policies and Court Cases in Canada and the US   xxiv

I.4   Immigrant Source Regions and Categories   xxviii

4.1   Mean Residential Segregation between Whites and Selected Non-European Immigrant Groups, US (2000) and Canada (2001)   77

4.2   Mean Residential Segregation between Whites and Selected Non-European Immigrant Groups and Their Socio-economic Characteristics, US (2000) and Canada (2001)   78

4.3   Mean Residential Segregation Rank and Socio-economic Status Rank for Selected Non-European Immigrant Groups, US (2000) and Canada (2001)   79

6.1   Educational Levels by Immigration Status   117

6.2   Official-Language Capability by Immigration Status   118

6.3   Industrial Participation among Different Population Groups, Canada and the US   120–1

6.4   Occupational Concentration of Immigrants Relative to Non-Immigrants and by Place of Birth   126

6.5   Characteristics of Selected Large and Medium-Sized Cities   128

8.1   Ethnic Variation in Health Status, Canada and the US   163

8.2   Ethnic Variation in Health-Care Use, Canada and the US   167

8.3   Health Status and Health-Care Use of Chinese in Canada and Hispanics in the US   169

9.1   Reported Voter Registration and Voting among US-Born and Naturalized Citizens in the US, by Region of Origin, 2002–8   186

9.2   Self-Reported Rates of Voter Participation by Selected Place of Birth, 2000 Canadian Federal Election   187

9.3   Self-Reported Rates of Voter Participation, Selected Visible Minorities by Place of Birth, 2000 Canadian Federal Election   187

9.4   Demographic Characteristics of the Foreign-Born Population, Selected US Cities, 2005–7   196–7

9.5   Demographic Characteristics of the Foreign-Born Population, Selected Canadian CMAs, 2006   198–9

10.1   Socio-economic Characteristics of Asian Immigrants in the US, 2005–7, Ages 15+   213

10.2   Socio-economic Characteristics of Asian Immigrants in Canada, 2006, Ages 15+   215

10.3   Labour Market Performance of Asian Immigrant Labour Force in the US, 2005–7 (%)   219

10.4   Labour Market Performance of Asian Immigrants in Canada, 2006 (%)   221

11.1   Characteristics of Foreign-Born Blacks and Blacks Born in the US and "
Canada   244–5

11.2   Characteristics of Foreign-Born Black Women and Men, US and Canada   249

12.1   Countries of Origin of the Latin American Foreign-Born Populations, US
and Canada   258

12.2   Top 20 Metropolitan Settlement Areas, Latin American Foreign-Born
in the US   262

12.3   Top 20 Metropolitan Settlement Areas, Latin American Foreign-Born
in Canada   263

12.4   Selected Demographic Characteristics of the Foreign-Born Latin American
Population by Region of Birth, US and Canada   265–70

12.5   Waves of Latin American Immigration to the United States and Canada   278–9

13.1   Top 20 American-Rich Canadian Cities (2006) and Canadian-Rich US Cities
(2000)   302

FIGURES

I.1   Permanent Resident Immigrants by Category, Canada, 1986–2008   xxvi
I.2   Immigrants by Class Admission, US, 1986–2008   xxvii
3.1   The Lower East Side Tenement Museum, New York City   50
3.2   A Typical Cartoon Depicting the Chinese in Vancouver, Early Twentieth Century   51
3.3   Post-War European Settlement, 1945–1960s   53
3.4   A Typology of Urban and Suburban Immigrant Settlements in US and Canadian Cities   59
5.1   Affordable Immigrant Housing Changes the Architectural Landscape of Small Centres: Row Houses in Steinbach, Manitoba   93
7.1   Gender Composition of Permanent Residents by Admission Class, Canada and the US, 2008   140
9.1   Naturalization Rates, Selected US and Canadian Cities   184
9.2   Foreign-Born and Ethnic and Racial/Visible Minority Representation on City Councils, Selected US and Canadian Cities   190–1
10.1   Changes in Number of Asian Immigrants in the United States and Canada   210
11.1   Foreign-Born Blacks in Counties in the US, 2000   236
11.2   Foreign-Born Blacks in Census Tracts in New York City, 2000   237
11.3   Geographic Concentration of Caribbean and African Immigrants in Metropolitan Toronto, 2006   241
12.1   Latin American–Born Immigrants in the United States, 2007   260
12.2   Latin American–Born Immigrants in Canada: 2006   261
13.1   Canadian-Born Population Distribution in the US, 2007, and Percentage Change in Canadian-Born Population in the US, 1970–2007   290
13.2   US-Born Population Distribution in Canada, 2006, and Percentage Change in US-Born Population in Canada, 1970–2006   291
13.3   Canadian Population in Atlanta MSA by Census Tract, 2000   304
13.4   American Population in Toronto by Census Tract, 2006   306

PREFACE

As co-editors, we are excited and proud to present this book to the academic and non-academic communities of Canada and the United States. It is the result of professional commitment, personal conviction, hard labour, and much pleasure.

Despite the long traditions of immigration to both Canada and the US (each of which has been dubbed a 'country of immigrants'), as well as their proximity and many similarities, this book marks the first systematic effort to compare and contrast the urban immigration geographies of the two nations.

Immigrant Geographies of North American Cities evolved out of an informal conversation between Katherine Skene, Oxford University Press acquisitions editor, who was visiting the University of British Columbia–Okanagan campus in Kelowna, BC, and Carlos Teixeira, one of our co-editors. Teixeira had expressed his frustration with the fact that, despite their geographical and cultural proximity, Canadian and American scholars seem to live with their backs against one another. He noted that the lack of comparative work between the two countries in the field of urban and social geography was truly remarkable.

Skene responded by offering Teixeira the challenge of designing a text for courses in urban/social geography with a Canada/US focus. As an immigrant himself and long-time resident of two of the most multicultural cities in Canada—Montreal and Toronto—and a current resident of British Columbia, Teixeira was an ideal candidate to take the leadership role on a project of such magnitude. In the face of such an opportunity, he immediately contacted a long-time collaborator and friend, Wei Li of Arizona State University, who unhesitatingly embraced the project with enthusiasm and passion. Later on, Professor Audrey Kobayashi of Queen's University, a leading social geographer in North America, was invited to join the team, which benefited tremendously from her impressive knowledge, constructive criticism, and thorough editing of the entire manuscript.

Beyond our professional commitment to the project, we all shared a strong personal conviction that this book would contribute in a positive way to immigrants' lives and advance equality both within and beyond Canada and the US. Fittingly, our two original co-editors are among some 50 million international migrants living in Canada and the US, and come from two main 'traditional' and 'non-traditional' source regions across the Atlantic and Pacific Oceans, Europe and Asia, respectively. They experienced the often painful decision-making process that compels countless international migrants to leave their families behind in order to settle, survive, and hopefully thrive in their adopted countries. As such, much of what is written in this book is 'up close and personal' for them. Our third co-editor, a Kelowna-born Canadian who is a member of a visible minority, has lived through racialization and racism in a liberal democratic society and has worked at the forefront of anti-racism and human rights theorization and practice. Our collective personal experiences have convinced us of the importance of telling immigrants' stories in a form that is accessible to students, academics, and general readers alike.

There have been numerous books on immigration, ethnic groups, or urban social geographies in the United States (e.g., Frazier and Tettey-Fio, 2006; Knox and Pinch, 2006; Miyares and Airriess, 2006; Singer et al., 2008) and in Canada (e.g., Bunting et al., 2010; Li, 2003; Reitz, 2003). The immigration dynamics in each country are not always well understood, and critical comparative examinations of such dynamics are scant. Very few works have considered immigrants in North American cities in a comparative context. This book is unique in that it draws together, for the first time, a variety of scholarly perspectives on a major topic—immigrant geographies—in two multi-ethnic countries, thus offering American *and* Canadian scholars and students a foundation text for their study and research. To reinforce the comparative yet integrated nature of the book and enhance its value as a scholarly yet accessible text for the widest possible readership, the volume is structured so that an American scholar and Canadian scholar are paired on each chapter.

This collection represents an attempt to address a gap in the scholarly literature by drawing together original research by leading and emerging American and Canadian scholars on immigration and the urban geography of North American cities. The critical perspective of this volume is thus at once both broad and particular, ranging from historical overview to contemporary analysis, and from national or group-specific discussions to urban or regional case studies, all within an interdisciplinary analytical framework. The comparative structure of the volume—which reveals similarities and differences between immigration in two distinct social and national contexts—is intended to give readers a deeper understanding of the rich, highly complex immigrant geographies in North American cities.

In this book, we have aimed to:

1. Develop a textbook suitable for academic programs in geography, immigration studies, ethnic studies, and North American studies.
2. Provide an overview of immigration in the major metropolitan areas of the United States and Canada.
3. Explore different stages of immigrant settlement and integration and how they affect our cities, suburbs, and neighbourhoods.
4. Describe examples of unique ethnic populations and immigrant groups and how they shape the urban social geography of our cities.
5. Inform students by offering insights into immigration and immigrants' impact on the internal structure of our cities and suburbs.
6. Suggest topics that researchers can consider in the areas of immigration, integration, and social geography.
7. Identify data and issues for policy-makers and analysts to consider with respect to furthering policy development in the areas of immigration, settlement, and service provision.

8. Provide ethnocultural communities with material to support their programs (e.g., settlement, housing).

Carlos Teixeira, Wei Li, and Audrey Kobayashi

REFERENCES

Bunting, T., P. Filion, and R. Walker, eds. 2010. *Canadian Cities in Transition: New Directions in the Twenty-First Century,* 4th edn. Toronto: Oxford University Press.

Frazier, J. and E. Tettey-Fio, eds. 2006. *Race, Ethnicity, and Place in a Changing America.* Binghamton, NY: Global Academic Publishing.

Knox, P., and S. Pinch. 2006. *Urban Social Geography: An Introduction.* New York: Prentice-Hall.

Li, P.S. 2003. *Destination Canada: Immigration Debates and Issues.* Toronto: Oxford University Press.

Miyares, I.M., and C.A. Airriess, eds. 2006. *Contemporary Ethnic Geographies in America.* Lanham, Md: Rowman & Littlefield.

Reitz, Jeffrey, G. 2003. *Host Societies and the Reception of Immigrants.* La Jolla, Calif.: Center for Comparative Immigrant Studies, University of California, San Diego.

Singer, A., C. Bretell, and S. Hardwick, eds. 2008. *Twenty-First-Century Immigrant Gateways: Immigrant Incorporation in Suburban America.* Washington: Brookings Institution.

ACKNOWLEDGEMENTS

A book of this magnitude, with 13 chapters plus substantive Introduction and Conclusion written by 28 contributors from both Canada and the US, is impossible to pull together without great commitment and hard work on the part of many people and institutions.

The editors wish to thank all the contributors, whose enthusiastic participation and dedication are central to this work. We are grateful to those contributors who participated in the extensive sharing of information that took place at the one-and-a-half day workshop at UBC–Okanagan in Kelowna, BC, and who made our end product much stronger. Thanks also to those who were unable to join us in person but were with us in spirit and contributed via PowerPoint presentations. All of these contributors kept up a spirit of co-operation and camaraderie, even when deadlines seemed impossible to meet. The paired contributors of some chapters are long-time collaborators, which made their collaborations both natural and seamless. Others were put into an 'arranged marriage' situation—having known each other only through reading one another's work prior to this project—and communicated largely via e-mail or phone during the writing process. In the end, all pairs worked well together and produced high-quality work.

We are also indebted to the many people and institutions in Canada and the United States who encouraged us to go ahead with this project and who helped us with advice, constructive criticism, and/or financial support. In particular we thank the Canadian Embassy in the US and the Dean's Office at UBC–Okanagan, which together provided the financing that made the Kelowna workshop possible. Thanks also to Irene Bloemraad of the University of California, Berkeley, and Daniel Hiebert of the University of British Columbia, whose strong recommendation letters contributed to our successful landing of the Canadian Embassy Conference Grant; John Oh and Raja Wariach of UBC–Okanagan and Wan Yu of Arizona State University assisted us in various stages of the project. We also appreciate the encouragement and input of the team from Oxford University Press, in particular, Katherine Skene, for believing in this project; Mark Thompson, Patti Sayle, and Phyllis Wilson for being such great liaisons between ourselves and the press; and Richard Tallman for his excellent editing skills. Seven book prospectus reviewers gave us the confidence that our ideas were worth exercising and three book manuscript reviewers provided critical comments and suggestions that improved the final outcome. We are truly grateful to them all. For errors, omissions, and weaknesses, however, we must take responsibility.

Wei Li expresses her deep gratitude to the Canada–US Fulbright Foundation, whose 2006–2007 Fulbright Visiting Research Chairship to Queen's University made it possible for her to live and work in Canada—the third country she has ever lived in as a resident—for nine and a half months. The experience contributed significantly to her understanding of Canada and Canadian immigration policies and practices, both of which have been a key inspiration to her ongoing comparative work between the two countries. She thanks the former director and current faculty head of Asian Pacific American Studies of the School of Social Transformation at Arizona State University, Karen Leong and Kathy Nakagawa, respectively, for their trust, guidance, and support. She also thanks her research associate and Ph.D. student, Wan Yu, whose key assistance eased much of her anxiety and contributed significantly to this book. Wei's gratitude also goes to the book's other two co-editors, Carlos, for his long collaboration, trust, and tolerance of her occasional lack of patience, and Audrey, for her guidance and hospitality during her Fulbright year, and, more importantly, for being a role model to Wei and to all female racial minority members struggling in academia. Wei is extremely grateful to have had the opportunity to work with you both and proud to call you good friends!

Audrey wishes to thank the Fulbright Foundation for a fellowship that allowed her to spend a year in Washington, DC, as an initial phase of developing a comparative understanding of immigration in the US and Canada, and between the US and Canada. She also thanks Carlos and Wei for their invitation to join the team, and for their enthusiastic, professional, and insightful approach to this important and innovative topic.

Carlos Teixeira, Wei Li, and Audrey Kobayashi

CONTRIBUTORS

James Allen
California State University–Northridge

Caroline Andrew
University of Ottawa

Thomas Boswell
University of Miami

Thomas Carter
University of Winnipeg

Elizabeth Chacko
George Washington University

Joe Darden
Michigan State University

Eric Fong
University of Toronto

Els de Graauw
Baruch College, City University of New York

Susan Hardwick
University of Oregon

Dirk Hoerder
Arizona State University

Audrey Kobayashi
Queen's University

Helga Leitner
University of Minnesota

Wei Li
Arizona State University

Lucia Lo
York University

Robert A. Murdie
York University

Valerie Preston
York University

Brian Ray
University of Ottawa

Damaris Rose
Institut national de recherché scientifique, Montreal

Emily Skop
University of Colorado, Colorado Springs

Heather Smith
University of North Carolina, Charlotte

Carlos Teixeira
University of British Columbia–Okanagan

Luisa Veronis
University of Ottawa

Domenic Vitiello
University of Pennsylvania

Scott Walker
Arizona State University

Lu Wang
Ryerson University

Qingfang Wang
University of North Carolina, Charlotte

Shuguang Wang
Ryerson University

Lindsay Withers
George Washington University

INTRODUCTION

IMMIGRANT GEOGRAPHIES: ISSUES AND DEBATES

Audrey Kobayashi, Wei Li, and Carlos Teixeira

Canada and the United States share a 5,527-mile/8,893-kilometre border.[1] The two countries also share a common history of long-established indigenous settlement many centuries before immigrants, first from Europe, then from the rest of the world, came to settle and establish the two modern nation-states. Canada has a larger land area than the US (about 158,000 square kilometres or 60,000 square miles bigger), but the US has about a 10 times greater population size and density and gross domestic product (GDP) than Canada. Both countries are widely known as major destinations for large numbers of international migrants,[2] who have played a role in shaping the social, economic, and political landscapes of the two countries in the modern era. In 2010, Canada and the US have international migrant populations of 7,202,340 and 42,813,281, respectively, comprising 21.3 per cent of Canada's total population and 13.5 per cent of that of the US (Table I.1).

This volume examines immigration in the context of the evolving urban geography of North American cities, particularly since World War II, when new patterns of immigration were linked to the evolving international economic system and to changing immigration policy regimes in both countries. We fill a gap in the existing literature on immigration, which is remarkable for its lack of sustained, comparative research, especially given the radical shifts in immigration admission and integration policies in both countries since the mid-1960s. Such shifts have facilitated large-scale heterogeneous immigrant flows from diverse and non-traditional source countries.

In this 'new age of migration', certain metropolitan areas have witnessed immense changes in their social, economic, and political landscapes, to become 'cities of nations', among the most multicultural places on the planet (Anisef and Lanphier, 2003). One of the common characteristics of recent immigration in both countries is the fact that immigrants prefer settling in major urban areas: almost 95 per cent of immigrants living in Canada in 2006 and close to 96 per cent of the foreign-born population in the US in 2005 are urban residents (Chui

Table I.1   Population and International Migrants in the World, Canada, and the US, 1990–2010

Year	Estimated Number of International Migrants at Mid-year			International Migrants as % of Total Population		
	World	Canada	US	World	Canada	US
1990	155,518,065	4,497,521	23,251,026	2.9	16.2	9.1
1995	165,968,778	5,047,093	28,522,111	2.9	17.2	10.5
2000	178,498,563	5,555,019	34,814,053	2.9	18.1	12.1
2005	195,245,404	6,304,024	39,266,451	3.0	19.5	13.0
2010	213,943,812	7,202,340	42,813,281	3.1	21.3	13.5

	Annual Rate of Change of Immigrant Stock (%)				Population at Mid-year (000)s		
	World	Canada	US	Year	World	Canada	US
1990-5	1.3	2.3	4.1	1990	5,290,452	27,701	254,865
1995-2000	1.5	1.9	4.0	1995	5,713,073	29,302	270,648
2000-5	1.8	2.5	2.4	2000	6,115,367	30,687	287,842
2005-10	1.8	2.7	1.7	2005	6,512,276	32,307	302,741
				2010	6,908,688	33,890	317,641

Source: Calculations based on esa.un.org/migration/p2k0data.asp.

et al., 2007; Singer, 2008). In Toronto, immigrants accounted for 44 per cent of the total population at the beginning of the twenty-first century; in Montreal and Vancouver the proportion was 37 per cent. In Miami, immigrants made up 59 per cent of the population and the figures were 41 per cent for Los Angeles and 36 per cent for New York (Hoernig and Walton-Roberts, 2006).

While large numbers of newcomers have been working-class immigrants or refugees in search of better lives, one of the interesting characteristics of immigration trends in recent decades has been that these cities have become magnets for highly educated and skilled international migrants, as well as those with substantial financial assets to invest in the North American economy (Ley, 2010). It is clearly imperative that scholarship address the new realities of immigration in the urban areas of both countries.

IMMIGRATION OVERVIEW: GLOBAL AND REGIONAL CONTEXT

Cross-national human migration is a centuries-old phenomenon. Millions of migrants seek to live and work outside their countries of birth, looking for better and safer living and working environments. Some migrant-sending countries, such as the Philippines, encourage out-migration with policy incentives or administrative assistance, in order to relieve population pressure or to tap into remittances sent back by migrants. In the meantime, migrant-receiving countries, in order to replenish the population base, resolve labour shortages, or address humanitarian issues, develop immigration admission policies to recruit certain types of international migrants while screening out those they deem 'undesirables'.

In the past century and a half, immigration to Canada and the US has fluctuated. Both countries received large numbers of immigrants just after the turn of the twentieth century, peaking just before World War I, then dropping to the lowest numbers of newcomers arriving between the two world wars, and rising again after World War II. Rates of immigration to the two countries have sometimes converged. For example, both countries received significant numbers of Irish immigrants during the potato famine of the mid-nineteenth century; both also received large numbers of Chinese immigrants during the era of transcontinental railway construction. But at other times, they have diverged. The US has seen a much larger influx of Latin American migrants, and Canada has welcomed a relatively larger number of Asian migrants in recent decades, and these two groups are very different in terms of cultural practices, language, education, and integration to the labour market.

With accelerated economic globalization, the pace of international migration has increased to both countries in recent decades. The UN estimates that international migrants reached almost 214 million in 2010, with the top 10 destination countries receiving more than half of the total (the US, 42.8 million; the Russian Federation, 12.7 million; Germany, 10.8 million; Saudi Arabia, 7.3 million; Canada, 7.2 million; France, 6.7 million; the United Kingdom, 6.4 million; Spain, 6.4 million; India, 5.3 million; and Australia, 4.7 million). In terms of percentage of international migrants in the total population, the top five countries in the world are Singapore (40.7 per cent), Israel (40.4 per cent), Luxembourg (35.2 per cent), Saudi Arabia (27.8 per cent), and Switzerland (23.2 per cent). The number of immigrants provides only one means of measuring the impact of immigration on these countries, however. Traditional immigrant countries such as Canada and the US are the result of cumulative permanent immigration over a long period; others, such as Saudi Arabia, are relatively recent destinations for labour migrants who do not take up permanent residence but return to their home countries after fulfilling their contracts.

At 21.3 per cent, Canada's percentage of international migrants in the total population is close to that of New Zealand (22.4 per cent), Australia (21.9 per cent), and Ireland (19.6 per cent). The US, at 13.5 per cent, is close to Germany, at 13.1 per cent (see Table I.2). Nevertheless, as Table I.1 demonstrates, both Canada and the US are important contemporary immigrant-receiving countries. With large and constantly increasing numbers and growing percentages of international migrants arriving between 1990 and 2010, the annual growth rates for the number of immigrants are almost constantly higher than the world average. Table I.2 shows selected OECD countries that received at least

Table I.2   Number and Percentage of International Migrants,
Selected OECD Countries

Selected OECD Countries	International Migrants (000s)			International Migrants as % of Total Population		
	1990	2000	2010	1990	2000	2010
Australia	3,581	4,106	4,711	21.0	21.0	21.9
Austria	793	996	1,310	10.3	12.5	15.6
Belgium	892	879	975	9.0	8.6	9.1
Canada	4,498	5,555	7,202	16.2	18.1	21.3
France	5,897	6,279	6,685	10.4	10.6	10.7
Germany	5,936	9,980	10,758	7.5	12.2	13.1
Greece	412	732	1,133	4.1	6.7	10.1
Ireland	228	385	899	6.5	10.1	19.6
Japan	1,076	1,687	2,176	0.9	1.3	1.7
Mexico	701	520	725	0.8	0.5	0.7
Netherlands	1,192	1,585	1,753	8.0	10.0	10.5
New Zealand	523	685	962	15.5	17.7	22.4
Poland	1,128	823	827	3.0	2.1	2.2
South Korea	572	568	534	1.3	1.2	1.1
Spain	829	1,753	6,375	2.1	4.4	14.1
Sweden	776	993	1,306	9.1	11.2	14.1
Switzerland	1,376	1,563	1,763	20.5	21.8	23.2
Turkey	1,150	1,263	1,411	2.1	1.9	1.9
UK	3,716	4,789	6,252	6.5	8.1	10.4
US	23,251	34,814	42,813	9.1	11.1	13.5

Source: esa.un.org/migration/p2k0data.asp.

a half-million international migrants in 2010; total numbers and percentages of international migrants have increased steadily in most cases in the past two decades. Among these 20 OECD countries, while the US has received the largest numbers of immigrants in these two decades, Canada ranks third in total number and fourth in the percentage of international migrants in the total population as of 2010.

What these figures do not show, however, is the range of citizenship prospects for migrants to these various countries. Some countries, notably Germany and most of the Eastern European nation-states, practise a policy of *jus sanguinis* (Latin: right of blood), meaning that citizenship is inherited within a common ethnocultural ancestry, excluding or severely limiting access to those considered to be from other 'blood' lines. Other countries practise *jus soli* (Latin: right of the soil) and consider to be citizens, with full and equal rights, all those who are born on national soil, regardless of their parentage, and all who have met the requirements for acquired citizenship. Both the US and Canada belong to the latter group. Both countries therefore have an obligation to treat all immigrants, and their children, by the same standards accorded to their non-immigrant populations. As the following chapters will show, however, equality in principle does not translate into equality in practice, for a wide range of reasons including the adoption of more complex temporary immigrant categories that do not confer full citizenship rights (although children born in Canada or the US have full citizenship rights no matter what their parentage), and structural discrimination and racism that result in marginalization of immigrant communities.

Both countries are also signatories to the United Nations Convention Relating to the Status of Refugees (1951, Article 1A), which defines a **refugee** as a person who 'owing to a well-founded fear of being persecuted for reasons of race, religion, nationality, membership of a particular social group, or political opinion, is outside the country of his nationality, and is unable to or, owing to such fear, is unwilling to avail himself of the protection of that country.' The countries from which the largest number of refugees sought refuge in 2009 are Afghanistan, Iraq, Somalia, the Russian Federation, and China (United Nations High Commission for Refugees, 2010). But the movement of refugees to North America does not always reflect the international numbers, because many refugees find themselves in close or adjoining countries (for example, some 3–4 million people are displaced within continental Africa), and because there are significant political, social, and economic elements to the refugee process. In Canada and the US, the largest number of requests comes from China, Mexico, Haiti, Colombia, and El Salvador. The US is the number-one refugee destination in the world, and received 49,000 asylum requests in 2009, 13 per cent of the world total. About a third of these were from the People's Republic of China. Canada is the third largest recipient of asylum requests internationally, with 33,900 (9 per cent of the world total) in 2009, a slight drop from recent years but still proportionally much higher than the US rate per capita. Refugees face significant challenges of resettlement: they are more often than not impoverished, have low levels of human capital, including language and education, and they are often fleeing extremely difficult circumstances that take a toll on individual and family lives. In many if not most instances, they are people who have been severely traumatized. At the same time, the line between refugees and economic migrants is often blurred.

The vast majority of migrants move for economic reasons. Low-skilled—and low-paid—migrant workers still account for the majority of labour migration flows, but increasing numbers of more highly educated people with professional skills are becoming permanent or temporary migrants. This trend is the result of global economic restructuring and especially the growth of the knowledge-based economy. Responding to demands for workers with high levels of knowledge-based human capital, receiving countries have developed favourable immigration admission policies towards this segment of international migrants. Data from the International Organization for Migration

(2008) indicate that 41 per cent of Asian and 29 per cent of Latin American/ Caribbean immigrants to the United States and 35 per cent of Asian and European immigrants to Canada have some post-secondary education, indicating a high level of self-selection among migrants with the highest levels of human capital.

UNDERSTANDING MIGRATION

There is no 'one-size-fits-all' approach to understanding the migration process or the complex array of social, cultural, and geographical consequences that attend the act of emigration from one place and the process of settlement and integration in a new country. Traditional discussions of migration tend to talk about 'push' factors, the conditions that bring people to the point of leaving, and 'pull' factors, the conditions that draw them to the new country. No one would deny that such factors exist. **Push factors** in the source countries include poverty, political, religious, or other forms of oppression, and events such as war and natural disasters that make life in the source country untenable. **Pull factors** include the promise of a higher standard of living, including better jobs, education, and health care, as well as the public policy and citizenship conditions that facilitate entry to the destination country. Historically, the public policy factors have included immigrant recruitment (usually by targeting source countries), policies that discriminate against certain groups but encourage others, a strong (or stronger) economy and labour market, and conditions that facilitate permanent settlement, including settlement programs and the potential for naturalization, as well as programs designed to encourage specific groups of people.

Although it is important to acknowledge such factors, they tell only part of the story. A fuller understanding of the immigration process requires detailed contextual understanding of the circumstances of departure, arrival, settlement, and integration, as well as an understanding of how variations occur within populations. We no longer think of immigration as a one-way process, for example, as many of today's migrants reflect what is called **transnationalism**, maintaining significant social, economic, and political ties with their countries of origin, their activities facilitated by modern transportation and communication. How does transnationalism affect the integration process, the second generation born as citizens of the destination country, and the ongoing political, social, and cultural practices of migrants? What factors influence who chooses to migrate? How do we understand how different ethnocultural groups, including differently **racialized** groups, fare upon arrival in their new countries? For that matter, how do we understand how individuals within a specific migrant population, who vary by gender, age, **human capital**, cultural traits, language, etc., may have very different experiences of settlement? And how do all these factors vary over time and from city to city within the destination country? Established scholarship, and the chapters in this book, shed light on some of these questions.

MIGRATION CATEGORIES AND ADMISSION POLICIES: CONVERGENCE AND DIVERGENCE

Both Canada and the United States have developed more complex migration categories over time. Although official categories did not exist historically, we can look back to see significant differences, especially between economic migrants and refugees. The earliest colonial migrants to both countries immigrated largely for economic reasons. The French immigrants to Canada, for example, established the basic, or staple, elements of the economy in the fur trade, forestry, and agriculture, and they were well organized within the French colonial administration. The earliest migrants also included refugees. Religious groups, such as the 'Puritans', who were among the earliest immigrants to the United States, were effectively refugees from religious persecution and established communities based on a fierce independence from Britain. The ramifications of the circumstances of immigration for just these two groups reverberate through the histories of the two countries. They also have significant regional effects. For example, the histories of Quebec, the Atlantic provinces, and the New England states have developed from the conditions established by these two groups of migrants.

Historically, Canada and the US have had similar immigration admission policies. Generally more open-armed during economic booms and more restrictive during economic busts, they have been traditionally more favourable towards immigrants from the United Kingdom, Northern and Western Europe, and gradually to immigrants from Eastern and Southern Europe. Immigrants from other continents, on the other hand, have often been the 'indispensable enemy' (Saxton, 1995); in other words, their presence was tolerated, if not welcomed, when their labour was needed, but they were restricted or even excluded from permanent settlement and nation-building (Lee, 2003; Li, 2003) even though, in the case of the Chinese, they were instrumental in completing the railway infrastructure required for the nation-building project. Every major immigrant group of colour—from the slaves brought from Africa starting in the seventeenth century, to the Asians brought to work on railway construction and in mines during the nineteenth century, to those who have come since World War II as farm labourers and domestic workers—has been deeply racialized, with ongoing implications for citizenship participation, social class distinctions, and residential segregation.

Discriminatory immigration laws were often based on 'race', gender, or nationality, and especially targetted non-white groups. Both countries set quotas, engaged in bilateral negotiation with source countries, and used other forms of policy manipulation to limit or completely exclude immigrants from specific parts of the world, or to prohibit them from gaining the full rights of citizenship after arrival. For example, the US passed legislation to exclude Chinese immigrants in 1882 and again in 1924. Similarly, Canada passed 'head tax' legislation, a prohibitive surcharge on each Chinese immigrant entering the country (see Chapter 2). Such public policy manipulation has significant knock-on effects.

When the US signed a 'Gentlemen's Agreement' with Japan to limit migration, the response was a shift in Japanese immigrants from the US to Canada. More than 4,000 Japanese immigrants entered Canada in 1907, where they became the major ethnic group in the west coast sawmill industry. The migration of Japanese immigrants to Canada was short-lived, however, as the Hayashi–Lemieux 'Gentlemen's Agreement' of later the same year severely curtailed any growth in their numbers until well after World War II. Other legal manipulations included various attempts to impose language tests on certain nationalities and, in the case of South Asian immigrants to Canada, the continuous passage policy that disallowed immigrants from disembarking unless they had travelled from their places of origin without stopping. All Asian immigrant groups were limited in their citizenship rights, including the franchise, landownership, and access to public goods such as education and professional accreditation. These imbalances began to change after World War II with the repeal of discriminatory laws and the extension of naturalization rights to previously disenfranchised immigrant groups in both countries. Similar manipulation of public policy occurred in the US, especially with respect to the Bracero Program, which encouraged large numbers of Mexican agricultural workers to enter the country between World War II and the mid-1960s, and with respect to the fate of the large number of undocumented workers, again mainly from Mexico, who now make up a signifi-cant if unofficial part of the US labour force.

It was not until the mid-1960s that immigration admission policies in both countries changed to eliminate discrimination based on race or national origin (1965 in the US and 1967 in Canada). This period also saw a marked divergence of their respective policies. In the United States, the Immigration and Naturalization Act (known as the Hart-Celler Act) of 1965 emphasized the principle of family reunification and, on a country-by-country basis, allocated to that category immigrant quotas of up to 80 per cent of total annual immigrants; the remainder were employment-based immigrants and refugees. Canada, in 1967, amended its Immigration Act but did not set any quotas; rather, potential immigrants were assessed largely on labour market criteria (educational attain-ment and official-language capability), as well as on the presence of an employ-ment offer or family connections in Canada. This system, which continues to the present, allocates 'points' to individual immigrant applicants based on their human capital. After the Vietnam War ended, the US passed the 1980 Refugee Act, taking refugees out of the immigrant category and giving the administra-tion the right to decide the annual level of entries. In Canada, the 1978 Canadian Immigration Act added a humanitarian category for refugees, and the propor-tion of refugees in the Canadian system has been significantly higher than that of the US ever since. On the other end of the immigrant spectrum, Canada introduced an entrepreneur immigrant stream in 1976 and an investor stream in 1986 to enlist wealthy immigrants to invest in Canada, even when their human capital (as measured by 'points') may not otherwise have gained them entry.

With the acceleration of economic globalization, both countries in recent years have revamped their immigration admission policies to join the global race for highly skilled migrants and to ensure their retention. The US passed the 1990 Immigration Act to triple its worldwide employment-based immigrant quota to 140,000 per year, to establish the H-1B visa for temporary migrants who have at least a bachelor's degree or equivalent and a US job offer, and to create the EB-5 visa for investor immigrants, similar to the Canadian investor stream but with a higher investment dollar threshold and different management mechanisms. The quotas for H-1B visas were initially set at 65,000 persons per year, although this number was significantly raised during the late 1990s and early 2000s, and another 20,000 have been added since 2005 for US-educated holders of a master's degree or higher. (Furthermore, academic institutions and non-governmental organizations are exempt from such quotas altogether.) The Canadian 2002 Immigration and Refugee Protection Act revised the points system to emphasize official-language capability, education, and work experience. A 'Canadian Experience Class' was introduced in 2008 to retain Canadian-educated international students, among others, with Canadian working experience—a category somewhat similar to the H-1B visa. Canada also has ramped up the number of temporary migrants at the lower end of the economic scale, through a **Seasonal Agricultural Workers Program** and a **Live-in Caregiver Program**. Temporary migrant workers in all categories now number closer to the number of permanent immigrants, while in the US the number of undocumented workers has grown to an estimated 12 million.

To sum up, while numerically the US continues to admit more immigrants than Canada, Canada still has a much higher per capita admission rate and a higher overall foreign-born population, especially in its largest cities. Administrative variations between the two countries, as well as geographical factors, have resulted in differences in immigrant source regions, types of immigrants, their labour market skills, and their status as temporary or permanent. The different attitudes towards economic migrants versus family reunification, in particular, affect who gets into each country. The differences that ensue from this policy discrepancy, however, are not straightforward. As shown especially in Chapters 2 and 6, other factors—including labour market demands, opportunities for self-employment, language and education levels, recognition of foreign credentials, residential location, and forms of racialization and prejudice—all complicate this picture immensely.

IMMIGRANT INTEGRATION IDEALS AND PRACTICES: INDIVIDUALISTIC ASSIMILATION VERSUS OFFICIAL MULTICULTURALISM

Immigration admission policies mark only one side of the immigration dynamics of who gets in and who is kept out. Once immigrants land, the issue becomes how to make them an integral part of the economic and socio-cultural fabric. In this regard, the two countries also exhibit historically convergent and currently divergent paths. Both countries espouse social democratic ideals, despite their different political systems, and both have been countries of immigrants since their inception. It is generally believed, however, that American society is more individualistic whereas Canada is more collectivist, and these approaches are evident in immigration integration ideals and practices. Whereas the US has continuously advocated **assimilation** and the personal responsibilities of immigrants themselves, Canada has enshrined **multiculturalism** as state policy since the 1970s and has provided government-funded immigrant integration programs of various kinds.

Historically, both countries advocated the assimilation of immigrants from different parts of the world to the white Protestant mainstream societies domin-ated by the Anglo-Saxon core, although in Canada there are two 'charter' groups (francophones, mainly in the province of Quebec, and anglophones, mainly in the rest of Canada). The famous Chicago School 'race relations cycle' (Park et al., 1925) and the classic and revisionist depictions of assimilation theories and ideals (Gordon, 1964; Alba and Nee, 2003) are deeply rooted in the minds of academics, the public, and governments (Anderson, 1991; Nagle, 2009). The two countries, however, are notably different in terms of integration ideals, and their respective implementation of these ideals in policies and practices have diverged in recent decades.

One of the most significant of these differences lies in Canada's adoption of multiculturalism as a federal policy framework in 1971. In 1982, multicultur-alism was officially enshrined in the Canadian Constitution, and six years later was legislated in greater detail in the Multiculturalism Act. Canadian multi-culturalism emphasizes 'pluralism' as the core value of Canadian identity, and policies at all levels of government are mandated to reflect the preservation and enhancement of the diverse cultural heritages of all citizens and immigrants to Canada, as well as the pursuit of anti-racism, employment equity, equal treat-ment of all groups, and redress for group discrimination in the past (Kobayashi, 1993; Ley, 2007). In contrast, in the US, the dominant paradigm of immigration has long been assimilation. In Canada, both the federal and provincial govern-ments devote considerable resources to immigrant settlement and integration, although in December 2010 the Conservative federal government announced cuts of over $50 million to funding for local settlement agencies, with most of the budget-slashing targeted on Ontario (Keung, 2010). In the US, on the other hand, admission policies are made at the federal level, but integration policies are based primarily at the state or local level. No federal policies are designed, and no resources are earmarked for immigrant integration purposes (refugee policies are separate and different, and are beyond the scope of this book); rather, integration is considered the responsibility of immigrants themselves. In general, multiculturalism can be seen as taking a distant second place in the American popular consciousness to the 'melting pot' as the defining metaphor for immigrant settlement and integration.

The melting pot/multiculturalism distinction should not be interpreted, however, as leading in any linear fashion to distinctive immigrant experiences in the two countries. As ideological concepts the two terms have been widely accepted in the popular discourses of their respective countries, and public opinion polling has shown that Americans tend to favour the concept of the melting pot, while Canadians generally support multiculturalism, although the sense of allegiance to both ideas has eroded in recent years. Also, the two concepts drive the adoption of public policy in significantly different ways, most notably in the array of immigrant integration programs that the Canadian state is obligated to deliver, compared to the laissez-faire approach to integration adopted in the US. On the other hand, and the chapters in this volume show, the distinctions between the two countries in terms of immigrant experiences are by no means so clear, and they are complicated by differences among source regions, human capital, and local conditions created by variations in the housing and labour markets. One of our major goals, therefore, is to be able to sort out the differences according to a range of variables.

DIFFERENTIAL IMMIGRANT PROFILES

Source Regions

Immigrant source regions in the two countries were very similar prior to the mid-1960s, with the majority of immigrants coming from Europe. The past four decades, however, have witnessed dramatic changes in source regions in both countries, but different *kinds* of changes in each country (Table I.3). The total numbers and percentages of immigrants from developing countries in Africa, Asia, and Latin America have increased rapidly for both countries, but the proportion of immigrants coming from each source region, and from countries within regions, varies significantly.

In Canada, the internationalization of immigration since the mid-1960s has been supported by Canadian immigration admission and integration policies that have facilitated heterogeneous immigration flows from diverse, non-traditional source countries. For example, Europe was the primary source for Canada's earlier immigrants (among those arriving in Canada before 1981),

Table I.3   Immigration Admission Policies and Court Cases in Canada and the US

United States	Canada
1. Initial Immigration—Labour Migration	
2. Exclusion Era	
1875: Page Act bars Chinese female immigrants	
1882: Chinese Exclusion Act	1885: Chinese head tax imposed
1907: Gentlemen's agreement restricts Japanese	1908: East Indians kept out of Canada by Order-in-Council
1917: Asiatic barred zone adds East Indians to the exclusion list	1909-13: Denial of Indian voting rights, exclusion from professions

1923: *U.S. v. Bhagat Singh Thind* reverses naturalization rights for Indians	1910: $200 in cash for 'Asiatic' immigrants
1934: Tydings-McDuffie Act restricts Filipinos	1923: Chinese Exclusion Act
1935: Repatriation Act sends back Filipinos/ Mexicans	

3. Transition Period (World War II to Mid-1960s)

1943: Repeal of Chinese Exclusion Act	1947: Repeal of Chinese Exclusion Act
1946: Luce-Celler Bill sets annual quota of 105 for Filipinos/Indians with rights to naturalization	1947: Citizenship Act allows all groups for naturalization except Japanese Canadians to vote
1952: Asia-Pacific triangle: 2,000 total annual immigrant quota for the entire region	1952: Immigration Act: British subjects and French citizens as preferred classes; 150 East Indian annual immigration quota

4. Open-Door Period (Mid-1960s to Late Twentieth Century)

1962: All discrimination eliminated	1967: Amended Immigration Act creates points system
1965: **Immigration and Naturalization Act** (Hart-Celler Act) abolishes discriminatory quota system; regulates family reunion (80 per cent of immigrants) and professional (20 per cent) classes; sets 20,000/country/year quota; permits citizenship after five years of permanent residency	
	1976: Immigration Act: citizenship after three years; provinces allowed some powers over immigration; four new immigrant classes

5. Selective Period (since Late Twentieth Century)

1990: Immigration Act: favours family-sponsored immigrants; employment-based: 140,000/year worldwide; diversity.	
1998: American Competitiveness and Workforce Improvement Act; H-1B visa increased from 65,000 to 115,000/year	
2000: H-1B visa entrants increased to 195,000/ year	2002: Immigration and Refugee Protection Act establishes immigrant classes: family class; economic class: skilled workers, business immigrants, provincial/territorial nominees, live-in caregivers, seasonal agricultural workers; refugees. Points system emphasizes French- and English-language proficiency, education, and work experience
2003: H-1B visa entrants reduced to 65,000/year	
2005: Additional 20,000/year H-1B visas for master's degrees obtained in the US; immigration admission policy at federal level	2008: Canadian Experience Class introduced to attract temporary foreign workers and students who graduated from Canadian universities; immigration policy at both federal and provincial/ territorial levels

Source: Adapted from Li and Lo (2009).

with Italy and the United Kingdom being the most common countries of birth, accounting for 31 per cent of immigrants. In recent years, however, this pattern has changed radically. Currently, Asia sends the most immigrants to Canada and China is the leading country of birth among immigrants (Justus, 2004). In the US, however, immigrants from Latin America make up the majority, followed by Asian immigrants. Immigrants from 'non-traditional' source regions often belong to the category of **racial minorities** in the US or **visible minorities** in Canada. How they are perceived and received in the two destination countries reflects changing racial dynamics, the racialization process, and the intricate complexity of race, class, gender, nativity, and religion, a topic that the following chapters will discuss at greater length.

Immigration by Categories

Figures I.1 and I.2 and Table I.4 illustrate the different immigrant categories that each country has admitted in recent years and the outcomes of different immigration admission policies. While the skilled class accounted for as much as 61.2 per cent of immigration to Canada in 2005, the family class accounted for almost 70 per cent in the US in 2005 and percentages of immigrants admitted in the family class increased steadily from 2005 to 2007. The lower percentage of skilled permanent immigrants admitted to the US each year tells only a partial story. In fact, since fiscal year 1992, when H-1B visas were first issued, the US has admitted as few as 93,069 H-1B visa holders (in 1993) and as many as 461,730 (in 2007).[3]

Thus, while Canada uses a points system to recruit skilled permanent immigrants, the US relies heavily on temporary H-1B migrants to fill the labour

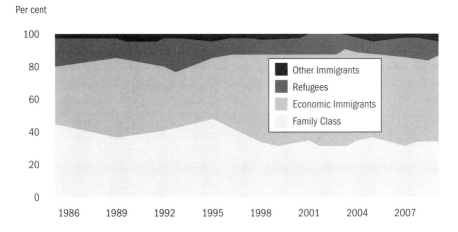

Figure I.1 Permanent Resident Immigrants by Category, Canada, 1986–2008

Source: Citizenship and Immigration Canada, 'Facts and Figures: Immigration Overview, Permanent and Temporary Residents' (2008), at: www.cic.gc.ca.

Per cent

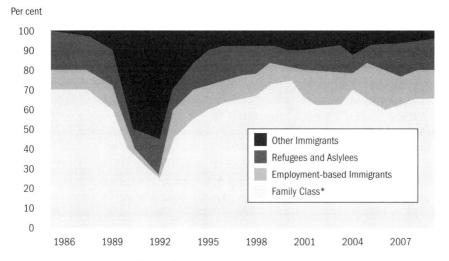

Figure I.2 Immigrants by Class Admission, US, 1986–2008

Sources: US Department of Homeland Security, *Yearbook of Immigration Statistics: 2008*, Table 6; US Department of Homeland Security, *Yearbook of Immigration Statistics: 2004*, Table 4.

*To be parallel with Canadian data, we incorporate U.S. 'Family-sponsored immigrants' and 'Immediate relatives of U.S. citizens' together into 'Family Class'.

needs of its knowledge-based economy. This pattern calls for a close examination of Canada's capacity to absorb immigrants admitted based on their skill and human capital levels versus the disenfranchisement of American H-1B visa holders. In Canada, immigrants face the ironic situation that notwithstanding their recruitment on the basis of achieving high points on the entry criteria scale, once landed they face discrimination in hiring, lack of recognition of their credentials (especially in professional occupations such as medicine and engineering), long waiting periods for retraining and re-accreditation, and other barriers such as linguistic difficulties. In the US, by contrast, despite their human capital, job offers, and comparable salary levels to Americans as stipulated by law, H-1B visa holders do not have equal rights and privileges as permanent immigrants, and their spouses are not permitted to seek employment inside the US, which results in problematic gender and family relations, and wastes education, talent, and experience when the spouses are skilled migrants themselves (see Box 6.1). Both systems, therefore, waste human capital and marginalize immigrants, but for significantly different reasons.

Settlement Patterns

While immigrants are largely urban-bound in both countries, the degree of concentration is different. In Canada, 66.3 per cent of all immigrants settle in the three largest metropolitan areas of Toronto, Vancouver, and Montreal (Statistics Canada, 2006). In contrast, the top three immigrant receiving metropolitan areas in the US—New York, Los Angeles, and Miami—were home

Table I.4   Immigrant Source Regions and Categories

Place of Birth	Canada (per cent of total)		United States (per cent of total)	
	1961	2006	1960	2005–7
Africa	0.5	6.1	0.7*	3.7
Asia	3.4	40.8	5.1	26.0
Europe	90.1	36.8	75.0	13.4
Latin America (including Caribbean)	1.5	11.3	9.4	53.4
North America	4.0	4.0	9.8	2.3
Oceania	0.4	1.0	n.a.	0.5

Canada	Skilled Stream/Class		Family		Refugees/ Protected Persons		Year Total
	Number	Per cent	Number	Per cent	Number	Per cent	Number
2005	156,310	61.2	63,354	24.8	35,768	14.0	255,432
2006	138,257	57.3	70,506	29.2	32,492	13.5	241,255
2007**	59,248	57.0	31,860	30.7	12,774	12.3	103,882
US							
2005	246,877	23.8	649,085	62.5	142,962	13.8	1,038,924
2006	159,081	13.5	802,577	68.1	216,454	18.4	1,178,112
2007	162,176	16.4	689,829	69.8	136,125	13.8	988,130

*Percentage includes all other areas.

**January to June only.

Sources: Canada: *Census of Canada, 1961* and *2006*; United States: *1960 Census*; 2005–2007 American Community Survey; IOM (2008: Table 11.1).

to less than one-third of all immigrants (31.1 per cent) during the 2005–7 period.[4]

Differential settlement patterns present both opportunities and challenges for the urban areas of both countries. For example, some groups concentrate spatially and form ethnic enclaves (such as Chinatowns in New York, San Francisco, Toronto, and Vancouver); others initially settle in immigrant reception areas close to downtown and later re-concentrate in the suburbs or, increasingly, they immigrate directly to suburban concentrations (e.g., Mississauga, Brampton, Markham, and Richmond Hill in the Toronto area or in Richmond and Surrey near Vancouver in Canada; and East New Orleans or the San Gabriel Valley near Los Angeles, Silicon Valley in the San Francisco Bay area, or the New York–New Jersey corridor in the US). Other immigrants disperse after acquiring a working knowledge of English (or French) and improving their socio-economic position. Still others disperse from the outset and never form spatial concentrations. The various settlement patterns and resulting urban immigrant community forms, which have been called ghettos, ethnic enclaves, ethnoburbs (Li, 1998), invisiburbs (Skop and Li, 2003), or heterolocalism (Zelinsky and Lee, 1998), represent differential formation dynamics and characteristics. With various theoretical underpinnings, these patterns are the main topic of Chapter 3 (Murdie and Skop, this volume).

Such different immigrant profiles and settlement patterns contribute to the different nature of immigration-related debates in the two countries, both in academia and in public policy, a subject that will be discussed in detail in later chapters.

MAJOR DEBATES: CONVERGENCE AND DIVERGENCE

Canadian Dreams: Immigrant Reality Check in Housing, Work, and Settlement

Housing and Poverty

Today's immigrants—particularly racialized minority immigrants—face significant challenges in Toronto's and Vancouver's expensive, tight rental housing markets. For example, pockets of concentration of Afro-Caribbean and African immigrants—often in public housing—occur in particular neighbourhoods of these cities (Mensah and Firang, 2007). The increasing number of new immigrants and racialized minorities in public housing can be seen as a consequence of low household income, compounded by supply, cost, and discriminatory constraints in Toronto and Vancouver rental markets. So far, none of these areas of concentration can be described as a ghetto, since they do not resemble the large-scale ghettos that characterize many US cities, but problems exist in the form of concentrated poverty (Ley and Smith, 1997; Murdie and Teixeira, 2006; Smith and Ley, 2008). Nevertheless, the historical treatment of racialized minorities and **First Nations** peoples in Canadian cities suggests that exclusion in almost all aspects of everyday life will likely remain a prominent feature of Canadian cities for many years to come (Murdie and Skop, this volume).

Immigrant Economies, Entrepreneurship, and Credentialism

While Canadian immigration policy has long acknowledged the importance of immigration as an engine of economic growth, the economic future is uncertain for many immigrant groups. Immigrants to Canada's major cities come from a vast range of cultures and social backgrounds, and represent a significant supply of human capital. Their diverse skills contribute positively to Canada's economy through participation in paid work and self-employment. For example, today immigration accounts for approximately 50 per cent of Canada's population growth and almost 70 per cent of its labour force growth (Jansen and Lam, 2003;

Teixeira et al., 2007). If current immigration rates remain constant, by 2011 immigration will account for virtually all labour force growth in the country (Hoernig and Walton-Roberts, 2006).

At the same time, many immigrants face significant barriers in finding employment or achieving success in the entrepreneurial sector. Moreover, many immigrants feel 'forced' to go into business for themselves because of the lack of other opportunities, which are constrained by discrimination and failure to recognize their credentials on the part of Canadian businesses. Canada's governments must facilitate the recognition of immigrant credentials and overseas experience to protect the future of the Canadian economy (Lo and Li, this volume).

One of the most significant economic impacts on Canada's major cities in recent years has been the increasing role of immigrant small businesses and entrepreneurs. Ever-larger numbers of immigrants have turned to self-employment, and today immigrants are more likely to be self-employed (15 per cent) than those born in Canada (12 per cent) (Teixeira et al., 2007). This trend is especially true for groups such as Chinese or Italian immigrants, who have constructed self-contained, institutionally complete communities and visible ethnic economies in Canada's major cities.

When immigrant economies are spatially concentrated, they are considered to be economic enclaves, which display important characteristics: their size (often small and family-oriented); their extensive use of family and co-ethnic labour; and in many cases their orientation to an ethnic consumer market. Most of the products sold and services provided by immigrant economies are largely oriented to a clientele that relies on imported products and services delivered in their own language. Such characteristics can result in heated internal competition and co-ethnic exploitation, with limited growth opportunities and little integration to mainstream economy. It is not clear, however, whether these enclaves will continue, as immigrants increasingly disperse to the suburbs of Canadian cities.

Spatial Concentration

The spatial concentration of some immigrant groups has had significant positive effects on their success in Canadian political, economic, and cultural life in Canada's cities. Residentially, some groups tend to concentrate (e.g., the Jews, the Chinese, the Portuguese, the Iranians) more than others, but their concentration appears to be due more to voluntary factors (e.g., retention of cultural, linguistic, and religious traditions) than to discriminatory practices by the receiving population. These groups also tend to live in owner-occupied dwellings in the city as well as in the suburbs.

One consequence of urban spatial concentration in ethnic enclaves is the maximizing of the political strength of immigrant groups, so that governments at all levels pay closer attention to their issues. Immigrant political gains have already occurred in Toronto and Vancouver and their suburbs, where the emergence of ethnic ridings for some groups (e.g., Chinese, Sikhs, Italians) is a recent phenomenon; however, the political involvement of minority communities has not been occurring at an equal pace for all groups in these cities (Bagga, 2007).

The Future of Immigrant Settlement in Canada

According to the best available projections, the annual volume of immigrants to Canada likely will remain at about 225,000, and the majority of this population will continue to settle in Toronto, Vancouver, and Montreal, although the numbers locating in smaller cities are currently increasing. Therefore, the racialized minority population of all cities will increase, in some cases dramatically. For example, Statistics Canada has estimated that by 2017 about half of the population of the Toronto CMA will belong to a visible minority group.

In terms of immigrant settlement, immigrants will probably continue to suburbanize, some re-segregating in the suburbs following initial settlement near the downtown core, others directly settling in the inner suburbs. The segregation levels between some immigrant groups and the rest of the population will remain high and the spatial outcome will be an increasingly fragmented and economically differentiated pattern of ethnic enclaves in many parts of the city (see Chapter 4, this volume). Observers of immigrant settlement in the major metropolitan areas of Canada emphasize, however, the multiple challenges that immigrants face upon arrival: from finding affordable and suitable rental housing, securing employment, and improving language skills/education, to securing access to health care and adapting to a new culture.

Notwithstanding the concentration in the largest cities, the immigrant population in nearly all Canadian cities continues to grow, and in some cities such as Calgary and Ottawa the numbers are increasing rapidly. While cities like Toronto and Vancouver clearly face substantial challenges managing the ever-increasing numbers of immigrants from a wider variety of source countries, smaller cities with limited infrastructure to assist newcomers are much further behind in adopting programs such as settlement, bridging, and language training. Some provinces, particularly Ontario, have recently ramped up efforts to expand integration programs outside the major cities. Most observers believe that the future of multicultural Toronto is bright, in large measure as a result of the recognition by all levels of government in Canada that large cities, particularly Toronto, are primary engines of the country's economic growth, and that immigration is a major contributing factor to this growth. The same may be the case for other, smaller cities, but it is too early to know. Thus, it is in the interests of all levels of government to ensure that this state of affairs continues. It remains to be seen whether the recently announced federal cutbacks to settlement services, aimed especially at Toronto and Ontario, are a sign of a less welcoming regime, or whether such cutbacks will have a negative effect on the integration of newcomers to the Canadian society (Hulchanski, 2010). Nonetheless, through wise policy development and implementation, it is likely that throughout the twenty-first century Canada's major port of entry for immigrants will be—more than ever—the 'World in a City' (Anisef and Lanphier, 2003).

Immigration Policy and Public Attitudes

While economics play a role in Canadian immigration policy, nation-building is its main goal. Since the introduction of official multiculturalism as federal policy in the 1970s, the balancing of economic interests and immigration levels to foster a more multicultural society has been a complex equation for a succession of governments. Over the past decade, public opinion surveys in Canada suggest that while Canadians may be less concerned about the numbers of immigrants entering the country, many are concerned about the process of immigrant integration. This separation of the two issues has given rise to what some regard as a set of contradictory attitudes among Canadians towards immigration (Sweetman and Warman, 2008). Recent polls also suggest that attitudes shift from one group to another concerning who is a 'good' immigrant and who is a 'bad' immigrant, that is, who can integrate 'successfully' and who cannot; although the discourse of successful integration is full of euphemisms that reflect common mythologies about the 'qualities' of certain groups, and the fact that some are more acceptable than others, with 'visible minorities' and, specifically, those of Muslim background being perceived as less able (or willing) to become integrated into Canadian society. In that respect, the Canadian multiculturalism discourse may be getting closer to the melting pot discourse.

American Dreams—A Candle Burning at Both Ends: Undocumented versus Skilled Immigrants

Attitudes towards Immigrants

One factor in the success of immigrant integration is the attitude of already established citizens towards them. Many Americans whose ancestors came as immigrants a generation or more ago have looked askance at newcomers, particularly those who seem most different from themselves. Other Americans have been more receptive, overlooking characteristics that seem out of place and recognizing that the processes of integration take time.

In the late twentieth century and the beginning years of the twenty-first century, anti-immigrant fears have arisen, so that Americans as a whole are highly conflicted and ambivalent regarding immigration. When immigrants have encountered hostility, their integration is presumably impeded. When they are treated in a friendly manner and with respect, they are encouraged to feel more at home and integrate more rapidly.

The mixed attitudes towards immigrants have several sources. In some cases animosity on the part of less-educated US citizens with fewer job skills sometimes results from their sense that immigrants, particularly Mexicans, are taking jobs away from them. Such concerns and debates accelerated during the global economic recession and prompted policies specifically targeted against this group (see Chapter 1). Meanwhile, highly skilled immigrants (e.g., H-1B visa holders, many of whom are South Asian or Chinese) are deemed to take jobs away from skilled American citizens. The arrival of newcomers from non-white racial groups is diminishing the proportion of whites in the population, a trend that many believe will weaken the country as they know it (Brimelow, 1995). A key issue in this debate concerns who are true 'Americans' (Li, 2009). Established concepts of American identity along ethnocultural and racial lines are therefore an integral part of attitudes towards immigration, although they are usually hidden or disguised. This, of course, leads to the enigma of how to change public attitudes, an issue that is beyond the scope of this book but that lies at the heart of why we study immigrant integration.

Services and Support

The need to provide health and educational support for less educated immigrants has been a large and growing concern, especially considering the many Mexican migrants living in the US without documentation. There are reportedly as many as 11 or 12 million undocumented migrants of various racial and ethnic backgrounds in the country, a statistic that has caused national, state, and local debates on immigration and led to the harsh SB1070 law on illegal immigrants being passed in Arizona in April 2010. Because the overwhelming majority of immigrants settle in urban areas and the enforcement of immigration laws has been increasingly localized, the contentious issues are more relevant and urgent for large cities (see Chapter 1). The cost of providing education to non-English-speaking children has also hit some localities hard. Moreover, large concentrations (*barrios*) where Spanish is the principal language, and Mexican immigrants in growing numbers also are settling beyond traditional gateways, so that the changing face of the US is a cause of concern on the part of some Americans. Their concern is fuelled by Manichaean notions of a 'clash of civilizations' (Huntington, 2004) between Americans—and those who can adapt to what is deemed 'American'—and the Others.

An Economic Necessity

On the other hand, most Americans recognize that immigrants contribute to the economy. Many are unskilled or low-skilled labourers willing to do difficult, unpleasant, and sometimes dangerous jobs at low wages that non-immigrant Americans do not want. Professionals such as physicians practise in small towns that previously had not had resident doctors. Other immigrants have opened new businesses, invested capital, and revitalized many neighbourhoods and cities, buying homes and businesses at higher prices than they would sell for if immigrants were fewer, and establishing trade and financial connections with their countries of origin that contribute to economic globalization.

Most Americans believe their country should continue to welcome immigrants, but in reduced numbers. A national survey in 2006 found that 75 per cent of Americans thought that immigration should be more restricted; support for restricted immigration ranged from 89 per cent of conservative Republicans to 54 per cent of liberal Democrats (Pew Research Center, 2007). On this and other aspects of immigration, surveys show much disagreement, with differences of opinion based not on social class or educational attainment but on political preferences and ethnicity. Asians, Latinos, Democrats, and liberals tend to see immigrants as a benefit rather than a burden, while whites, Republicans, and conservatives are more likely to stress problems associated with immigration (Myers, 2007: 137). In this context, US immigration debates—in the forms of anti-immigrant sentiment and racialization of certain groups, especially in the post-9/11 context, and recent failed efforts at comprehensive immigration reform by the federal government—differ significantly from those in Canada.

CONCLUSION

In this introduction we have highlighted similarities and differences between the United States and Canada regarding immigration histories, admission policies, integration ideals and practices, the resulting differential immigrant profiles, and major national debates related to international migration. We have not intended to offer an exhaustive discussion on everything to do with immigration— that would be impossible within a single volume—but have, instead, focused on selected topics of particular significance. We therefore have left out some important issues associated with immigration, such as the economic impacts of immigrants, return migration and remigration, and specific immigrant groups such as LGBTs and retirees. We hope our work represents a beginning, not an end, of comprehensive and comparative studies of international migration in Canada and the US within the contexts of globalization and changing geopolitics. Such global forces have profoundly altered the world order as we know it, along with international migration dynamics and the daily lives of immigrants and non-immigrants alike.

NOTES ON LANGUAGE

We faced significant linguistic challenges in compiling this set of essays. Some of the issues concern differences in terminology between the two countries. Canadians uniquely use the term 'visible minorities' while Americans tend to say 'racial minorities'. Canadians refer to the 'First Nations' and 'Aboriginal peoples' while Americans use 'Native American'. And, of course, Canadians for many years have seen their society of immigrants as a 'cultural mosaic' compared to the American 'melting pot'. These differences reflect profoundly distinct ideological, historical discourses from which the meanings of these terms cannot be decontextualized; they are not mere technical differences in definition. We also faced the issue of how to be respectful of the many immigrant groups of whom we write when the terms used in popular discourse, public documents, or even academic literature are not necessarily those used within communities. For example, many permutations and nuanced differences surround the terms 'black', 'African American', 'African Canadian', and 'Caribbean', just as a range of loaded meanings cluster around the term 'white'. One of the most difficult terms refers to individuals with origins in the large number of countries referred to as 'Latin America'. Canadian officials use the term 'Latin American' regardless of racialization, while American officials use 'non-Hispanic white' to divide the 'white' population between those of Latin American and European origins. Community names such as 'Hispanic', **'Latino/a'**, and 'Chicano/a' reflect a range of political and cultural identities within those communities. We take the position that all of these terms are social constructions, that is, products of historical discourse, that carry huge power to classify, segregate, and in many cases oppress specific groups. After lengthy discussions and consultations among all the authors (the majority of whom are themselves immigrants and/or racialized as non-white) we decided on a few terms that we use consistently throughout the book, with full recognition that we cannot capture all of the nuanced differences our terminology implies:

- We use 'black' and 'white' (uncapitalized) to refer to the large number of groups who are historically racialized into those designations, rather than use terms that refer to specific origins such as Europe, Africa, the Caribbean.
- We use the terms 'Latin American' and 'Latino/a' interchangeably, with explanations to avoid terminological differences that occur in the official censuses and use 'Latinos' and 'Filipinos' as non-gendered collectives for these two groups of people, as is the common colloquial usage, reserving the feminine 'Latinas' and 'Filipinas' for specific reference to those who are gendered female.
- We refer to people by country of origin (Chinese, Italian, Mexican) where specificity is required, but only in the sense of national or ethnic origin, not nationality, since the majority of immigrants (more in Canada than the US) have acquired full citizenship status, and therefore are fully Canadian or American.
- 'Undocumented' refers to those who do not have authorized immigrant status, particularly in the US.
- Various terms refer to the experience of new immigrants—'acculturation', 'assimilation', 'integration', 'incorporation'—each of which carries implications about the relationship between the newcomer and the long-settled. We use the term 'integration' wherever possible because it carries the least connotation of an immigrant's obligation to change to become more like a normative, dominant population. We recognize, however, that issues around these terms are fraught both politically and theoretically.
- We avoid the assumptive use of the term 'race', especially as a category that essentializes human difference, instead referring where possible to

'racialized minorities' in order to indicate that racial categorization is a social process, deeply embedded in histories of colonialism and racism.

No doubt, some ambiguities and assumptions remain embedded in the language throughout the book, but we have endeavoured wherever possible to clarify, and the Glossary at the end of the book provides further clarification.

QUESTIONS FOR CRITICAL THOUGHT

1. Discuss why Canada and the United States are known as major destinations ('ports of entry') for large numbers of international migrants. Comment on immigrants' impact in shaping the social, economic, and political landscapes of these two countries, using examples where possible.

2. What do the main conclusions of this chapter reveal about the similarities and differences between the two countries with regard to immigration and integration policies?

3. According to most projections, the Canadian and US populations are expected to become more heterogeneous in the future, both culturally and linguistically. Discuss the principal challenges that new immigrants and refugees face in their settlement/integration process in each country.

4. Does the distinction between the 'melting pot' in the US and the 'cultural mosaic' in Canada represent a fundamental difference between the two countries, or is that difference overrated? Explain.

SUGGESTED READINGS

1. Hirschman, C., P. Kasinitz, and J. DeWind. 1999. *The Handbook of International Migration: The American Experience.* New York: Russell Sage Foundation. This general reference volume comprehensively reviews issues related to international migration, mainly from a distinct US perspective.

2. Kobayashi, Audrey, and Valerie Preston. 2007. 'Transnationalism through the Life Course: Hong Kong Immigrants in Canada', *Asia Pacific Viewpoints* 48, 2: 151–67. This study of one of the largest groups of migrants in history reveals that family decisions regarding migration revolve around life-course stages, including education, entry to the labour market, and retirement.

3. Ley, D. 2010. *Millionaire Migrants: Trans-Pacific Life Lines.* West Sussex, UK: Wiley Blackwell. This book provides a detailed overview on wealthy immigrants' transnational work and lives in East Asia and Canada, and discusses the impacts of Canadian immigration policy initiatives.

4. Li, W., ed. 2006. *From Urban Enclave to Ethnic Suburb.* Honolulu: University of Hawai'i Press. This book documents the changing immigrant settlement patterns in eight metropolitan areas in four Pacific Rim immigrant

countries, and how the intertwining of globalization, racialization, national policies, and changing immigrant profiles and lives facilitates such settlement and community changes.

NOTES

1. We thank all contributors who participated in the workshop for this book, which took place at the University of British Columbia–Okanagan in November 2009, for their inputs and suggestions for this introduction chapter, especially James Allen, whose earlier draft has been adapted and used in the subsection of 'American Dreams—Candle Burning at Both Ends: Undocumented versus Skilled Immigrants', and Wan Yu at Arizona State University for her assistance in compiling data and creating graphics.
2. We are keenly aware that 'immigration' and 'international migration' have different legal meanings and ramifications. For the sake of simplicity, however, we use the two terms interchangeably in this chapter.
3. Data extracted from the US Citizenship and Immigration Service immigration statistical yearbooks.
4. Calculation based on the 2006 Canadian census and the 2005–7 American Community Survey.

REFERENCES

1. Alba, R., and V. Nee. 2003. *Remaking the American Mainstream: Assimilation and Contemporary Immigration.* Cambridge, Mass.: Harvard University Press.
2. Anderson, K.L. 1991. *Vancouver's Chinatown: Racial Discourse in Canada, 1875–1980.* Montreal and Kingston: McGill-Queen's University Press.
3. Anisef, P., and M. Lanphier. 2003. 'Introduction: Immigration and the Accommodation of Diversity', in P. Anisef and M. Lanphier, eds, *The World in a City.* Toronto: University of Toronto Press, 3–18.
4. Bagga, G. 2007. 'From the *Komagata Maru* to Six Sikh MPs in Parliament: Factors Influencing Electoral Political Participation in the Canadian-Sikh Community', in K. Graham, ed., *Our Diverse Cities.* Ottawa: Metropolis, 161–5.
5. Brimelow, P. 1995. *Alien Nation: Common Sense about America's Immigration Disaster.* New York: Random House.
6. Chui, T., K. Tran, and H. Maheux. 2007. *Immigration in Canada: A Portrait of the Foreign-born Population, 2006 Census.* Ottawa: Statistics Canada.
7. Gordon, M.M. 1964. *Assimilation in American Life: The Role of Race, Religion, and National Origins.* New York: Oxford University Press.
8. Hoernig, H., and M. Walton-Roberts. 2006. 'Immigration and Urban Change:

National, Regional, and Local Perspectives', in T. Bunting and P. Filion, eds, *Canadian Cities in Transition: Local Through Global Perspectives.* Toronto: Oxford University Press, 408–18.
9. Hulchanski, J.D. 2010. *The Three Cities within Toronto: Polarization among Toronto's Neighbourhoods, 1970–2005.* At: www.urbancentre.utoronto.ca/pdfs/curp/tnrn/Three-Cities-Within-Toronto-2010-Final.pdf.
10. Huntington, S.P. 2004. *Who Are We? The Challenge to America's National Identity.* New York: Simon & Schuster.
11. International Organization for Migration (IOM). 2008. *World Migration 2008: Managing Labour Mobility in the Evolving Global Economy, 2008.*
12. Jansen, C., and L. Lam. 2003. 'Immigrants in the Greater Toronto Area: A Sociodemographic Overview', in P. Anisef and M. Lanphier, eds, *The World in a City.* Toronto: University of Toronto Press, 63–131.
13. Justus, M. 2004. 'Immigrants in Canadian Cities', in C. Andrew, ed., *Our Diverse Cities.* Ottawa: Metropolis, 41–57.
14. Keung, Nicholas. 2010. 'Funding Axed for Immigrant Services', *Toronto Star,* 23 Dec., A1, A4.
15. Kobayashi, A. 1993. 'Multiculturalism: Representing a Canadian Institution', in

J.S. Duncan and D. Ley, eds, *Place/Culture/Representation*. London: Routledge, 205–31.

16. Lee, E. 2003. *At America's Gates: Chinese Immigration during the Exclusion Era, 1882–1943*. Chapel Hill: University of North Carolina Press.

17. Ley, D. 2007. 'Multiculturalism: A Canadian Defense'. At: mbc.metropolis.net/assets/uploads/files/wp/2007/WP07-04.pdf. (6 Nov. 2010)

18. ———. 2010. *Millionaire Migrants: Trans-Pacific Life Lines*. West Sussex, UK: Wiley-Blackwell.

19. ——— and H. Smith. 1997. *Is There an Immigrant 'Underclass' in Canadian Cities*. Vancouver: Vancouver Centre of Excellence, Research on Immigration and Integration in the Metropolis.

20. Li, P.S. 2003. *Destination Canada: Immigration Debates and Issues*. Toronto: Oxford University Press.

21. Li, W. 1998. 'Anatomy of a New Ethnic Settlement: The Chinese Ethnoburb in Los Angeles', *Urban Studies* 35, 3: 479–501.

22. ———. 2009. *Ethnoburb: The New Ethnic Community in Urban America*. Honolulu: University of Hawai'i Press.

23. ——— and L. Lo. 2009. 'New Geographies of Migration? A Canada–US Comparison of Highly-Skilled Chinese and Indian Migration', *Journal of Asian American Studies* (under review).

24. Mensah, J., and D. Firang. 2007. 'The Heterogeneity of Blacks in Ontario and the Racial Discrimination Boomerang', in K. Graham, ed., *Our Diverse Cities*. Ottawa: Metropolis, 20–5.

25. Murdie, R.A., and C. Teixeira. 2006. 'Urban Social Space', in T. Bunting and P. Filion, eds, *Canadian Cities in Transition: Local Through Global Perspectives*. Toronto: Oxford University Press, 154–70.

26. Myers, D. 2007. *Immigrants and Boomers: Forging a New Social Contract for the Future of America*. New York: Russell Sage Foundation.

27. Nagle, C.G. 2009. 'Rethinking Geographies of Assimilation', *Professional Geographer* 61, 3: 400–7.

28. Park, R.E., E. Burgess, and R.D. McKenzie. 1925. *The City*. Chicago: University of Chicago Press.

29. Pew Research Center. 2007. 'The Immigration Divide'. At: pewresearch.org/pubs/450/immigration-wedge-issue. (11 June 2010)

30. Saxton, A. 1995. *The Indispensable Enemy, Labor and the Anti-Chinese Movement in California*. Berkeley: University of California Press.

31. Singer, A. 2008. 'Twenty-First-Century Gateways: An Introduction', in A. Singer, C. Bretell, and S. Hardwick, eds, *Twenty-First-Century Immigrant Gateways: Immigrant Incorporation in Suburban America*. Washington: Brookings Institution, 3–30.

32. Skop, E., and W. Li. 2003. 'From the Ghetto to the Invisoburb: Shifting Patterns of Immigrant Settlement in Contemporary America', in J.W. Frazier and F.L. Margai, eds, *Multi-Cultural Geographies: Persistence and Change in U.S. Racial/Ethnic Geography*. Binghamton, NY: Global Academic Publishing, 113–24.

33. Smith, H., and D. Ley. 2008. 'Even in Canada? The Multiscalar Construction and Experience of Concentrated Immigrant Poverty in Gateway Cities', *Annals, Association of American Geographers* 98, 3: 686–713.

34. Statistics Canada. 2006. *Census of Canada, 2006*. Ottawa.

35. Sweetman, A., and C. Warman. 2008. 'Integration, Impact, and Responsibility: An Economic Perspective on Canadian Immigration Policy', in J. Biles, M. Burstein, and J. Frideres, eds, *Immigration and Integration in Canada in the Twenty-First Century*. Montreal and Kingston: McGill-Queen's University Press, 19–44.

Teixeira, C., L. Lo, and M. Truelove. 2007. 'Immigrant Entrepreneurship, Institutional Discrimination, and Implications for Public Policy: A Case Study of Toronto', *Environment and Planning C* 25: 176–93.

United Nations High Commission for Refugees. 2010. *Asylum Levels and Trends in Industrial Countries 2009*. At: www.unhcr.org/4ba7341a9.html. (31 July 2010)

Zelinsky, W., and B.A. Lee. 1998. 'Heterolocalism: An Alternative Model of the Sociospatial Behaviour of Immigrant Ethnic Communities', *International Journal of Population Geography* 4, 4: 281–98.

PART I

THE INTERNATIONALIZATION OF NORTH AMERICAN CITIES AND SUBURBS

CHAPTER 1

GOING LOCAL: CANADIAN AND AMERICAN IMMIGRATION POLICY IN THE NEW CENTURY

Helga Leitner and Valerie Preston

INTRODUCTION

Canada and the United States are countries of immigration that define themselves as nations of immigrants (Castles and Miller, 2009; Stalker, 2008). Each has developed an elaborate institutional apparatus to regulate immigration. **Control policies** concern the numbers and characteristics of those who will be allowed to enter national territory and how the state polices entry; **incorporation policies** set the terms by which immigrants become full members of the national community. But there are also significant differences, in particular with respect to the perceived role of the state in promoting immigrant incorporation. In Canada the state takes an active role in integrating immigrants whereas in the US incorporation is seen mainly as an individual responsibility.[1] This chapter compares recent tendencies in immigration policies of these two **nation-states** within the context of changing characteristics of immigration as well as larger transformations in governance and geopolitical developments. Of special interest in this regard are the rise of **neo-liberalism**, a political philosophy that emphasizes market rationality, a self-sufficient, self-interested citizen, and a declining role for the nation-state (Peck and Tickell, 2002; Leitner et al., 2007), and **neo-conservatism**, a moral-political rationality favouring government intrusion to address security and morality issues (Brown, 2006). We contend that the ascendancy of neo-liberal and neo-conservative immigration policies has been influenced by the growing securitization of both societies, which began well before the events of 11 September 2001, and that the distinct and different histories of federalism in the two countries also have affected how these ideologies play out in current immigration policies.

We focus on the changing geographies of immigration governance and control. For most of the past century, border control in Canada and the US has been under the purview of the national governments.[2] This monopoly has eroded recently with the devolution of some social responsibilities from the national to local levels and the efforts of some American states to enact their own immigration control measures. We will describe and explain commonalities and differences in these new spaces of immigration control and governance in the US and Canada.

We begin with a brief description of their shared experiences as countries of immigration to provide background for a discussion of common policy tendencies in the two national jurisdictions. The next two sections explore how these tendencies are localized in the Canadian and American contexts. We conclude by discussing how we may make sense of the variations and distinct geographies of local immigration policies.

COMMON EXPERIENCES, RESPONSES, AND POLICY TENDENCIES

Canadian and American national identities are founded on the myth that each nation has welcomed migrants to settle and build the nation, even though the welcome was always selective (Stasiulis and Jhappan, 1995). In both countries, immigration policy is increasingly motivated by the imperatives of economic development and growth. As a result, economic migrants and their families now account for more than 60 per cent of all **permanent residents** admitted annually to Canada (Citizenship and Immigration Canada, 2009). In the US, first priority is still given to family reunification with well over half of all immigrants being admitted through family sponsorship; however, the need to admit more skilled immigrants is also under discussion (US Senate, 2006).

Growing numbers of **temporary** and **undocumented migrants** are of major concern in both countries.[3] Temporary populations have increased significantly in Canada since 1998, growing by 10 to 20 per cent annually in recent years (Fudge and McPhail, 2009). The annual number of temporary visa holders admitted to Canada now exceeds the number admitted as permanent residents (Alboim, 2009; Citizenship and Immigration Canada, 2009), largely as a result of rapid expansion of the temporary foreign worker program, under which employers who have demonstrated that there are no qualified Canadians available for a job may apply for permission to hire a foreign worker who receives a visa valid for two years (Alboim, 2009). As visas expire, many temporary visa holders become undocumented, a circumstance that is likely to fuel anti-immigrant sentiments in Canada. The number of temporary residents in the US also has increased and is now estimated to total approximately 2,000,000 people (Papademetriou et al., 2009); however, this figure pales in comparison to the number of undocumented immigrants, which according to some estimates ranged between 8 and 12 million in 2008 (Passel and Cohn, 2009). Heightened enforcement along the border with Mexico has neither stopped immigrants from entering the US nor reduced the presence of undocumented immigrants. To the contrary, the number of undocumented immigrants has increased; border crossings have become more difficult and costly, preventing the previously common circular movement between Mexico and the US.

The discourse of illegality that views immigrants as a threat to the security of the nation has been intensifying in both countries. Anti-immigrant sentiments have been on the rise, and at the centre of these passions is the figure of the undocumented migrant. In Canada, the resurgence of anti-immigrant discourse was one reason for the **Bouchard-Taylor Commission** (Bouchard and Taylor, 2008) that examined Quebec residents' views about immigration and its impacts. The recent detention of Sri Lankans, who arrived in British Columbia without visas on a battered freighter, resonates with a long history of Canadian panic about illegal entry at the border (Hier and Greenberg, 2002; Mountz, 2004). In the US, right-wing and conservative civil organizations have often demonized undocumented immigrants by presenting them as a security threat. Organizations such as The Minuteman Project, a self-declared vigilante group, patrol borders and construct border fortifications to keep out undocumented immigrants.

In the twenty-first century, the two states also share several policy tendencies concerning immigration. In both countries, there have been active efforts to restrict the **social rights** of immigrants, particularly their claims on the welfare state. In the US, immigrants' access to social programs designed for low-income Americans has been reduced since passage of the Personal Responsibility and Work Opportunity Reconciliation Act of 1996 (US Congress, 1996). For example, most legal permanent residents who entered the US on or after 22 August 1996 are ineligible for Medicaid for five years. In Canada, the government has expanded temporary immigration that by definition provides limited access to social rights. Temporary foreign workers are not eligible for federally funded language training offered free to permanent residents. Without fluency in one of Canada's official languages, temporary foreign workers are often unaware of their rights and at risk in the workplace. The recent deaths of two Chinese workers in Alberta have been attributed in part to their limited knowledge of English that left them unable to understand safety training and unaware of health and safety regulations (Alboim, 2009; Elgersma, 2007; Fudge and McPhail, 2009).

As indicated above, **border control** has intensified in both Canada and the United States, a change often attributed to the events of 11 September 2001; however, 9/11 provided an opportunity to introduce restrictions on the entry of refugees that had been contemplated for more than 10 years, and to further step up border controls that had been introduced in the mid-1990s in both countries. The Third Safe Country Agreement signed in December 2002 dramatically cut the number of refugee claimants being admitted to Canada since individuals must now claim refugee status in the first safe country of arrival (Canadian Council for Refugees, 2007). For example, those who arrive at an airport in New York City and take a bus to the Canadian border at Niagara Falls to claim refugee status will be returned to the US on the grounds that the claims must be adjudicated in the first safe country where the claimants arrived. In December 2001, only four months after 9/11, Canada and the US issued the Canada–US Smart Border Declaration that called for greater collaboration and co-operation on border security issues. The Smart Border initiative, like the Illegal Immigration Reform and Immigrant Responsibility Act (IIRIRA) of 1996, was intended to reduce the influx of undocumented immigrants by increasing border control. In both countries increased border control has shaped and been shaped by increasing anti-immigrant sentiment, blaming immigrants of colour for numerous social ills. In the US, provocative neo-conservative talk-show hosts have zeroed in on undocumented workers—'illegal aliens'—repre-senting them as a threat to national security, and as evidence that immigration is beyond government control and is undermining the power and control of the white majority.

Last but not least, there has been a tendency towards increasing **localization** of immigration policies, which will be the focus of the second part of this chapter. Complex forces have been driving this localization. First and foremost, neo-liberal restructuring has resulted in a downscaling of central/federal state authority to lower tiers of the state. Second, neo-conservative tendencies have reinforced and promoted the drive to devolve state authority over immigration policies.

UNDERSTANDING COMMON POLICY TENDENCIES

Recently, a number of authors (Abu-Laban and Gabriel, 2002; Varsanyi, 2009; Coleman, 2009) have argued that the rise of neo-liberalism as a policy regime and form of governmentality has been instrumental in shaping immigration and citizenship policies. With others (Brown, 2006; Leitner et al., 2007), we argue that neo-liberalism also has been articulated with neo-conservatism. Neo-liberalism's market rationality promotes open borders—not just for commodities but also for people—a self-sufficient, self-interested citizen-subject. In contrast, neo-conservatism promotes the policing of borders, justified by discourses of patriotism, protection of the homeland, national order and the rule of law, and favouring government intrusion to address security and morality issues. In terms of cultural and racial differences, neo-liberalism promotes the erasure of cultural and racial boundaries (colour-blindness) and the commodi-fication of difference through the market. Diversity is promoted as a means of increasing productivity and expanding the market for goods and services. In contrast, neo-conservatism, as a moral-political rationality, promotes policing of national and cultural borders, identifying itself as the protector of a potentially vanishing past, and opposing the loss of national identity and core values. In the United States, these two rationalities collide all the time, in what many have framed as the impossibility of the Republican Party trying to be both the party of moral values and the party of big business, as evident in immigration policy debates (Brown, 2006).

Similar debates are occurring in Canada where neo-liberal tendencies are evident in current and past decisions to control the federal deficit by shrinking social programs rather than raising taxes (McBride and Shields, 1997; Brodie, 2002). With a greater role for government provision of services—from health care to post-secondary education—the neo-liberal shift in Canada has represented a more profound change than in the US, while simultaneously being more limited in scope. Neo-conservatism also takes slightly different forms in Canada than in the US. Despite some progressive social legislation, such as the legalization of same-sex marriages and the recognition that same-sex partners can be sponsored as immigrants, neo-conservative social policies have gained more traction since 2005. The minority Conservative government has promoted a traditional view of the nuclear family headed by a male breadwinner and has heightened surveillance and control at the border as part of its neo-conservative 'get tough on crime' agenda; however, the policy landscape in Canada and the US is complex and varied; the neo-conservatism advocated at the federal scale is often contradicted at state, provincial, and local scales.

While on the surface neo-liberal and neo-conservative rationalities are contradictory and collide all the time, we suggest that a closer look reveals that they have come to work symbiotically, frequently enabling one another. For example, illegality—primarily a neo-conservative discourse—works in the service of a neo-liberal agenda. Nancy Hiemstra (2007) has shown how discourses of illegality maintain immigrants as an unorganized, cheap, and flexible labour force, suggesting that neo-conservative discourses of illegality, the rule of law and order, and national security are supportive of neo-liberal policies and processes.

Our argument recognizes the multi-faceted nature of **immigration policies**, which are concerned with the admission, selection, and incorporation of migrants. We emphasize that these different facets of immigration regulation are tightly interlinked. The state's views about the likelihood and desirability of incorporating migrants affect policies concerning admission and selection and, in turn, these policies affect immigrant incorporation (Massey, 1999; Freeman, 2004). We also recognize that the state is concerned with migrants of all types—temporary, permanent, and undocumented, refugees and asylum seekers, etc. Migrants' plans to stay permanently at their destinations vary from person to person and for the same person, may change over time. People who arrive as sojourners sometimes find themselves staying permanently (Leitner and Ehrkamp, 2006), while others who intend to settle permanently end up leaving to pursue opportunities elsewhere (Kobayashi and Preston, 2007). Refugee claimants who enter legally, but whose claims are refused, may find themselves suddenly without legal rights to residency. The complexity of individual migration histories complicates state efforts to police and secure borders and disrupts policy narratives that assign a fixed legal status.

Neo-liberalism and neo-conservatism are associated with increasing decentralization of administrative responsibilities, often referred to as downscaling or downloading (Peck and Tickell, 2002; Leitner et al., 2007). Lower tiers of government are charged with implementing and delivering policies and programs subject to regulation and oversight by the central government. This process has applied to immigration policies in both countries, although the content and mechanisms by which decision-making and implementation occur are different. Processes of localization are path dependent, reflecting the long and complex history of struggle over jurisdictional power and authority between different tiers of government in Canada and the US.

LOCALIZATION OF IMMIGRATION POLICIES IN CANADA

The localization of Canadian immigration policy has occurred within the context of a highly decentralized federation in which control of natural resources, provision of social welfare, and many other state functions are predominantly under provincial jurisdiction. Since its inception in 1867, Canada's immigration policies have been a shared jurisdiction between the federal and provincial governments (Hawkins, 1988). Provincial jurisdiction is limited insofar as the provinces' actions must not be 'repugnant to any Act of the Parliament of Canada' (Becklumb, 2008b). While the federal government took responsibility for immigration control and largely retained authority of selection policies, its involvement in settlement policies has waxed and waned while provincial interest in immigration has increased since 1991 when the province of Quebec acquired the power to determine the number of immigrants admitted each year and to control selection, admission, and settlement services for all immigrants to Quebec (ibid.).[4] With substantial increases in federal funding for immigration and settlement services, per capita funding for selection and settlement is higher in Quebec than in other provinces.

Provincial Nominee Programs

Since the Quebec–Canada Accord, more limited and less remunerative agreements have been reached with each of the other provinces and the Yukon Territory. **Provincial nominee programs** that allow provinces to select applicants for permanent residence according to provincial rather than federal criteria are central aspects of devolution (ibid.). The programs have proven very popular, with the numbers of provincial nominees increasing from '477 in 1999 to 17,095 in, 2007', 7.2 per cent of all immigrants (Becklumb, 2008a: 18).

The provincial nominee programs illustrate how the localization of immigration policy in Canada is part and parcel of processes of neo-liberalization. The programs respond directly to employers' needs for workers by allowing employers to nominate applicants for permanent residence. Employers fill vacancies by recruiting migrants directly and skirt the demanding requirements for language fluency, educational attainment, and skills inherent in the points system by which applications from economic migrants are judged. In some provinces, employers may even recruit temporary foreign workers who are then allowed to apply for permanent residence under the provincial nominee program. In addition to enabling the recruitment of a foreign-born workforce tailored to employers' needs, the programs admit socially desirable immigrants who are economically self-sufficient. Despite their shared goals, each provincial nominee program operates slightly differently, as the following examples from Manitoba and Ontario demonstrate.[5]

The Manitoba provincial nominee program that is now the dominant means of recruiting immigrants, accounting for more than 70 per cent of those landed in Manitoba in 2007, is closely linked to the recruitment of temporary foreign workers (Manitoba, 2009). The province encourages temporary foreign workers to apply for permanent residence as provincial nominees after working for a Manitoba employer for at least six months (Manitoba, 2010). Under this scheme, employers specify who they want to hire, temporary foreign worker visas are processed quickly,[6] and speedy access to permanent residence ensures that employers retain workers, avoiding additional recruitment and training costs.

The racial backgrounds of the workforce recruited through the Manitoba provincial nominee program are consistent with the neo-conservative desire for a white, Christian Canada (Abu-Laban, 1998; Abu-Laban and Gabriel, 2002). Several streams of the temporary migrant program promote chain migration by requiring social links to Manitoba, resulting in higher numbers of Europeans to Manitoba than to other provinces. Germany was the second largest source country for immigrants in 2007, accounting for 13 per cent of those who landed that year (Manitoba, 2009), when approximately 1.1 per cent of all immigrants admitted to Canada originated in Germany (Citizenship and Immigration Canada, 2009).

In Ontario, Canada's largest province and the destination for more than half of all permanent residents settling in the country annually, the provincial nominee program is much smaller and more focused on highly skilled workers (Ontario, 2010). In 2009, the annual target for nominees increased from 500 to 1,000, many fewer than the number of nominees in Manitoba and only 1.7 per cent of all economic immigrants admitted to Ontario in 2008 (Citizenship and Immigration Canada, 2009). The Ontario program also allows those who invest at least $3,000,000 in Ontario and create at least five permanent, full-time jobs to recruit key professional, managerial, and skilled trades employees (ibid.). Thus, although both provincial nominee programs are employer-driven, the Ontario program only admits workers in highly skilled occupations and does not provide a path to permanent residence for temporary foreign workers.

Provincial and Municipal Involvement in Settlement Services

Increasing provincial involvement in settlement services, such as language training and job training, is heightening local variations in immigration policies.

Designed to facilitate the speedy and successful integration of newcomers, **settlement services** for all legal immigrants are a government responsibility in Canada, unlike in the US where the government only provides settlement services for refugees. The federal government funds many settlement services, but the money often flows through the provinces to local non-governmental organizations that deliver the services. The devolution has two contradictory effects. On the one hand, the federal government mandates the types of settlement services that can be funded and sets eligibility rules. Only legal landed immigrants are eligible for many settlement services, and growing numbers of temporary migrants and undocumented migrants are excluded from services. On the other hand, federal funds nominally intended for settlement services go directly into each province's general revenue stream, which the province may use as it sees fit. The diversion of these federal funds to other provincial priorities, such as health care and education, is blamed for a growing gap between funding for settlement service and demand (OCASI, 2009). The funding gap disadvantages immigrants who need settlement services to establish themselves, while favouring immigrants who can draw on their own resources to settle, that is, self-sufficient immigrants who quickly find secure positions within Canada's neo-liberal economy (Mitchell, 2004). In this funding context, settlement organizations also are increasingly hard-pressed to offer any unfunded services to temporary and undocumented migrants. The divisions between desirable, self-sufficient immigrants and those who legitimately need services to settle are often racialized, with French- or English-speaking immigrants from European origins more likely to settle independently than are many immigrants of colour from Africa and Latin America.

In Canada, municipalities are eager to influence the provision of settlement services. A recent report (Federation of Canadian Municipalities, 2009) called for predictable and equitable long-term funding for settlement services, a federal–municipal roundtable on immigration issues that would allow municipalities to contribute to planning immigration settlement policy, programs, and service delivery, and a cost-shared pilot program to support community efforts to attract and retrain more newcomers. The goal is to enhance municipal involvement in planning for successful settlement of immigrants rather than to ascertain the status of newcomers and enforce deportation and other regulations. The City of Toronto is a pioneer in this respect, the only municipality represented on federal–provincial committees and working groups concerned with settlement services and language training (Canada, 2006). The Toronto municipal government neither endorses nor seeks any responsibilities or powers for enforcing immigration legislation. To the contrary, local Toronto agencies have implemented 'don't ask, don't tell' policies that protect undocumented immigrants. Other cities and towns can be more exclusionary. For example, the municipal charter proposed recently in Hérouxville, Quebec, a town of approximately 1,300 people, explicitly bars women from being veiled (Abu-Laban and Abu-Laban, 2007). The resulting public controversy forced the municipality to rescind the proposed change in the charter.

The recent rise of provincial control over the selection of immigrants and the provision of settlement services has deep historical roots in Canada, where the federal and provincial governments jockey endlessly over shared jurisdictions (Brodie, 2002; McBride and Shields, 1997). The current round of provincial devolution that began with the Quebec–Canada Accord in 1991 can be seen as a continuation of historical trends that favour regional autonomy. Some provinces have seized the opportunity to ensure their economic futures by using the power to select migrants to respond to the labour needs of local employers. In Alberta and British Columbia, the provincial governments have allowed employers to recruit temporary foreign workers for low-skilled jobs; Manitoba has encouraged less-skilled temporary workers recruited by employers to become permanent residents; and in Ontario employers may recruit only highly skilled foreign workers.

Localization also contributes to a geographically uneven distribution of migrants. Increasing numbers who are not eligible for language courses or job training are concentrated in provinces that admit large numbers of temporary migrants, while a continuing emphasis on permanent residents in other provinces such as Ontario and Quebec exacerbates the impacts of funding shortages for settlement services (Alboim, 2009). The increasing geographical variability in eligibility for settlement services and the variability in funding that results from localization have the potential to undermine popular support for government involvement in settlement services, a desirable outcome from a neo-liberal perspective.

LOCALIZATION OF IMMIGRATION POLICIES IN THE UNITED STATES

Throughout the twentieth century, the US federal government held exclusive power over the admission of migrants into its territory and society. Some states have intermittently restricted immigrants' access to public benefits and employment, and/or introduced penalties for employers who knowingly hire undocumented immigrants. By and large, federal authority on immigration remained unchallenged, but this situation has changed during the past decade. In 2007, US state legislatures alone considered 1,000 pieces of legislation regulating immigrants and immigration (Rodriguez et al., 2007). In addition, local ordinances barring undocumented immigrants from working, obtaining housing, or using public benefits have proliferated across American counties and municipalities. Some states and municipalities have also introduced measures that expand immigrants' access to services and make them feel welcome irrespective of their legal status.

Devolution of Federal Responsibility for Immigration Control

Currently, the tendency with the greatest momentum in terms of promoting local immigration control is the federal **287(g) program**, which originated in the

1996 Illegal Immigration Reform and Immigrant Responsibility Act (IIRAIRA). It was designed to enrol states and localities in immigration law enforcement by authorizing local law enforcement officials to arrest and detain undocumented immigrants (US Immigration and Customs Enforcement, 2006). As of 2009, 67 state and local agencies have participated in this program, from Los Angeles County, California, to Mecklenburg County, North Carolina, to Morristown, New Jersey. Devolution has effectively 'pushed the border inward' from national to local scales, making immigration status an increasingly salient local issue (Coleman, 2007; Varsanyi, 2008, cited in Walker and Leitner, 2011).

Devolving and Restricting Immigrant Access to Public Benefits

The year 1996 also saw the passage of the Personal Responsibility and Work Opportunity Act (PRWORA), commonly referred to as the 1996 Federal Welfare Reform Act. This Act barred documented and undocumented immigrants from accessing public benefits such as food stamps and Supplemental Security Income (SSI)[7] and constituted a sharp break from past practices in the United States as well as in other countries of immigration. The provisions of the PRWORA followed the 1994 passage of Proposition 187 in California, which denied undocumented immigrants access to social, educational, and health services.[8] Since then other states, such as Arizona, Colorado, and Georgia, have passed similar laws that require proof of citizenship for voting and access to a range of public services, and deny in-state tuition to those without documents (for details, see Rodriguez et al., 2007). It is important to point out, however, that the majority of the proposals to restrict immigrants' access to public services that were considered by state legislatures (e.g., Indiana, Oklahoma, Virginia) failed to pass or were vetoed, and some that had been enacted were halted (for further details, see Broder, 2007).

Local Anti-immigration Measures and Ordinances

The presence of **day labourers** gathering in public squares, on street corners, and in parking lots seeking work (primarily in construction and landscaping) has become an issue of contention since the early 1990s, brilliantly chronicled in the documentary *Farmingville*, set in a suburban community in the New York metropolitan area. Many suburban and exurban localities across the United States have been raising concerns about the presence (or the prospect of the presence) of day labourers, and of undocumented immigrants more generally. Some have responded by opening worker centres and hiring halls to accommodate day labourers, while others have taken hostile actions against immigrants. Local governments have introduced a variety of local ordinances. Anti-socialization ordinances are designed to prevent individual workers from congregating on public streets to solicit work (e.g., Glendale, California, in 2004). Other ordinances make it illegal for landlords to rent to persons without proper documentation. Hazelton and Bridgeport, Pennsylvania, and Valley Park, Missouri, have passed an 'Illegal Immigration Relief Act Ordinance' that seeks to penalize landlords and employers who rent to or hire undocumented immigrants. In May 2008, the Suffolk County, NY, council passed an ordinance requiring that all 17,000 licensed contractors in the county prove their employees' working status. Other measures include English-only ordinances that make English the 'official' language of a particular city or town, e.g., Bridgeport, Pennsylvania, and Nashville, Tennessee. Still other localities have introduced ordinances that would make it illegal to display a foreign flag—unless an American flag is flown above it (e.g., Pahrump, Nevada, in 2006). Many of the measures and ordinances launched by county and municipal governments have been challenged successfully by local citizens and civil rights organizations on the streets and in court on the grounds that they interfere with federal law and violate due process (*ACLU Newsletter*, 2008).

Improving Access to Public Services and Local Pro-Immigration Ordinances

Not all states and localities have sought to exclude immigrants or prevent them from accessing public benefits. Indeed, many states and some counties, with long histories of immigrant settlement and large immigrant populations, have expanded access to services. For example, the states of Illinois and New York extended health coverage to children of all income scales regardless of immigration status, and some California counties have continued to expand access to health coverage for immigrant children (Broder, 2007). According to Broder (ibid., 15):

> more than half of the states currently spend their own funds to provide services to at least some immigrants who are ineligible for federal services. A growing number, recognizing that providing preventive care is an effective public health strategy, have eliminated immigration status as a prerequisite for public health coverage in programs serving children and/or pregnant women.

In addition, local governments have advanced policies that protect immigrants against discrimination and promote their integration into the larger society (see Box 1.1). According to Walker and Leitner (2011), nearly 100 cities and counties across the US have proposed or established sanctuary ordinances. These ordinances have their roots in the **Sanctuary Movement** of the 1980s, when churches provided refuge to Central Americans fleeing civil wars in their home countries. Contemporary sanctuary ordinances protect immigrants by declaring that local authorities will not participate in the federal 287(g) program and thus will not check residents' immigration status, will extend local voting rights to non-citizens to promote their integration, and will commit the local government to supporting the rights of undocumented migrants (ibid.). One such example is the sanctuary ordinance of the city and county of San Francisco, first passed in 1989 and reaffirmed in 2007:

This ordinance prohibits city employees from helping Immigration and Customs Enforcement (ICE) with immigration investigations or arrests unless such help is required by federal or state law or a warrant. . . . In 2007, Mayor Gavin Newsom reaffirmed San Francisco's commitment to immigrant communities by issuing an Executive Order that called on City departments to develop a protocol and training on the Sanctuary Ordinance. (www.sfgov.org/site/sanctuary_index.asp?id=80999).

BOX 1.1   Pro-Immigration Localities

In the United States and Canada, some states and municipalities have enacted legislation and regulations that promote the rights of all migrants. In the following two examples, we high-light the positive aspects of localization, discussing how public agencies in the City of Toronto in Ontario and the municipality of Takoma Park in Maryland, a suburb of Washington, DC, have acted to guarantee rights to undocumented migrants. When evaluating these examples, it is important to remember the variegated geographies of municipal responses in both countries. While our examples highlight the possibilities for pro-immigration policies at the local level, we recognize that municipalities in Canada and the US also have implemented anti-immigration bylaws and ordinances.

In the City of Toronto there has been an active 'don't ask, don't tell' campaign calling on mu-nicipal officials and all local agencies to explicitly forbid requests for information about immi-gration status. The city has not yet implemented this policy; however, in accord with provincial legislation that mandates that every child between the ages of 6 and 16 is required to attend elementary or secondary school and 'shall not be refused admission because the person or the person's parent or guardian is unlawfully in Canada' (Education Act, s. 49.1), the Toronto District School Board does not ask for, report on, or share information about whether a stu-dent is undocumented (CBC, 2007). The Toronto Police also have stated that police services should be available to all members of the community regardless of their immigration status. The Toronto Police now specify that victims and witnesses of crime should not be asked their immigration status unless there are 'bona fide reasons' (Toronto Police, 2009). Although the implementation of these policies is still contested, with proponents of immigration publiciz-ing cases where the policies appear to have been contravened, the pro-immigration policies themselves contribute to an inclusionary climate for all migrants.

In the US, where many municipalities have introduced exclusionary immigration policies, other municipalities have sought to ensure the rights of undocumented migrants through sanctuary policies. Takoma Park, an inner-ring suburb of Washington, has a sanctuary policy similar to that in San Francisco. This ordinance states that 'Takoma Park has a long and strong tradition of embracing and valuing diversity and respect[ing] the civil and human rights of all residents regardless of their race, sex, ethnicity, sexual orientation, national origin, or immigration status' (Takoma Park, 2008: 1). In 2008 the city reaffirmed and amended this ordinance, stipulating that no city employee (including local law enforcement officers) shall assist the US Immigration and Customs Enforcement in 'the investigation or arrest of any persons for civil or criminal violation of the immigration and nationality laws of the United States' (ibid., 2). In other words, it prevents any city employees from participating in the federal 287(g) program and thus helping to enforce federal immigration laws. In addition, the ordinance prohibits inquiries into the citizenship status of individuals, for example, when registering to vote in local elections or when stopped by a police officer. Finally, it bans discrimination based on citizenship and immigration status: 'No agent, officer or employee of the City, in the performance of official duties, shall discriminate against any person on the basis of citizenship or immigration status' (ibid.).

Geographies of Localization

Walker and Leitner (2011) have identified distinct regional and intra-metropolitan variations in the localization of immigration policies in the US. Anti-immigration policies are highly concentrated in the South. 'In contrast, the West is the only region where a majority of municipalities have adopted pro-immigration policies (52 per cent). 57 per cent of municipalities in the Midwest and 74 per cent in the Northeast have introduced local immigration ordinances designed to keep out or deter the settlement of immigrants' (ibid., 6). There are also distinct intra-metropolitan variations, with suburban and rural municipalities being more likely to introduce anti-immigration ordinances, whereas central cities are more likely to adopt sanctuary policies (ibid.).

Several explanations have been proposed for the emergence of local immigration policies. Some have suggested they are a response to rapid increases in the number of migrants in a local community and local residents' fear of being overrun (Esbenshade, 2007; Esbenshade and Obzurt, 2008). Fears that the influx of immigrants will negatively affect property values also may lead local residents to support anti-immigration ordinances (Light, 2006). Others have highlighted socio-demographic characteristics (Wilkes et al., 2008) as well as political persuasion (Ramakrishnan and Wong, 2007) to explain residents' attitudes towards immigration and immigrants and the local variation in immigration policies.

In their analysis of local immigration policies in the US, Walker and Leitner (2011) find that areas with a high percentage of Republican voters and owner-occupied housing are more likely to introduce exclusionary policies, whereas places with better-educated populations or with high unemployment are less likely to pass exclusionary policies. In addition, based on a textual analysis of local ordinances and supporting documents, they find that local communities that value and respect cultural and racial diversity in both their place of residence and the national community are more likely to reject anti-immigration ordinances and/or favour pro-immigration ordinances. By contrast, local communities that value cultural homogeneity are more likely to support anti-immigration ordinances. This suggests that the anti- or pro-immigration ordinances are rooted in different conceptions of community (national as well as local) and of place.

CONCLUSION

In this chapter, we have identified common tendencies but also differences between immigration policies in Canada and the US. Processes of neo-liberalization and the ascendency of a neo-conservative political ideology have left their imprint on immigration policies in both countries, but have been mediated by the distinct histories of federalism, divisions of power and authority among different tiers of the state, and different conceptions of citizenship rights and of the national community. We have focused on one common tendency in particular, the localization of immigration policies within both decentralized federal systems, showing how it takes a different form in each nation-state.

In Canada, localization occurs primarily but not exclusively at the provincial scale. Canadian provinces have been given more power in the area of immigration than have American states. The recent willingness of the Canadian federal government to cede more of its power over the admission and selection of immigrants is consistent with its neo-liberal commitment to 'build local solutions to local problems' (Brodie, 2002: 391). With aging populations, it is not surprising that provincial and municipal officials have seized this opportunity to control immigrant selection and incorporation to try to ensure that immigrants who settle in their jurisdictions will be economically successful. Localization in Canada also occurs at the provincial scale because Canadian municipalities are creatures of the provinces. There is much less direct federal involvement in Canadian cities than in the US, where the federal government directly funds urban services such as mass transit and neighbourhood improvements.

In contrast to Canadian provinces, American states have less power and authority to control immigration and facilitate immigrant incorporation. In the mid-1990s, the US federal government introduced the 287(g) program, which encourages lower tiers of government to become involved in immigration law enforcement at the local scale; however, this program does not constitute a devolution of power but a devolution of responsibilities to lower tiers of the state. The limited power and authority of state and local governments in the area of immigration in the US has intermittently resulted in intense struggles with the federal government over joint jurisdictions. We suggest the contemporary period is another such moment when state and local governments in the US are challenging the power and authority of the federal government in the area of immigration. State governments (most recently the state of Arizona) have introduced their own immigration laws, and local governments have introduced a range of ordinances, some designed to deter and keep out unwanted immigrants, others intended to expand immigrants' access to services and make them feel welcome irrespective of their legal status.

Our analysis has shown how local jurisdictions at the state, provincial, and municipal levels can resist neo-liberal and neo-conservative policy tendencies. In Canada, such resistance to neo-liberalism is evident in the continued provision of social services and social welfare by all levels of government (Brodie, 2002; Adams, 2004), although welfare state provisions do not always achieve equity, and Canada, like the US, is an increasingly unequal society despite government involvement in the welfare state. Both provinces and municipalities seek to plan and provide settlement services, and municipal governments often oppose efforts to devolve responsibility for enforcing immigration law to local jurisdictions. Some municipal and state governments in the US, and some municipal governments in Canada, have resisted federal efforts to download responsibilities for enforcing immigration policies, sometimes going so far as to enact sanctuary laws. The implementation of such sanctuary laws also suggests that the local is not necessarily parochial; it offers important and sometimes overlooked opportunities for progressive immigration politics. Immigrant organizations and their allies have been successful in challenging local anti-immigration ordinances in the courts and on the streets.

Despite the optimism associated with such resistance, exclusionary local immigration policies have been on the advance. For example, in May 2010, the state of Arizona passed a new immigration law that makes failure to carry immigration documents a crime and provides broad powers to the police to detain, prosecute, and deport anyone suspected to be present in the country without documents. Opponents have branded the law as an open invitation for racial profiling, irrespective of the individual's citizenship status. In Canada, exclusionary local policies have been rare, with the municipal charter proposed in Hérouxville as the best-known and most controversial example; however, renewed debates about veiling are associated with the rise of exclusionary policies at the provincial level (Abu-Laban and Abu-Laban, 2007). The unanimous February 2011 vote in Quebec's National Assembly to ban from the province's legislative buildings the kirpan—the ceremonial dagger worn by Sikh men—suggests the extent to which provinces are sometimes intent on pursuing such policies.

The increasingly divided landscape of state and local immigration policies in the US to some extent mirrors opposing viewpoints in national immigration policy debates, between proponents of more restrictive immigration policies and access to citizenship rights and proponents of open borders and easier access to citizenship. Underlying these opposing viewpoints are very different understandings of the American nation, who belongs and does not belong, and who has a place in the American community. An inclusive conception that celebrates and values cultural diversity and an open, constantly emerging nation, community, and place exists alongside and frequently in fierce competition with an exclusive conception that values and appreciates cultural/racial homogeneity—a white America—and a clear bounding of place, community, and nation (Walker and Leitner, 2011). Similar divisions are evident in Canada. Exclusionary policies are debated in Quebec, where the cultural diversity associated with contemporary immigration must be reconciled with ongoing efforts to preserve and promote the French language and culture. Simultaneously, in Quebec and other parts of Canada, inclusive policies at the municipal and provincial levels are justified on the grounds that Canada is a multicultural nation built by immigrants (Abu-Laban and Abu-Laban, 2007).

An important area for future comparative research is the impact of the localization of immigration policies on immigrants and the communities involved. To date no systematic study of these impacts exists. Some journalistic accounts and anecdotal evidence suggest that exclusionary ordinances have spurred the out-migration of documented and undocumented immigrants from certain counties and municipalities, and have created homelessness and other hardships within the immigrant community (Belson and Capuzzo, 2007). On the other hand, exclusionary policies also have catalyzed the mobilization of immigrants. For example, California's Proposition 187, which sought to deny a range of state services to undocumented immigrants, resulted in increased mobilization and electoral participation by Latinos (Garcia Bedolla, 2005). A research program comparing and contrasting local responses within these two similar but distinct national contexts would enhance our understanding of the divided landscapes that immigrants face and the future prospects for them and the communities in which they reside.

QUESTIONS FOR CRITICAL THOUGHT

1. How have neo-conservatism and neo-liberalism influenced the localization of immigration policies in Canada and the United States?

2. What factors contribute to the divergent approaches to immigration by provincial, state, and municipal governments in Canada and the United States?

3. Do you agree with the authors that a progressive immigration politics is possible at the local level? Illustrate your argument with examples.

SUGGESTED READINGS

1. Abu-Laban, Yasmeen, and Christina Gabriel. 2002. *Selling Diversity: Immigration, Multiculturalism, Employment Equity and Globalization.* Peterborough, Ont.: Broadview Press. This book examines recent developments in immigration policy, multiculturalism policy, and employment equity in Canada that have emphasized efficiency, markets, and the interest of business.

2. Coleman, Mathew. 2009. 'What Counts as the Politics and Practice of Security, and Where? Devolution and Immigrant Insecurity after 9/11', *Annals, Association of American Geographers* 99: 904–13. Coleman examines the devolution of immigration enforcement in the United States after 11 September 2001, which has intensified local risks for undocumented immigrants.

3. Fudge, Judy, and Fiona MacPhail. 2009. ~~The Temporary Foreign Worker~~ *Program in Canada: Low-skilled Workers as an Extreme Form of Flexible Labour.* Melbourne: University of Melbourne, Centre for Employment and Labour Relations Law, Working Paper no. 45. The authors focus on the legal regime that regulates the entry and exit of temporary foreign workers and these workers' rights and terms and conditions of employment while in Canada.

4. Varsanyi, Monica W. 2008. 'Rescaling the "Alien", Rescaling Personhood: Neoliberalism, Immigration, and the State', *Annals, Association of American Geographers* 98: 877–96. This article discusses the contemporary constitution of neo-liberal subjects via the devolution of immigration powers to state and local governments by the US federal government.

NOTES

1. In the US, the federal government does provide settlement services for refugees but other immigrants are responsible for finding and funding their own settlement services.
2. At the beginning of the twentieth century, the Canadian government continued to authorize the Canadian Pacific Railway to act as its agent in recruiting, transporting, and settling immigrants to Canada.
3. Temporary migrants have the right to reside in Canada or the US legally for a fixed period, after which they are expected to leave the country. Undocumented migrants are all those people who are living in Canada and the United States without legal rights to residence. They also are known as illegal migrants, unauthorized immigrants, and people without status.
4. Quebec's selection rules must accord with those set by the Parliament of Canada concerning criminality, security, and health (Becklumb, 2008b).
5. The Manitoba and Ontario provincial nominee programs illustrate the extremes of current programs in terms of their size, emphasis on skilled workers, and provision of a path to permanence for temporary foreign workers. Other provincial nominee programs fall between these two extremes.
6. Processing times vary, but there is a reported backlog of more than 900,000 applications for permanent residence from people living outside Canada.
7. SSI is a federal income supplement program designed to help aged, blind, and disabled people with little or no income. For further details on the impact of PRWORA, see Levinson (2002).
8. Proposition 187 was passed by voters on 8 November 1994, but implementation of its provisions was challenged in court. After years of legal and political debate the proposition has been voided.

REFERENCES

1. Abu-Laban, Yasmeen. 1998. 'Welcome/ Stay Out: The Contradiction of Canadian Integration and Immigration Policies at the Millennium', *Canadian Ethnic Studies* 30: 190–211.
2. ——— and Baha Abu-Laban. 2007. 'Reasonable Accommodation in a Global Village', *Policy Options* (Sept): 28–33.
3. ——— and Christina Gabriel. 2002. *Selling Diversity: Immigration, Multiculturalism,* *Employment Equity and Globalization.* Peterborough, Ont.: Broadview Press.
4. *ACLU Newsletter.* 2008. 'Civil Rights Groups Urge Appellate Court to Overturn Arizona Employer Sanctions Law', 12 June. At: www. aclu.org/cpredirect/35641. (24 Feb. 2010)
5. Adams, Michael. 2004. *Fire and Ice: The United States, Canada and the Myth of Converging Values.* Toronto: Penguin Canada.

6. Alboim, Naomi. 2009. *Adjusting the Balance: Fixing Canada's Economic Immigration Policies*. At: www.maytree.com/wp-content/uploads/2009/07/adjustingthebalance-final.pdf. (24 Feb. 2010)

7. Becklumb, Penny. 2008a. *Canada's Immigration Program*. Ottawa: Parliament of Canada, BP–190E.

8. ———. 2008b. *Immigration: The Canada–Quebec Accord*. Ottawa: Parliament of Canada, BP–252E.

9. Belson, Ken, and Jill Cappuzo. 2007. 'Towns Rethink Laws against Illegal Immigrants', *New York Times*, 26 Sept.

10. Bouchard, Gérard, and Charles Taylor. 2008. *Building the Future: A Time for Reconciliation*. Québec City: Commission de consultation sur les pratiques d'accommodement reliées aux différences culturelles.

11. Broder, Tanya. 2007. 'State and Local Policies on Immigrant Access to Services. Promoting Integration or Isolation?', National Immigration Law Center, May. At: www.nilc.org/ immspbs/sf_benefits/statelocalimmpolicies06-07_2007-05-24.pdf. (24 Feb. 2010)

12. Brodie, Janine. 2002. 'Citizenship and Solidarity: Reflections on the Canadian Way', *Citizenship Studies* 6, 4: 377–94.

13. Brown, Wendy. 2006. 'American Nightmare: Neoliberalism, Neoconservatism, and De-democratization', *Political Theory* 34: 690–714.

14. Canada. 2006. Canada–Ontario Immigration Agreement. At: www.cic.gc.ca/english/department/laws-policy/agreements/ontario/can-ont-toronto-mou.asp. (24 Feb. 2010)

15. Canadian Council for Refugees. 2007. *Safe Third Country*, Brief to the Standing Committee on Citizenship and Immigration. At: www.ccrweb.ca/SafethirdbriefSC.pdf. (26 Feb. 2010)

16. Castles, Stephen, and Mark Miller. 2009. *The Age of Migration: International Population Movements in the Modern World*. Houndsmills, UK: Palgrave Macmillan.

17. CBC News. 2007. 'Toronto School Board Pushes "Don't Ask, Don't Tell" Policy on Immigration Status'. At: www.cbc.ca/canada/toronto/story/2007/05/03/to-school-board-immigrant.html#ixzz0gell351W. (26 Feb. 2010)

18. Citizenship and Immigration Canada. 2009. *Facts and Figures: Immigration Overview Permanent and Temporary Residents*. Ottawa: Research and Evaluation Branch, Citizenship and Immigration Canada. At: www.cic.gc.ca/english/pdf/ research-stats/facts2008.pdf. (21 Jan. 2010)

19. Coleman, Mathew. 2007. 'Immigration Geopolitics beyond the Mexico–US Border', *Antipode* 39: 54–76.

20. ———. 2009. 'What Counts as the Politics and Practice of Security, and Where? Devolution and Immigrant Insecurity after 9/11', *Annals, Association of American Geographers* 99: 904–13.

21. Education Act (Ontario, s. 49.1). 1993. 'Persons Unlawfully in Canada'. Toronto: Province of Ontario.

22. Elgersma, Sandra. 2007. *Temporary Foreign Workers*. Ottawa: Library of Parliament, PRB 07–11E.

23. Esbenshade, Jill. 2007. *Division and Dislocation: Regulating Immigration through Local Housing Ordinances*. Washington: Immigration Policy Center, American Immigration Law Foundation.

24. ——— and Barbara Obzurt. 2008. 'Local Immigration Regulation: A Problematic Trend in Public Policy', *Harvard Journal of Hispanic Policy* 20: 33–47.

25. Federation of Canadian Municipalities. 2009. *Immigration and Diversity in Canadian Cities and Communities: Quality of Life in Canadian Communities*, Theme Report No. 5. Ottawa: Federation of Canadian Municipalities. At: www.fcm.ca/CMFiles/QofL%20Report%205%20En1JPA-3192009-2422.pdf. (26 Feb. 2010)

26. Freeman, Gary P. 2004. 'Immigrant Incorporation in Western Democracies', *International Migration Review* 38: 945–69.

27. Fudge, Judy, and Fiona MacPhail. 2009. *The Temporary Foreign Worker Program in Canada: Low-skilled Workers as an Extreme Form of Flexible Labour*. Melbourne: University of Melbourne, Centre for Employment and Labour Relations Law, Working Paper no. 45.

28. Garcia Bedolla, Lisa. 2005. *Fluid Borders: Latino Power, Identity, and Politics in Los Angeles*. Berkeley: University of California Press.

29. Hawkins, Freda. 1988. *Canada and Immigration: Public Policy and Public Concern*. Montreal and Kingston: McGill-Queen's University Press.

30. Hiemstra, Nancy. 2007. 'Immigrant "Illegality" as Neoliberal Governance in Leadville, CO', paper presented at Association of American Geographers annual meeting, San Francisco, Apr.
31. Hier, Sean P., and Joshua L. Greenberg. 2002. 'Constructing a Discursive Crisis: Risk, Problematization and Illegal Chinese in Canada', *Ethnic and Racial Studies* 25, 3: 490–513.
32. Kobayashi, Audrey, and Valerie Preston. 2007. 'Transnationalism through the Life Course: Hong Kong Immigrants in Canada', *Asia Pacific Viewpoint* 48, 2: 151–67.
33. Leitner, Helga, and Patricia Ehrkamp. 2006. 'Transnationalism and Migrants' Imaginings of Citizenship', *Environment and Planning A* 38, 9: 1615–32.
34. ———, Eric Sheppard, Kristin Sziarto, and Anant Maringanti. 2007. 'Contesting Urban Futures—Decentering Neoliberalism', in Helga Leitner, Jamie Peck, and Eric Sheppard, eds, *Contesting Neoliberalism—Urban Frontiers*. New York: Guildford Press, 1–25.
35. Levinson, Amanda. 2002. *Immigrants and Welfare Use*. Washington: Migration Policy Institute, Aug. At: www.migration information.org/USFocus/display.cfm?ID= 45. (24 Feb. 2010)
36. Light, Ivan. 2006. *Deflecting Immigration: Networks, Markets, and Regulation in Los Angeles*. New York: Russell Sage.
37. McBride, Stephen, and John Shields. 1997. *Dismantling a Nation: The Transition to Corporate Rule in Canada*. Halifax: Fernwood.
38. Manitoba. 2009. *Manitoba Immigration Facts, 2008 Statistical Report*. Winnipeg: Manitoba Labour and Immigration.
39. ———. 2010. *Manitoba Provincial Nominee Program*. Winnipeg: Manitoba Labour and Immigration. At: www2. immigratemanitoba.com/browse/howto immigrate/pnp/index.html. (26 Feb. 2010)
40. Massey, Douglas. 1999. 'International Migration at the Dawn of the Twenty-First Century: The Role of the State', *Population and Development Review* 25: 303–23.
41. Mitchell, Katharyne. 2004. *Crossing the Neoliberal Line: Pacific Rim Migration and the Metropolis*. Philadelphia: Temple University Press.
42. Mountz, Alison. 2004. 'Embodying the Nation-State: Canada's Response to Human Smuggling', *Antipode* 23: 323–45.

43. Ontario. 2010. *About Opportunities Ontario: Provincial Nominee Program*. At: www. ontarioimmigration.ca/english/PNPabout. asp. (26 Feb. 2010)
44. Ontario Coalition of Agencies Serving Immigrants OCASI. 2009. *Canada–Ontario Immigration Agreement (COIA): Crafting the Vision for the Sector*, OCASI Discussion Paper. Toronto: OCASI. At: www.ocasi.org/index. php?qid=1005. (26 Feb. 2010)
45. Papademetriou, Demetrios G., Doris Meissner, Marc R. Rosenblum, and Madeleine Sumption. 2009. *Aligning Temporary Immigration Visas with US Labor Market Needs: The Case for a New System of Provisional Visas*. Washington: Migration Policy Institute.
46. Passel, Jeffrey, and D'Vera Cohn. 2009. 'A Portrait of Unauthorized Immigrants in the United States', Report. Pew Hispanic Center, 14 Apr.
47. Peck, Jamie, and Adam Tickell. 2002. 'Neoliberalizing Space', *Antipode* 34, 3: 380–404.
48. Ramakrishnan, S. Karthick, and Tom Wong. 2007. *Immigration Policies Go Local: The Varying Responses of Local Governments to Low-Skilled and Undocumented Immigration*. Working Paper. Berkeley: University of California, Berkeley, School of Law. At: www.law.berkeley.edu/files/Ramakrishnan Wongpaperfinal.pdf. (24 Feb. 2010)
49. Rodriguez, Cristina, Chishti Muzaffar, and Kimberly Nortman. 2007. *Testing the Limits: A Framework for Assessing the Legality of State and Local Immigration Measures*. Washington: National Center on Immigrant Integration Policy, Migration Policy Institute.
50. San Francisco, City and County. At: www. sfgov.org/site/sanctuary_index.asp?id= 80999. (24 Feb. 2010)
51. Stalker, Peter. 2008. *The No-Nonsense Guide to International Migration*. Toronto: New Internationalist.
52. Stasiulis, Daiva, and Radha Jhappan. 1995. 'The Fractious Politics of a Settler Society: Canada', in Daiva K. Stasiulis and Nira Yuval-Davis, eds, *Unsettling Settler Societies: Articulations of Gender, Race, Ethnicity and Class*. London: Sage, 95–131.
53. Takoma Park, City of. 2008. 'Ordinance No. 2008-7. An Ordinance Making Technical Amendments to the City of Takoma Park's Immigration Sanctuary Law'.

54. Toronto Police. 2009. *Victims and Witnesses without Legal Status*. At: www.torontopolice.on.ca/publications/files/victims_and_witnesses_wthout_legal_status.pdf. (26 Feb. 2010)

55. US Congress. 1996. Public Law 104–193, Personal Responsibility and Work Opportunity Reconciliation Act of 1996.

56. US Immigration and Customs Enforcement. 2006. Section 287(g) Immigration and Nationality Act. Fact Sheet, 15 Aug., Office of Public Affairs, US Department of Homeland Security. At: www.ice.gov/partners/287g/Section287_g.htm. (11 May 2010)

57. US Senate. 2006. 'Examining the Value of a Skills-based Point System', Hearing of the Committee on Health, Education, Labor, and Pensions. Washington: US Senate, One Hundred Ninth Congress, 14 Sept. At: frwebgate.access.gpo.gov/cgi-bin/getdoc.cgi?dbname=109_senate_hearings&docid=f:30005.pdf. (24 Feb. 2010)

58. Varsanyi, Monica. 2008. 'Rescaling the "Alien", Rescaling Personhood: Neoliberalism, Immigration, and the State', *Annals, Association of American Geographers* 98: 877–96.

59. Walker, Kyle, and Helga Leitner. 2011 (forthcoming). 'The Variegated Landscape of Local Immigration Policies in the United States', *Urban Geography*.

60. Wilkes, Rima, Neil Guppy, and Lily Farris. 2008. '"No Thanks, We're Full": Individual Characteristics, National Context, and Changing Attitudes toward Immigration', *International Migration Review* 42: 302–29.

CHAPTER 2

IMMIGRATION TRENDS IN THE UNITED STATES AND CANADA: A HISTORICAL PERSPECTIVE

Dirk Hoerder and Scott Walker

INTRODUCTION

By the early twenty-first century, years of sustained immigration from Mexico and Latin American countries were changing the United States. As many as 12 million undocumented migrants lived in the country. US media barraged the public with stories of the 'Border War' to keep out Spanish-speaking immigrants arriving in large numbers. In Canada, after decades of a policy of multiculturalism, a nationalist revival criticized allegedly unintegrated immigrants. Canadians, proud of the longest undefended border in the world, saw their border with the US policed from the south, much as Mexico saw its northern border turned into a fortification. In Europe, internal borders were being dismantled; in North America, they began to resemble the Iron Curtain that once had divided East and West.

This chapter looks back at US and Canadian migration and immigration, seeking to put the ongoing changes into perspective. We address:

- national policies of both the US and Canada;
- demographic characteristics of the new arrivals; and
- evolving reactions of the citizens of these two countries.

Few of the present migrations, and the debates or fears about them, lack precedent. Although in both countries the total numbers of new arrivals have indeed increased, the percentage of immigrants or newcomers is on par with that of the early twentieth century. ('Immigrants' is standard US terminology referring to people who stay. '**Newcomers**' is the Canadian word.)

THE ROOTS OF THE AMERICAN AND CANADIAN PEOPLES: A HISTORY OF IMMIGRATION

The first 'immigrants' to the Americas were the Aboriginal peoples, who came by land across the Bering Strait over several centuries, and quite possibly by water, too. However they arrived, by 10,000 years ago these peoples populated the width and breadth of North America and had reached the tip of South America. In some cases they were organized into relatively densely populated city-states; other groups lived in smaller, sedentary communities; and in many instances they thrived in small semi-nomadic groups. When European explorers, soon followed by settlers, came to the western hemisphere beginning at the end of the fifteenth century, they brought with them their own assumptions regarding the peoples they encountered and the vast territories they discovered. Because the Native peoples' societies were unlike their own and because they were scattered across such a wide territory and lived by a different ethos and with different technologies, the European concept of *terra nullius*—uninhabited land—was applied to all the new discoveries. Consequently, the land was claimed in the name of European powers and monarchs. In effect, over a period of centuries as European discovery and settlement spanned the continent, what had been Aboriginal homelands ceased, in large measure, to be so. The story of a second immigration, first from Europe and then from throughout the world, had begun, and that is our focus in this chapter and in this book.

Immigration to what would become both the United States and Canada was strongly influenced by European colonial priorities, starting in the sixteenth century. Both governments—American and British—began to discuss policies and to pass legislation in the late eighteenth century. The nineteenth century was marked by a long period of open access. Entry papers—what is now called the *passport* system—were invented in the decades between the French Commune in 1871 and World War I, when fear of radicalism reached unprecedented heights (Torpey, 2000).

Early immigration to the colonies that would become the US had involved English, Dutch, Scots, Germans, Irish, Swedes, and others of several religious confessions. To the north, early immigration to New France had come mainly from one region of France, and was Catholic. After the British Empire annexed Acadia and renamed it Nova Scotia in 1713, Scottish and continental European Protestants came. The next annexation, in 1763, added Quebec to the British Empire, with a Catholic French-language population under a French legal system. Mainly for reasons of cost, but also in an enlightened mode, the imperial government decided to permit the French Canadians to live according to their own history and culture. Thus bilingual and bicultural British North America—what would become Canada in 1867—came into being.

With the independence of the United States achieved in the Peace of Paris of 1783 following the American Revolution, a south–north mass migration of United Empire Loyalists (called anti-revolutionary Tories in the US) brought perhaps 46,000 US refugees to Canada. Included among them were those of many European cultural backgrounds, along with free and enslaved blacks. In 1789, the French Revolution began in Europe. In reaction to both, the first immigration and exclusion legislation was passed. In British North America, an Act Respecting Aliens, 1794, required a check of the US refugees' loyalty to the British Crown. In the US, a conservative government fearful of French revolutionary ideas used the Alien and Sedition Act of 1798 to exclude migrants who might criticize the administration. These measures were soon abolished and in both countries immigration remained unrestricted.

The popular myth sees the nineteenth century as a time of almost exclusively agrarian migration to settle white farmers in the West. Native peoples were expelled by war or, in Canada, by treaties. Actually, in the mid-nineteenth century, only one-third of the immigrants to the US were rural settlers. All others went to cities and towns. In Canada, a far higher percentage continued to rural regions since settlement occurred later. While US settlement in the West ended in the 1890s, that of Canada continued into the 1930s. (Burnet with Palmer, 1988; Takaki 1993; Daniels, 2002; Hoerder, 1999).

In the 1880s, open access and legislative developments were similar in the two countries. But the composition of migrants differed, and so did the timing of concern about their racial and cultural backgrounds. Starting in the 1840s, US nativists expressed anti-immigrant sentiments. But this became an issue for the Canadian federal government only in the 1880s and 1890s. In the nineteenth century, the responsibility for immigration shifted from states or provinces to the federal governments. In Washington, the process extended from 1864 to 1891; in Canada, the Immigration Act of 1869 vested the power in the Ottawa government. Both governments were concerned with immigrants becoming public charges, the inscription on the pedestal of the US Statue of Liberty notwithstanding:

> . . . Give me your tired, your poor,
> Your huddled masses yearning to breathe free,
> The wretched refuse of your teeming shore.
> Send these, the homeless, tempest-tossed to me,
> I lift my lamp beside the golden door! (Emma Lazarus, 1883)

In the 1870s, when national discourses focused on settling the frontier/the prairies, both countries also needed workers for the building of a nationwide infrastructure, especially the railroads. In response, along the Pacific coast, men from the Canton province of China began to arrive from the late 1840s. On the Atlantic coast, workers arrived from rural Eastern and Southern Europe, starting in the mid-1880s. All of these groups were considered racially inferior: Italians and Slavs as 'dark, olive, and swarthy', 'Orientals' as the 'yellow race' (see Box 2.1). The majority of these transatlantic and transpacific migrants were illiterate. The work for which they were needed did not require literacy. These mass moves were complemented by a small migration from Mexico to the US. Of the racially unwanted, the Chinese were singled out; Japanese also came, as did Filipinos, Punjabis, and Koreans in small numbers. Although numbering only around

BOX 2.1   Two Immigrants

In a small village in 1880s industrializing northern Italy clear borderlines separated the poor from the rich and the village from nearby urban worlds. They forced labouring people to move in extended and multiple migratory spaces. At a young age, girls were sent to work in nearby communities' silk factories, boys migrated with their fathers as construction workers further afar. Young men went to France for seasonal or multi-annual labour and adult men, single or married, travelled to Missouri iron mines. Age-group-specific, and intergenerational migrations thus extended in several directions. The wives of some miners followed to run boarding houses, doing commodified household labour for many men. Rosa, who arrived in the Missouri community one evening, had to cook coffee for the boarders the next morning—coffee at home was for the rich only. Then she had to shop with a German farming couple, pick up mail from an English-speaking postmaster, and by noon had to have ready a meal that was palatable to men from many Italian regions, not only from her Lombardy. Her route was an established, well-travelled one; adjustment immediately after arrival required high adaptive skills. A secondary migration would bring her to Chicago after separation—in Italy unthinkable—from her abusive husband.

Source: Marie Hall Ets, ed., *Rosa: The Life of an Italian Immigrant* (Minneapolis: University of Minnesota Press, 1970), 76–88, 120–2.

From another village, in China's Guangdong province, Chan Sam left in 1913 for a job in an internationalized labour market segment open to 'Chinamen' in Vancouver. He donned Western-style clothes but left behind his wife Huangbo and their daughter. On his trajectory between past and future he crossed stages of economic development and mentalities: the peasant became a shingle mill worker. He had to negotiate two identities, respected villager and cheap 'Oriental' labourer. Huangbo had to negotiate the position of a woman left behind in a society ruled by older men and had to improve their plot with his remittances. This couple and their children and grandchildren accommodated the frame of two states and two local societies in their material and mental worlds when two world wars, the Great Depression, and the change from Nationalist to People's Republic created havoc around them.

Source: Denise Chong, *The Concubine's Children: Portrait of a Family Divided* (Toronto: Penguin, 1994).

100,000 in the US and 17,000 in Canada in 1901, their presence was pronounced in mining and subsequently in railroad construction as well as in lumbering in Pacific Canada.

Whites in the US West were particularly attuned to the powerful rhetoric of Manifest Destiny and the ideal of free labour; those in British Columbia to 'white men's standard of living'. In this line of thinking, white miners personified free labour and family wages. The frugal Chinese were considered the antithesis. They were called 'coolies' though indentured labour, the 'coolie system' as practised in the plantation belt of the British and French empires, never reached North America's Pacific Rim. Chinese men came under credit ticket arrangements—as European indentured servants or 'redemptioners' had done during the 1820s.

In both countries, the unwanted migrants spurred a complex debate juxtaposing national economic needs against ideologies of race and immigrant cultural inferiority, agrarian settlers against industrial workers. Both countries passed legislation to keep some people out: migrants from Asia; people from Europe who were poor, held radical political beliefs, or were suspected of moral defects. Laws phrased in gender-neutral terms had a particularly restrictive effect on women: when Asian exclusion was imposed, few Asian women were in either country. When fear of immorality among white immigrants mounted, immigration officials suspected arriving single women of becoming prostitutes. From 1875 to the 1920s, a series of laws closed the doors to the US and, from 1885 to the 1930s, the same was true in Canada.

WHAT KIND OF IMMIGRANTS FIT THE NATION? THE ECONOMY? THE RACE?

The moment the US and Canada completed their intercontinental railway systems, and the US and Mexico were connected by south–north lines, a violent debate began about the workers who built it—mainly Chinese in the West, Mexicans in the South, and Eastern and Southern Europeans in the East. The focus was, first, on the 'Orientals', deemed racially inferior, morally depraved, and economically without needs, thus undercutting wages. This discourse, however, was but a trial run for exclusion of dark Europeans against whom US author Madison Grant railed in 1916 in his *The Passing of the Great Race, Or the Racial Basis of European History*. Immigrant workers, required for economic growth, were seen as subverting the American people's racial composition. In Canada, attitudes were similar, but terminologies more cautious: migrants from the British Isles were 'preferred'.

Exclusion of those deemed racially inferior would have slowed national economic development, hurting citizens' well-being and each country's position in the international arena. Debates were charged. For example, an American in 1910 might have waxed nostalgic about the 'good old days' of the 1860s, when the country was more homogeneous, but reminiscence without solid historical data often paints badly distorted pictures. In 1860, the 4.2 million immigrants accounted for 13.2 per cent of the population. By 1910, immigrant totals had risen to 13 million, but absolute numbers are misleading. In 1920, the *percentage* of foreign-born people levelled out to exactly what it had been in 1860: 13.2 per cent. Nativist charges were as uninformed as they were biased.

In Canada, the debates were different, but policies were similar. The prairies and the Northwest had only begun to be incorporated into the nation in 1870, with the establishment of the at-first tiny province of Manitoba and the acquisition of extensive territories that had been under the control of the Hudson's Bay Company and the British Crown. While British Columbia and eastern Canada from the Maritimes to Ontario were settled societies, the new country's middle section was 'empty' because nation-building or nationalist statesmen did not consider the First Peoples or the Métis a part of Canada. At this time, though, Canada felt threatened by the US, which was strongly engaged in continental expansion and had just bought Alaska. The region needed to be populated.

By the 1880s, industrialization provided job options in Western and Northern Europe. Immigration declined. The Canadian Secretary of the Interior, Clifford Sifton, forestalled a US-style racist debate: he called Ukrainian peasants 'men in sheepskin coats', making them appear similar to desired and preferred British yeomen. The image and garment hid the alleged darker skin on which racializing debates focused. Male historians, for decades, did not cite the rest of his gendered and intergenerational view—Sifton wanted men 'with stout wives and a half-dozen number of children'. Farming in the Prairie provinces (Saskatchewan and Alberta became provinces in 1905) involved family labour—and procreation. Little debate ensued from the census data. In the decade from 1901 to 1911, 28.0 per cent of Canada's population was foreign-born, and this remained high, at 20.2 per cent, from 1911 to 1921—almost double the US rate.

In the US, what many feared most was the difference between the 'new immigrants' of Slavic or Mediterranean 'race' and the 'old-stock Americans' and 'old immigrants'. Of the 778,000 immigrants who arrived in 1882, 87 per cent were from Northern and Western Europe, mainly of British, Irish, German, and Scandinavian background. But by 1907, of the 1,285,350 immigrants in that year, only 20 per cent came from Northern and Western Europe. The others were 'swarthy', almost all male, unskilled, and with no intention of staying permanently or of owning property (Gabaccia, 1997). Rather than work the land like their immigrant predecessors, they 'chose' to live in crowded urban slums among their own. Moreover, few of the new arrivals were Protestant. The 'Russians', fleeing religious and ethnic oppression, were Jews and Catholic Poles.

To those born in the United States, the motley new immigrants seemed alien to American values and unwilling and unfit for assimilation. In Canada, in contrast, the Ukrainians, Poles, and Russian Mennonites were considered skilled agriculturalists. In both countries, **Homestead Acts** encouraged ownership of 'free' land—land freed of First Peoples. But the mass of urban workers came from unfree countries and entered dependent wage labour (Bumsted, 1992: vol. 2, 68–142).

The nativists refused to acknowledge that 'new' immigrants entered a new economic context. Cities and mass-producing factories held the economic opportunities. In Canada, the Ottawa government's analysis fell short of this reality. Sifton did not want an industrial proletariat of single men whom he saw as rootless, unwilling to work hard, and worst, ready to organize trade unions.

Migrants, by contrast, showed more intelligence. They understood the options that industrialization and raw material sectors like lumber and minerals offered. They and North American employers knew that they were needed to sustain economic development; both understood this context far better than statesmen and administrators. The new migrants were accused of being unskilled, illiterate, and of Catholic or Jewish faith. Yet similar accusations had greeted the earlier-arriving Irish and southern Germans who, at the time of the new immigration, were assimilating rapidly. The newcoming men and women faced an economic stage of development that no longer demanded their agrarian skills. It was only for factory work that they were unskilled—like those born in the country. Their migration involved deskilling processes that fit neatly with Taylorization, industry's deskilling of factory work.

RURAL MIGRANTS IN PROLETARIAN URBAN NEIGHBOURHOODS

Around 1850, two-thirds of the immigrants to the US had been farmers. By 1900, 95 per cent went to the cities, though many came from the peasant classes in Europe or Asia. In Canada, the proportion of men and women heading for urban jobs was slightly lower.

Emma Lazarus's invitation to the poor of the earth was not what either of the two governments wanted. Those likely to become a public charge were rigorously excluded. From 1910 on, men and women arriving in Canada had to prove that they came with $25 in addition to a ticket to their final destination—a sum higher than even many of the preferred British could produce. In the US, data from 1910 indicated that the poorest migrants, Russian Jews, arrived with an average of $12; German-speakers, as the 'richest', landed with $41. On average, new arrivals owned $21.50. They had been impoverished by conditions and structures of their societies of origin. They came with a will to improve their lot by hard work and education for their children. By the 1920s and 1930s, scholars like Manuel Gamio (1971 [1931]) found the same to be true for the increasing number of migrants from Mexico.

Hardship pushed migrants from their countries of birth. They left to escape wars, political upheaval, natural disasters, and religious or cultural oppression, as migrants in the early 2000s do across the globe. Still, the overarching reason was economic: a shortage of land caused by population growth and modernization, rising birth rates and decreasing mortality, and a pace of industrialization that did not provide jobs for the rural displaced. Dislocated peasant families had to join a landless, mobile proletariat 'at home' or afar. Since the home did not feed them, their term for migration was going 'to bread'. Around 1900, 40 per cent of the European migrants were women (Gabaccia, 1996; Harzig et al., 2009).

Migrants could select between destinations, and many sought work within Europe. But in the 1870s steamships made transatlantic migration faster and much cheaper. Irish men or women could migrate to New York or Montreal for just under $10 in the 1890s, those from the continent for $30. Arranging the trip was easy. Shipping company agents across Europe and local brokers in some regions of Asia advertised low rates and fast travel—comparable to modern arranged group tourism. While the US advertised itself as 'land of opportunity', most migrants had reliable sources of information among acquaintances and kin who had migrated earlier (Wyman, 1993: 21–5).

The data point to extended local and familial networks—as is the case for present-day migrants. In the decade prior to 1914, 94 per cent of the US-bound went to kin and friends. They moved transregionally and transculturally rather than transnationally—a discourse prevailed about 'going to America' (not the US in particular) and the job options (or earlier, the land) at a very local destination counted for the actual decision. The immigrants departed from and arrived in particular localities according to information flows and friendship spaces that extended across continents and oceans (Dillingham Commission., 1911, vol. 3: 358–9, 362–5). The detailed pre-departure information level of migrants is evident from migrant letters (Kamphoefner et al., 1991: 1–35; Hoerder, 1986).

Similarly, people in the Pacific Rim's Chinese diaspora moved between locations about which they had information. 'The Chinese', as they were labelled, came from only particular locations in two of the Empire's provinces (Pan, 1999; Wang, 2000). Mexicans of this period could also be traced to particular departure regions. The men and women were neither generic nationals nor undifferentiated imperial subjects of Europe's dynasties. It was immigration officers who classified them by country of origin, and scholars who invented the nation-to-ethnic-enclave cliché. Migrants moved along trajectories of specific subnational cultural affinity and emotional ties. Thus the recently fashionable term 'transnational' demands specification. At least from the early twentieth century, the data indicate translocal and transregional belongings or, with open spatial extent, transcultural ones.

SCIENTIFIC MANAGEMENT AND THE UNSKILLING OF IMMIGRANT LABOUR

Around 1900, the industrializing countries in the expanding global capitalist system competed for migrant or immigrant workers. For those willing to cross the Atlantic, the US was one option but Canada, Brazil, and Argentina presented others. Chinese could move to Southeast Asia or the Americas. In some periods, national governments or, in the case of the US, individual states subsidized transportation and helped with accommodations upon arrival and facilitated naturalization of some European groups.

Although the United States offered little help, the fast expansion of industrial jobs attracted more migrants than any other industrializing region. From 1906 to 1915, Canada received three million immigrants, Argentina two million,

Australia and New Zealand around 900,000, while the US received 9.4 million. At this time, so many southern Italians had family in the US that many 'felt more confident about going to New York than Rome' (Wyman, 1993: 25).

In the US, Frederick Winslow Taylor measured labour input along the production lines and broke complicated processes into many small steps. This so-called 'scientific management' implied 'scientific' displacement: American-born or Canadian-born, highly skilled workers lost their jobs to unskilled and semi-skilled immigrants. A foreman would watch over workers executing menial, repetitive tasks at stopwatch speed. On the positive side, the economies of scale translated into wages higher than in Argentina, for example.

Rural migrants from Europe became urban proletarians. From their point of view, they could feed themselves and, they hoped, their families. Still, living conditions were as gruelling as among migrants in the service sector at the beginning of the twenty-first century. Propaganda of steamship agents hyped 'America'—the US—into a promised land. Ideological billboards proclaimed unlimited opportunities for all who submitted to Taylorized work routine and factory discipline. While most men and women, through letters or oral information from relatives and friends, knew better, those who had great hopes quickly came to understand the realities. This was also true for migrants from Asia. Since the California and British Columbia gold rushes of 1848 and 1858, the Chinese name for North America was 'Gold Mountain'—but lives were spent in segregated accommodations and exploitative working conditions.

ADMISSION REGULATIONS AND ENTRY MYTHS

The contradictory experiences and discourses are reflected in the entry myths and experiences particular to the US, centralized from 1892 at **Ellis Island**. The New York skyline and the Statue of Liberty—donated to the US by France in 1886—beaconed to steerage passengers. Then came inspection: head tax, poking, prodding, washing, spraying, and being investigated. Although few were rejected, a total of perhaps 20 per cent were detained for various lengths of time.

The focus on Ellis Island conceals the multiplicity of immigrant admission stations. Since the early 1900s migrants from the Caribbean arrived in the ports of the Gulf of Mexico, migrants from Asia in Pacific coast ports—especially, since 1910, at the detention (rather than admission) station on **Angel Island** in San Francisco Bay. Furthermore, the land borders to Mexico and to Canada were open. US citizens emigrated. Mexicans and Canadians moved to the US.

In Canada, no entry myth emerged. Most migrants from Europe debarked in Montreal. During the Irish famine a quarantine station had operated on an island below Quebec City. From the early twentieth century on, migrants also debarked in Halifax. Those heading for the prairies remembered the mass accommodation in Winnipeg's railway sheds—grey buildings that never assumed landmark status in immigrants' minds or in national lore.

ANTI-IMMIGRANT NATIVISM AND ANGLOCENTRISM

It is sadly ironic that the deplorable conditions with which immigrants had (and have) to put up with only drew more ire and loathing from native society. As urban centres accommodated ever more immigrants, **nativism** in the US and an emphasis on **Britishness** in Canada grew more intense. Such hostility was not new (or restricted to North America). Its first US phase, from the 1840s, had been anti-Catholic, directed mainly against the Irish and Germans. The second phase, against the Chinese, Japanese, and other Asians, was particularly virulent in California and British Columbia. The third phase began in the US when increasing numbers of strikes by immigrant workers protested exploitation and miserable living conditions. In Canada, the Secretary of the Interior's fear of British (and other) trade unionists pointed in the same direction. When, in Chicago in 1886, a bomb detonated during an anarchist rally at Haymarket Square, politicians and the media blamed immigrants, German-background socialists in particular. Show trials fuelled further nativism and ignited fears across the country, mainly of Eastern and Southern European 'aliens'.

In Canada, government investigating commissions labelled Chinese in British Columbia and Italians in Quebec as unfit for civilized society. Since the 'aliens' were working-class men and women, anti-immigrant propaganda involved aspects of class and anti-radicalism. During 'Red scares' the resident middle- and working-classes vilified the 'Others' who asserted their right to decent living conditions. Alleged radicals were deported. John Higham and others have pointed to the anti-Semitic positions of nativism. Segments of the Anglo-Canadian and, in Quebec, French-speaking resident majorities feared that their societies would be culturally, socially, and racially polluted should immigration continue unabated.

The immigrant experience has been heavily debated among US historians and social scientists: Some have emphasized ghetto squalor and the oppressively racist society, others rapid upward mobility. Many scholars fell for the slogans, too, and never asked about downward mobility. Reality was diverse and complex: fast and easy access to jobs as well as physically debilitating work conditions and poverty. Scholars agree that hard work and harsh living conditions were the rule for most immigrants, whether of European, Asian, or Mexican background. Many had planned to come only temporarily or did not like the conditions—around 1900 an estimated one-third of the transatlantic *migrants* returned to Europe. Contemporary scholars who talked about '*immigrants*' and 'opportunities' never understood this fact, although the data were available. Many Europeans came as 'sojourners', a label affixed by white scholars to Chinese (Kazal, 1995; Bodnar, 1985).

MIGRANTS AS AGENTS

The migrants came as fully socialized, gendered individuals, mostly in their late teens, twenties, or early thirties. Traditional immigrant historians of the 1940s and 1950s, such as Oscar Handlin (1951), saw them as in limbo between old and new, as 'uprooted', at loss; Caroline Ware (1994 [1935]), who provided a sophisticated study of gendered and intergenerational immigrant enculturation in Greenwich Village, New York, was the only exception (Hoerder, 2006a). More recently, historians have shown that immigrant communities held on to cultures that mediated between their social skills and the new environment. John Bodnar (1980) argued that the spread of global capitalism meant that immigrants had already encountered capitalism before they left home and thus had already dealt with many of the challenges it posed to their often-rural customs.

In both the US and Canada, immigrants responded to their relative powerlessness by forming **ethnic enclaves** in which mutual aid could be provided and viable aspects of the culture of origin (e.g., food habits) could be retained while other, less viable aspects, such as social hierarchies, could be discarded over time or, sometimes, quickly. They certainly did not simply accept and unconditionally assimilate to middle-class American values of individualism, economic mobility, and education, or to English-Canadian ways of life or to Quebecers' French-Canadian customs (Hoerder, 1996; Lucassen and Lucassen, 1997; Hoerder, 1994; Hoerder, 2010: chs 3, 5, 8).

In the early 1900s, the vast majority of migrants from Europe (and the smaller numbers from Mexico and Asia) lived in harsh conditions—as labour migrants from across the globe do in the early 2000s. While countries of origin and cultural backgrounds have changed, most still form the bottom layer of the workforce, whether farm workers from Mexico, domestic and caretaking women workers from the Philippines, or motel keepers from South Asia. In the past as in the present, people leave their countries of origin because they see few if any economic opportunities if they stay. They know about a slightly broader range of options in North America (or elsewhere). As around 1900, many of today's migrants come for a few years to support family at home, to permit survival of a marginal agricultural, small-town, or lower-class urban life in their country of origin. Migration has always separated and truncated families; migrants support the receiving society by their work and taxes, their societies and families of origin by remittances.

RACIALIZATION AND EXCLUSION IN THE US AND CANADA

In the racializing 1880s, both the Canadian and American governments had to frame laws acceptable to a majority of voters without hurting the economy or losing to international competition, for at this time Britain, France, and Germany needed immigrant workers for their expanding industries and mines. Both countries began with Asian exclusion, first against the Chinese in 1875/1882 (US exclusion) and 1885 (Canada's **'head tax'** or entry fee). Within the British Empire, the Queen's subjects could migrate freely and, to exclude South Asians from 'British India', Canada circumvented imperial law by administrative subterfuge, a regulation established by Order-in-Council in January 1908 requiring that immigrants from Asia must arrive by continuous journey from their country of origin without intermediate stops, but steamships needed to stop for re-coaling to traverse the distance and, in any case, no steamship lines went directly from India to Canada. A year earlier, in 1907, Japanese and US diplomats concluded a 'gentlemen's' agreement to the detriment of Japanese commoners. Under US pressure, Japan agreed to stop emigration. Exclusion was also directed against Koreans, although Filipinos could not so easily be excluded after American occupation of the islands following the war of 1898.

After exclusion, the predominantly male immigrants pursued culturally different gender strategies: brides were selected by acquaintances of both sides for Japanese men, who legally brought in their wives so that a gendered and inter-generational community emerged; for Chinese women, following Confucian prescripts of women's role as subservient to their husband's parents and male migration patterns (sons were expected to return to care for the memory of deceased parents), departure remained nearly impossible, and in the US as early as 1875 the US Page Law had made their entry difficult. Consequently, the Chinese communities remained predominantly male. Exclusion also placed new requirements on those in the country who planned to visit home. They had to obtain certification to re-enter. In Canada, the entrance fee for Chinese immigrants, known as the 'head tax', was raised until it reached $500 in 1903. Students and merchants could continue to come since they were useful for trade and cultural relations.

Ironically, the exclusion measures were victories of the growing North American labour movements, which consisted of earlier arriving white immigrant workers. Industrious and frugal Chinese workers felt they were persecuted not for their vices but for their virtues. While images of Chinese 'opium dens' abounded, recent scholarship has validated the view of hard-working immigrants who never accepted the hard-drinking practices of some white workers (Ward, 1978; Hoerder, 2006b).

Concerning Europeans, in contrast, only narrowly defined 'prohibited classes' were excluded from 1882 on: criminals and felons, paupers likely to become a public charge (a clause assumed to apply in particular to single female migrants), 'insane' and physically disabled persons, those with contagious diseases, women suspected of prostitution (the 'white slave' paranoia), polygamists (i.e., Mormons in the US), and, from 1903 in the US, anarchists and subversives.

The exclusion of 'contract labourers' was problematic. US industrial unions had fought pre-migration contracting of European workers in order to prevent the arrival of cheap labour, but farm and domestic workers and others were exempted from the provision. How was a potential migrant to know whether he or she could have a contract or not and, for persons intent on providing for themselves, what better guarantee could there be than having a job waiting upon arrival?

Exclusion provisions in Canada, not substantially different from those in the US, were codified into comprehensive laws in 1906 and 1910. Still, since 1880, some 20 million immigrants had arrived in the US and over 4 million in

Canada—when mass migration came to a sudden stop in 1914 with the beginning of World War I.

RACIALIZING 'BROWN' PEOPLE: MEXICANS

The emphasis on migrants from Asia obscures the similar **racism** directed against 'brown' people, i.e., Mexicans. Those annexed with the territories gained by the US after the war of 1846–8, as well as in Texas after its 1845 annexation, had lost influence and property to Anglos. After 1900 a confluence of factors brought the first great Mexican migration to the US (hardly any reached Canada). The causes for migration date back to the mid-nineteenth century when a first assault by the Liberal Party in Mexico on the landholding peasantry, Native people in particular, abolished the *ejido* or communal lands of the indigenous peoples. By 1892, private companies held about 20 per cent of Mexico's land; by 1910, companies had acquired an additional 27 per cent of public lands. The dictatorship of Porfirio Díaz, beginning in 1876, offered favourable conditions for foreign investment and US capitalists opened mines and drilled oil wells. To transport the extracted minerals to the US, they built south–north railroads. These facilitated migration from central Mexico to its northern states, where the jobs in mining and infrastructure were expanding.

When population growth increased the labour pool, wages fell. By the early twentieth century, growing discontent led to the revolution against the 'Porfiriato'. As Lawrence Cardoso puts it, 'the initial impetus for emigration resulted from attempts to achieve capitalist modernization in the countryside'; the dictatorship's anti-labour policies pushed further tens of thousands to depart, and the revolutionary violence sent about 200,000 fleeing (Cardoso, 1980: 73; Ruiz, 1998: 27; Hart, 2002).

Mexicans' reasons to depart their native land were thus starkly similar to those of Europeans. Their experience in the US, however, was markedly different. In the early decades they remained in the Southwest in seasonal agricultural work, in building and maintenance of railroad tracks, and in Los Angeles brickyards (see Box 2.2).

LEGISLATIVE MEASURES AND DEPORTATION

Perhaps the most substantive force in the early restrictionist movement was the US Congress's joint **Dillingham Commission**, formed in 1907. Its 42-volume report of 1911 provided abundant data, still used by historians. Often contradicting the data, the conclusions upheld the biased view that the new immigrants lacked the moral, physical, and emotional fortitude of earlier Northern Europeans. The commissioners recommended a literacy test and an admittance quota based on an ethnic group's percentage of the population. Such measures had been discussed in Congress repeatedly for a quarter-century. In 1896, President Grover Cleveland vetoed a law to that intent, since working-class immigration was indispensable for US economic growth (as it was for Canada, government

BOX 2.2   Migration of the Mexican Peasantry

The first assault on Mexico's peasantry came with the liberal constitution of 1857, which abolished the *ejido* system, or communal lands inhabited largely by indigenous residents. By 1910, private companies had acquired 27 per cent of public lands. The peasantry and the vanguard middle class rebelled, ousting dictator Porfirio Diaz in 1911 and creating a power vacuum. Violent political upheaval became the norm, which continued until the 1930s.

Repression in the United States was less overt. The Woodrow Wilson administration nurtured growing fears of immigrants and radicalism to promote the war effort. The Committee for Public Information (CPI) spread wartime propaganda, scapegoating immigrants, especially Germans. Headed by muckraking journalist George Creel, the committee launched an energetic, sometimes lurid disinformation campaign against German immigrants. Though disbanded in 1918, the CPI left an indelible impression and was partly responsible for subsequent restrictive immigration laws.

The immense railroad buildup made it easier to migrate to Mexico's northern states, where the jobs were moving. As the labour pool in the northern region grew, wages plummeted. The highest-paid workers in Mexico made about 50 cents a day working on the railroad. Lower-paid workers in the south made as little as 15 cents a day. Migration from southern to northern Mexico paralleled migration to the US, and for the same reasons.

Mexican experiences in the United States were unique. Avoiding the crowded eastern cities, they became an enclave population in the US Southwest, working in the fields rather than in the factories. Their presence in these hinterlands left them less noticeable, and their immigrant experience less recorded than that of those who landed on Ellis Island. This anonymity helped in some ways. They largely evaded the nativist hostility that met the earlier Chinese and Europeans. Supported by a powerful agribusiness lobby, Mexicans stayed under the restrictionist radar—at least for a time.

opposition notwithstanding). Immigrants from Southern and Eastern Europe were far more likely to be illiterate than earlier ones. In 1900 only 2.7 per cent of the Irish and 1 per cent of the British were illiterate—but 54 per cent of Italians who in Canada became the focus of governmental investigations were illiterate (Hoerder, 1999: chs 11–15).

World War I provided the climate to increase the push for restriction. In the US, the Committee of Public Information distributed millions of pamphlets in several languages 'explaining' the dangers posed by immigrants and radicals, illustrated these by expositions in dozens of cities, and added a barrage of some 6,000 press releases. In consequence, a 1917 law established a literacy test, doubled the head tax to eight dollars for all immigrants, and created a barred zone that kept out all Asians except Japanese and Filipinos; however, in its practical application, only a fraction of the illiterate immigrants from

Europe were turned away. The border to Mexico, like that to Canada, remained open.

The victory of 1918 did not ease the fears. The Bolshevik Revolution and the high visibility of internal radical organizations like the Industrial Workers of the World induced a Red scare in the US. In 1919–20, hundreds of immigrants were deported. Canada also passed deportation measures. The 1919 Winnipeg General Strike, an attempt to alleviate miserable living conditions, created fears of radicalism among the political elites and much of the press, and led to the deportation of a number of foreign workers and labour activists. In the 1920s and 1930s radicals, unemployed people, and single women who were pregnant were sent back. Nor did the return to economic normalcy in the early 1920s ease nativist fears. The rise of the second, urban Ku Klux Klan in the US—with its anti-Catholic, anti-immigrant, and anti-radical rhetoric in addition to anti-black racism—continued.

This sustained, government-supported paranoia brought the Dillingham Commission's recommendations to legislative fruition. The US **Emergency Quota Act** of 1921 limited the number of immigrants to 3 per cent of their population in the census of 1910, and consequently the cap of 357,800 annually was weighted towards the 'old-stock' groups.

Only three years later, in 1924, the **Johnson-Reed Act** further inscribed racist rhetoric into law. In Congressman Albert Johnson's words: 'the United States is our land . . . we intend to maintain it so.' The Act limited the total number of immigrants to 150,000 per year and lowered the national origin quota to 2 per cent of the 1890 census—in 1890 few new immigrants had arrived. Western and Northern European countries received generous quotas: Britain's 65,000 compares to a cap on Italian migration at 5,803, Polish at 6,524, and Jews at 2,784. Due to extensive lobbying by western agribusiness, the 'western hemisphere', meaning Mexico, was exempted from the provisions (Kraut, 1982: 177).

By the time the law took full effect in 1929, the Great Depression drastically reduced migration and Mexicans, too, faced discriminatory action, especially in the Southwest. With a surplus of native labour supply, including dust bowl refugees who migrated west from prairie states to California, agribusiness ceased to defend Mexican migration. Deportations sent perhaps 300,000–400,000 Mexicans back across the border. This repression often did not differentiate between Mexican nationals and US citizens of Mexican origin. To racist police and sheriffs' deputies they were all the same. Uneasy about the loss of labour power through emigration, the Mexican government co-operated and provided land for settlement.

World War II-related production pulled the US and the Canadian economies out of economic depression. Only about 120,000 migrants arrived in the United States from Europe during the war years. With men in the army, both countries feminized their factory labour force. But the US faced a tremendous shortage of labour for seasonal agriculture, and through the government's Bracero Program, initiated in 1942, temporary Mexican workers came. Despite resistance from agricultural unions, the program continued until 1964.

RETHINKING IMMIGRATION, 1945–1965

After World War II and two decades of extremely filtered immigration—the period of closed doors—attitudes towards immigrants began to shift. Those who had been feared so much in the 1910s and 1920s had largely become integrated into the American and Canadian societies. Italians, Jews, and Slavs were virtually indiscernible from white society. In the 1930s and 1940s, both countries had kept their doors closed to refugees from Fascism, Jews in particular. Because China had become an ally in the war, the US ended Chinese exclusion in 1943. The quota was only 105 persons annually, about equal to that for the tiny Baltic states. Canada ended Chinese exclusion in a step-by-step process beginning in 1947 (Alba and Nee, 2003; Reimers, 1985; Foner, 2005, 2009).

After 1945 both countries admitted displaced persons from Nazi forced labour camps, responding more to labour force needs than to humanitarian concerns. Those recruited for Canada had to leave surviving family members and work in assigned, usually disagreeable, jobs for a year before they could apply for family reunification. After years of labour in Fascist camps this added insult to injury. The two economies were booming and needed workers.

Still, racial and political nativism endured. Prime Minister Mackenzie King assured Canadians in 1947 that further immigration would be encouraged—as long as it did not change the racial character of the society. A new comprehensive Immigration Act (effective in June 1953) combined openness to newcomers with exclusionary powers. According to a 1952 regulation, Asian men and women could immigrate only if bilateral agreements existed, and Canada did sign such treaties with China, Japan, India, Pakistan, and Sri Lanka (Hawkins, 1988).

In the US, the beginning of the Cold War fuelled anxieties about Communists and other radicals. The restriction and exclusion of certain nationalities were a diplomatic embarrassment: The war had been fought against racist Fascism, yet segregation in the South recalled the treatment of German Jews in 1930s. Despite a growing challenge, the nativist-racist forces won out. Congress passed the **McCarran-Walter Act** over the veto of President Harry Truman in 1952 but the law gave the President the power to admit immigrants in an emergency. President Truman's open-minded stance began a period when hundreds of thousands of refugees arrived outside the quota system. From 1953 to 1965, for the first time, immigrants from Europe did not dominate admission, falling to 43 per cent of the total.

Internationally, the post-war years brought the end of old-style colonialism and stateside imperialism, though not of economic domination. States with 'coloured populations' became independent and the white, free, capitalist Western world had to come to terms with this. Both the US and Canada were Pacific as much as Atlantic powers. In both countries, political and economic elites assumed that consumer demand for North American industrial products would be high in populous Asian countries. Discrimination provided no basis for good commercial relations.

In 1962, Canada ended discrimination in immigrant admission based on race or ethnicity. Five years later, in 1967, it introduced the **points system** to select men and women who would fit into Canadian society and be an economic advantage. Education, training, and English or French language were assets. Potential immigrants from the many societies in Asia with their highly developed educational systems could easily fulfill these criteria. Canada also recognized the increasing role of the service sector. As early as 1955, the government began to admit women from the Caribbean, especially the British West Indies, as domestic workers and caregivers. Racists wanted to admit these women only with temporary work visas, but policy-makers decided to grant them immigrant status provided they had worked for a year (or longer, depending on the regulation in force) in domestic jobs. After four years in Canada, all immigrants could apply for citizenship as a right rather than as a favour at the discretion of bureaucrats. As citizens, they could sponsor relatives who did not have to fit the points criteria, as long as the sponsor assumed financial responsibility for them (Harzig, 1999).

Canada's policy change came when the Immigration Act of 1962 finally ended race-based immigrant selection. In the US, the watershed came three years later when the new Act, conceived by the Kennedy administration, was guided through Congress in President Lyndon Johnson's Great Society program. Supporters of the law assumed that immigrant composition would not change drastically and did not believe the law would affect the lives of Americans. The law seemed moderate, indeed. It set an annual global immigration cap to the US of 290,000, with 120,000 from the western hemisphere and 170,000 from the eastern (i.e., Europe, Asia, and Africa), with no more than 20,000 immigrants coming from any particular country annually. Some noted that the family preference system would retain the integrity of the previous immigration composition, since Africans, Indians, and Asians were only a small percentage of the population (US Congress, 1964: 418).

The actual numbers before and after passage of the law far exceeded quota limits since the Act retained the presidential power of discretion and, more importantly, a hierarchical system of preferred immigrants existed within the law. Direct family members of US citizens had always been exempted from the quota system. Parents of naturalized immigrants held the highest priority; immigrants with exceptional talent were the second; other family members followed. Designed to enable **family reunification**, the provisions inadvertently encouraged a migration pattern that resembled a pyramid more than a chain. A new immigrant could receive citizenship in five years and thus establish preferential entrance for relatives and children. Once naturalized, these in turn could bring over their families.

Proponents of the law gave scant consideration to demographic changes in source countries. The push factors that sent Europeans abroad were forcefully operating in Latin America and Asia. In the 1950s, population growth in the poorer nations outpaced that of the richer nations for the first time. Nascent industrialization and a new capital-intensive agriculture in these regions displaced peasantry, creating a landless proletariat. This process was reflected in immigration figures: the annual cap of 290,000 was almost instantly surpassed by arrivals. By the 1970s, 4.4 million legal immigrants had entered the US. The number grew to more than 600,000 entering the country annually by 1978. By the 1980s, the number had increased to 7.3 million for the decade. The ethno-cultural change in immigration was clear: European immigration dropped from 52 per cent in the 1950s, to 33 per cent in the 1960s, to around 12 per cent by the 1980s. Canadian immigrants to the US, largely of European descent, similarly dropped from 22 per cent in the 1930s, to 2 per cent in the mid-1980s. The new immigrants came from Asia and Latin America.

The most dramatic change involved migrants from Asia. Given the previous restriction, arrivals rose from a trickle to six million Asian immigrants to the United States from 1970 to 1995. Asian Americans had accounted for around 0.6 per cent of the population in the 1960s, and Asians comprised only about 5 per cent of the total immigrants between 1931 and 1965. These numbers rose to 12 per cent during the 1960s, 34 per cent during the 1970s, and 48 per cent of total legal immigrants by the 1980s, surpassing even Latin Americans. In total, they accounted for over 40 per cent of arrivals from 1970 to 1995. The US currently has 14.9 million inhabitants claiming Asian heritage.

The proximity of Latin American countries, especially Mexico, transformed Latino immigration. Latin America accounted for 15 per cent of immigrants over the 1931–60 period but for 38 per cent from 1961 to 1969 (the same percentage as Europeans), and 41 per cent in the 1970s, dropping to 35 per cent in the 1980s. What had been a Southwest borderlands population with a few clusters in northern cities became the largest minority group in the country.

In Canada, the federal government, regardless of whether the Liberals or Conservatives have been in power, has pursued a policy of immigration that brings in about 1 per cent of the current total population per year, 250,000 in the 1980s, at present 300,000 annually with a population close to about 34 million. Again, the increase in migrants came from Asia, along with those from the Caribbean. Latin Americans came only in small numbers, but migration from North Africa and sub-Saharan Africa also grew even if total numbers remained small. Asian newcomers—as in the US—included ever larger numbers from the Philippines, as well as South Asians and Southeast Asians. Chinese came to be differentiated as originating from the People's Republic of China, Hong Kong, and Taiwan.

DEBATES ABOUT MIGRATION AND ACCULTURATION SINCE THE LATE 1960S

Although racial criteria had finally been deleted from the immigration laws, the North American societies continued to experience racialization, ghettos, and other discrimination. The impetus to address these issues came from larger forces. The decolonization movement worldwide undercut the claims of white superiority. In the US, **Chicanos** added their voice to that of African Americans in the civil rights movement. In Canada, a debate about the historic treatment of Asian immigrants, Chinese exclusion, and the wartime internment and deportation of Japanese Canadians pointed to injustices. Mass migration from the anglophone and francophone Caribbean to the big cities added another voice to the Canadian chorus. Puerto Rican and Cuban arrivals bolstered the US debate.

Regional differences are part of all countries of the Western world, the US included. Canada, however, was a special case. In addition to pronounced regionalism, two self-styled founding nations, British and French, claimed status within Canada and, in the country's original constitutional document, the British North America Act, both English and French were recognized as the languages of the federal Parliament. To address the cleavage, the federal government in 1963 created an independent Royal Commission on Bilingualism and Biculturalism. When the 'ethnics'—long-resident descendants of primarily European immigrants, notably Ukrainians and Germans—emphasized their status as a third force, the Commission, in its report, recommended in 1969 equal status and services for **francophones** in Quebec and in other provinces and for anglophones in Quebec, and called for a policy to support the 'other ethnic groups'. These groups were integrated into Canadian society and largely spoke English, but in various ways they continued to identify with their countries of origin and with the cultures they and their parents and grandparents had brought with them to Canada. Canadian society was depicted as a 'mosaic' of peoples, but historic power relations had made it a 'vertical mosaic' with the most recent newcomers at the bottom. Implementing the recommendations, the Liberal government of Pierre Trudeau, with support of all other parties, announced a policy of multiculturalism in 1971. Society had changed to such a degree, especially among the younger generations, that this caused almost no debate. Rather, it was viewed as a step towards bringing governmental policies in line with societal practices.

Contrary to later misconstructions, the policy did not encourage retention of immigrant culture in insular groups. It explicitly furthered full participation of all and a sharing of cultural traditions. It sought to reduce dividing lines and cultural hierarchies, guaranteeing francophones outside of Quebec and Canadians from minority ethnocultural groups equality in funding for cultural affairs as compared to the majoritarian, dominant British Canadians.

In the US, student dissent, the civil rights movement, and the Chicano movement had already dislodged the racial status quo. In addition, official support for right-wing dictatorships in Latin America set in motion a critical chain of events. Refugees from the dictatorships headed north in large numbers, providing an anchor point for even greater Mexican, Central American, and South American migration. The refugees were not admitted legally but also, because of the death squads at home, were not deported. About three-quarters of undocumented migrants since 1965 have been Hispanic. Congress attempted to curb the movement and reduce legal problems with the Immigration Reform and Control Act (IRCA) in 1986, which gave amnesty to migrants who had been in the country since 1982. The law also augmented the Border Patrol and penalized employers who knowingly hired migrants without papers.

In view of economic and political conditions south of the US border (often abetted by US capital and conservative governments), the ineffectiveness of the IRCA became quickly evident. Immigration continued to log record numbers. In the 1990s, 10 million arrived, augmented by an estimated 286,000 illegal entrants annually. Southwest states, especially California, New Mexico, Texas, and Arizona, experienced massive demographic changes. New Mexico became the first state to have a non-white majority; California and Texas followed with 36 per cent Hispanics by 2006. In the US, the combined Mexican-American and Mexican (legal and unauthorized) immigrant population grew from 6.4 per cent in 1980 to over 10 per cent in 1997. In 2005 and 2006, those of Mexican birth and ancestry accounted for almost half the population growth in the US—1.4 million of the 2.9 million increase. By 2006, people of Latin American descent outnumbered African Americans as the largest minority group. Refugees in particular but also, in small numbers, seasonal Mexican workers crossed the border from the US to Canada under legal and administrative provisions.

The low number of temporary work visas and, as in the case of Chinese more than a century earlier, the difficulties of re-entering the US after a visit home led to large-scale undocumented presence and brought illegals and the border war to the forefront of the US debate. A 2006 law increasing the total quota to 675,000 did nothing to curtail it. Despite the militarization of the border, only the economic crisis emerging from the collapse of Wall Street banks in 2008 curbed border crossings in 2009. The inability to wall off and militarize the border has been a humbling lesson for the United States. Efforts at restrictive legislation, such as Arizona's 2010 immigration law, SB1070, aimed at identifying and deporting illegals through racial profiling, and vigilante citizen groups patrolling the US–Mexico border chiefly have succeeded only in polarizing public opinion and in pitting one level of government against another and private citizens against government officials.

With immigration to Canada ever more diverse in terms of culture, language, religion, gender relations, parent-imposed intergenerational hierarchies, and everyday practices, critiques of multiculturalism began to mount in the 1990s. Many of them came from monolingual old-style British Canadians unwilling to adjust to 'mosaic madness', but other criticisms of the policy came from successful 'ethnics' who wanted to be Canadians rather than 'ethnics' or 'visible minorities'. Some results of multicultural policy had undercut multicultural practices. Ethnocultural and ethnoreligious spokespersons (rather than whole groups) had advocated a problem-laden distinctiveness and separation. On the other hand, some major institutions, the Canadian Broadcasting Corporation for example, remained closed and Anglo-Canadian until public criticism attacked the 'whiteness' component of its hiring practices. To provide a frame situated above 'national' or 'ethnic cultural rights', many people today speak for a 'Canadianness' that provides human rights to every citizen. The debate over ethnicity continues, and, in an unholy alliance, recent Conservative governments have curtailed some aspects of multicultural policies while immigrant fundamentalists undermined its societal acceptance. On the level of everyday life, however, cultural interaction, *métissage*, and fusion continue.

Always present, as subtext on the borders of immigration policy and in discussions of migrants, have been the first migrants to North America, the Aboriginal peoples. Their needs, demands, and disposition often have been framed by policies designed for immigrants and newcomers, and determined by those who arrived in later waves of immigration.

CONCLUSION

This survey of more than a century of immigration and legislation indicates that present debates and patterns of arrival and acculturation do not present a radical change from the past. The cultural backgrounds and skin colours of those arriving have changed, however, and so have the contexts for migration, for incorporation, and for discrimination. Old immigration cities, often coastal port cities, have receded in importance, particularly in the United States. New 'gateway cities' in the interior, such as Atlanta and Calgary, attract migrants in large numbers, and their human, social, and financial capital have induced economic growth and urban expansion. In Canada, however, a large majority of immigrants continue to settle in the three largest cities, Toronto, Montreal, and Vancouver.

Human agency remains paramount in shaping people's lives and providing their children with increased options, either where they were born, or by short-distance migration, or by settlement in a different and distant country. Wherever they settle, they add their capabilities to the residents—and they subtract their capabilities and economic input from the societies of origin that paid for their educations and bore the cost of their childhoods. Economists have called migration 'development aid' from less developed societies to more developed ones. Political frames may change fundamentally—in periods of open doors there can be no 'illegals' because everybody can come; in the period of civil rights certain types of discrimination became illegal. But 'free' North American societies today, as in the past, are making some human beings 'illegal'. In the long view, hierarchies of cultures, at present often cast as Christian West over Muslim, have always existed and need to be addressed in the present as much as in the past. Although 'clashes of civilization' have been re-imagined by academics like Bernard Lewis (1990) and Samuel Huntington (1997)—who, unable to deal with diversity, turned to prophets of doom—societies are not one-way streets and certainly not monocultural dead-end alleys. They negotiate difference, albeit in power hierarchies.

QUESTIONS FOR CRITICAL THOUGHT

1. In the US the recent arrival of large numbers of immigrants has drawn criticism from opponents, who have characterized it as 'unprecedented' in numbers and origin. Are these sentiments accurate? May similarities with past immigration be detected? What are the differences between reactions in Canada and the US?

2. The high levels of immigration began in the 1880s and slowed to a trickle by the end of the twenties in the US, and in Canada in the thirties. What factors led to this decline in newcomers and which were the most important ones?

3. By the 1960s, 30 years of low immigration ended and a new surge of arrivals began in both the US and Canada. What factors led to this revitalization?

SUGGESTED READINGS

1. Daniels, Roger. 2002. *Coming to America: A History of Immigration and Ethnicity in American Life*, rev edn. New York: HarperCollins. Daniels presents a chronological survey of immigration with emphasis on all groups involved, reference to major changes in immigration law, and acculturation patterns.

2. Hoerder, Dirk. 1999. *Creating Societies: Immigrant Lives in Canada*. Montreal and Kingston: McGill-Queen's University Press. Hoerder provides a chronological, region-specific survey of immigration to Canada from the 1830s, paying particular attention to the agency of immigrant men and women.

3. Knowles, Valerie. 1997. *Strangers at Our Gates: Canadian Immigration and Immigration Policy, 1540–1997*, rev. edn. Toronto: Dundurn Press. Knowles concisely summarizes Canadian immigration policies and laws, and the debates about them, and sets the context within which immigrants could make their decisions.

4. Spickard, Paul. 2007. *Almost All Aliens: Immigration, Race and Colonialism in American History and Identity*. New York: Routledge. Spickard provides the most recent, issue-oriented summary of debates about immigration, race, and identity in US society from 1600 to the present.

REFERENCES

1. Alba, Richard, and Victor Nee. 2003. *Remaking the American Mainstream: Assimilation and Contemporary Immigration*. Cambridge, Mass.: Harvard University Press.

2. Bodnar, John. 1985. *The Transplanted: A History of Immigrants in Urban America*. Bloomington: Indiana University Press.

3. Bumsted, J.M. 1992. *The Peoples of Canada*, 2 vols. Toronto: Oxford University Press.

4. Burnet, Jean R., with Howard Palmer. 1988. *"Coming Canadians": An Introduction to a History of Canada's Peoples*. Toronto: McClelland & Stewart.

5. Cardoso, Lawrence. 1980. *Mexican Immigration to the United States*. Tucson: University of Arizona Press.

6. Chong, Denise. 1994. *The Concubine's Children: Portrait of a Family Divided*. Toronto: Penguin.

7. Daniels, Roger. 2002. *Coming to America: A History of Immigration and Ethnicity in American Life*, rev edn. New York: HarperCollins.

8. Dillingham Commission, US Senate. 1911. *Reports of the Immigration Commission (1907–1910)*, 41 vols. Washington: Government Printing Office.

9. Ets, Marie Hall. 1970. *Rosa: The Life of an Italian Immigrant*. Minneapolis: University of Minnesota Press.

10. Foner, Nancy. 2005. *In a New Land: A Comparative View of Immigration*. New York: New York University Press.

11. ———, ed. 2009. *Immigrant Families in America*. New York: New York University Press.

12. Gabaccia, Donna. 1996. 'Women of the Mass Migrations: From Minority to Majority, 1820–1930', in Dirk Hoerder and Leslie Page Moch, eds, *European Migrants: Global and Local Perspectives*. Boston: Northeastern University Press, 90–111.

13. ———. 1997. 'The "Yellow Peril" and the "Chinese of Europe": Global Perspectives on Race and Labor, 1815–1930', in Jan Lucassen and Leo Lucassen, eds, *Migration, Migration History, History: Old Paradigms and New Perspectives*. Bern: Lang, 177–96.

14. Gamio, Manuel. 1971 [1931]. *The Life Story of the Mexican Immigrant—Autobiographic Documents*. New York: Dover.

15. Handlin, Oscar. 1951 *The Uprooted: The Epic Story of the Great Migrations that Made the American People*. Boston: Little, Brown.

16. Hart, John M. 2002. *Empire and Revolution: The Americans in Mexico since the Civil War*. Berkeley: University of California Press.

17. Harzig, Christiane. 1999. '"The Movement of 100 Girls": 1950's Canadian Immigration Policy and the Market for Domestic Labour', *Zeitschrift für Kanada Studien* 19: 131–46.

18. ——— and Dirk Hoerder, with Donna Gabaccia. 2009. *What Is Migration History?* Cambridge: Polity.

19. Hawkins, Freda. 1988. *Canada and Immigration: Public Policy and Public Concern*,

2nd edn. Montreal and Kingston: McGill-Queen's University Press.

20. Hoerder, Dirk, ed. 1986. *"Struggle a Hard Battle": Essays on Working Class Immigrants*. DeKalb: Northern Illinois University Press.

21. ———. 1994. 'Changing Paradigms in Migration History: From "To America" to Worldwide Systems', *Canadian Review of American Studies* 24, 2 (Spring 1994): 105–26.

22. ———. 1996. 'From Migrants to Ethnics: Acculturation in a Societal Framework', in D. Hoerder and L.P. Moch, eds, *European Migrants: Global and Local Perspectives*. Boston: Northeastern University Press.

23. ———. 1999. *Creating Societies: Immigrant Lives in Canada*. Montreal and Kingston: McGill-Queen's University Press.

24. ———. 2006a. 'Historians and Their Data: The Complex Shift from Nation-State Approaches to the Study of People's Transcultural Lives', *Journal of American Ethnic History* 25, 4: 85–96.

25. ———. 2006b. '"Of Habits Subversive" or "Capable and Compassionate": Perceptions of Transpacific Migrants, 1850s–1940s', *Canadian Ethnic Studies* 38, 1: 1–22.

26. ———. 2010. *"To Know Our Many Selves Changing Across Time and Space": From the Study of Canada to Canadian Studies*. Edmonton: Athabasca University Press.

27. Huntington, Samuel. 1997. *The Clash of Civilizations and the Remaking of World Order*. New York: Simon & Schuster.

28. Kamphoefner, Walter D., Wolfgang Helbich, and Ulrike Sommer, eds. 1991 [German orig. 1988]. *News from the Land of Freedom: German Immigrants Write Home*, trans. Susan Carter Vogel. Ithaca, NY: Cornell University Press.

29. Kazal, Russell A. 1995. 'Revisiting Assimilation: The Rise, Fall and Reappraisal of a Concept of American Ethnic History', *American Historical Review* 100 (Apr.): 437–71.

30. Kraut, Alan M. 1982. *The Huddled Masses: The Immigrant in American Society, 1880–1921*. Arlington Heights, Ill.: Harlan Davidson.

31. Lewis, Bernard. 1990. 'The Roots of Muslim Rage', *Atlantic Monthly* 266, 3 (Sept.): 47–60.

32. Lucassen, Jan, and Leo Lucassen, eds. 1997. *Migrations, Migration History, History: Old Paradigms and New Perspectives*. Bern/New York: P. Lang.

33. Pan, Lynn, gen. ed. 1999. *The Encyclopedia of the Chinese Overseas.* Cambridge, Mass.: Harvard University Press.

34. Reimers, David M. 1985. *Still the Golden Door: The Third World Comes to America.* New York: Columbia University Press.

35. Ruiz, Vicki L. 1998. *From Out of the Shadows: Mexican Women in Twentieth-Century America.* New York: Oxford University Press.

36. Takaki, Ronald. 1993. *A Different Mirror: A History of Multicultural America.* Boston: Little, Brown.

37. Torpey, John. 2000. *The Invention of the Passport: Surveillance, Citizenship and the State.* Cambridge: Cambridge University Press.

38. US Congress, House of Representatives.

1964. *1964 Hearings.* Washington: Government Printing Office.

39. Wang, Gungwu. 2000. *The Chinese Overseas: From Earthbound China to the Quest for Autonomy.* Cambridge, Mass.: Harvard University Press.

40. Ward, Peter. 1978. *White Canada Forever: Popular Attitudes and Public Policy toward Orientals in British Columbia.* Montreal and Kingston: McGill-Queen's University Press.

41. Ware, Caroline F. 1994 [1935]. *Greenwich Village, 1920–1930: A Comment on American Civilization in the Post-war Years.* Berkeley: University of California Press.

42. Wyman, Mark. 1993. *Round Trip to America: The Immigrants Return to Europe, 1880–1930.* Ithaca, NY: Cornell University Press.

PART II

THE IMPRINT OF IMMIGRATION IN NORTH
AMERICAN CITIES AND SUBURBS

CHAPTER 3

IMMIGRATION AND URBAN AND SUBURBAN SETTLEMENTS

Robert A. Murdie and Emily Skop

INTRODUCTION

The US and Canada have experienced major shifts in the source countries, socio-economic characteristics, and settlement patterns of immigrants since World War II. This chapter focuses on these changes, including: (1) a change in immigrant origins from Britain and continental Europe to countries in the Caribbean, Latin America, Asia, and Africa; (2) an increase in post-industrial restructuring and labour market segmentation, which prompts the arrival of migrants with varying skill levels; and (3) a trend towards the suburbanization of the immigrant population. The first change coincided with the removal of discriminatory immigration policies in the mid-1960s, the second with the emergence of the global political economy, and the third with changes in urban form that began about the same time and affected the spatial distribution of immigrants.

Although the spatial patterns of immigrant settlement are similar in both countries, the details vary, partially as a result of differences in integration ideals and practices between the US and Canada. As noted in the Introduction to this volume, attitudes towards immigrant settlement patterns in US cities have been based on the ideal of individualistic assimilation whereas in Canada, at least since the early 1970s, multiculturalism is an ideal that fosters the retention of cultural traditions. Peach (2005) argues that these divergent ideals and practices will likely produce different immigrant settlement patterns, with US assimilationist tendencies resulting in spatial diffusion and suburbanization as the economic conditions of newcomers improve (a **spatial assimilation** model) and Canadian multiculturalism leaving immigrant groups more spatially concentrated, regardless of economic progress or central-city/suburban location (a **spatial mosaic** model). Peach concludes, however, that the differences do not appear to be as extreme as might be expected. In this chapter, we evaluate this conclusion by examining the geography of post-World War II immigrant settlement in the US and Canada.

We begin with a discussion of the changing geography of immigrant settlement in the two countries, focusing on two time periods: 1945 to 1969 and 1970 to the present. Next, we present a typology with examples of contemporary urban and suburban ethnic settlements in the two countries. Then, we evaluate the consequences of spatial concentration and dispersion in terms of **immigrant integration**. In the conclusion, we summarize the major themes and highlight similarities and differences in the immigrant geographies of the US and Canada.

BOX 3.1   Immigrant Ghettos in the US Industrial City

New York City's Lower East Side Tenement Museum (www.tenement.org), located at 97 Orchard Street, incorporates elements of the visual landscape—factories, transportation lines, and tenements—to illustrate the lived experience of immigrant enclave dwellers and the impacts of racism in everyday immigrant life. A six-storey brick tenement houses the museum (Figure 3.1) and is one of the few surviving buildings that once accommodated immigrants to New York during the great wave of US immigration in the late nineteenth and early twentieth centuries. Erected in 1863–4, it represents the first rush of tenement building in New York City. The top two floors of the museum contain rooms, wallpaper, plumbing, and lighting preserved as they were left almost 60 years ago, when they were boarded up and sealed until their discovery in 1988.

Something of an urban time machine, the building conveys a vivid sense of the deplorable living conditions experienced by its tenants who, during its 72-year tenure as housing, may have numbered as many as 10,000 residents in total. During the height of immigration, the industrial city's large, ugly, and appalling immigrant **ghettos** often experienced crime and other social ills that led to the popular stereotype that 'unsavoury' European immigrants suffered from pathological behaviours resulting from their inferior racial status. In this environment, East European Jews, Italians, Poles, Ukrainians, and Lithuanians, as well as many other non-Northern European immigrants, experienced strikingly harsh conditions of work and residence. These degraded conditions of life only further confirmed the racist views of those who saw Southern and Eastern European immigrants as innately degenerate. Such racism continued well into the twentieth century and still has not completely disappeared, although it has largely been replaced by discrimination and racist actions towards more recently arrived immigrant groups, especially racialized minorities.

Figure 3.1 The Lower East Side Tenement Museum, New York City: building in centre of photograph.

Source: Photograph by William Kimber.

BOX 3.2   Vancouver's Chinatown

In the second half of the nineteenth century, large numbers of Chinese arrived in British Columbia as contract labourers working under dangerous and exploitive conditions laying track for the Canadian Pacific Railway and in various resource industries. Continued discrimination forced the Chinese at the turn of the twentieth century into older and poorer parts of Canada's cities, where the familiar 'Chinatowns' emerged. Vancouver's Chinatown was the largest, and the Chinese immigrants who lived there experienced a variety of racist policies and actions. Indeed, Chinatown itself was a response to the climate of racial hostility. The Chinese immigrants were not legally required to live apart from people of European background, but the 'unfriendly feelings' in the rest of the city made it seem the wiser course. They were viewed as 'others' by white European Vancouverites, who perceived themselves as the superior race and Chinatown as a place that deviated dramatically from their values and moral standards. From the late nineteenth century into the early twentieth century, anti-Chinese rallies helped to raise the ire of the nativist white population and at times led to vigilante action and riots against the Chinese.

As Figure 3.2 demonstrates, discrimination took many forms, from disparaging cartoons in local newspapers to systematic harassment by city inspectors against Chinese immigrant-operated laundries, Chinatown's overcrowded and poor living conditions, and violations of acceptable public morality, including gambling, prostitution, and drugs. The Chinese immigrants were not allowed to vote and at one time or another they experienced restrictions in landownership, housing, and public accommodations. These discriminatory regulations, along with other prohibitive social practices and immigration restrictions, resulted in declining Chinese immigration, a predominantly male population, limited occupation choices, and spatial segregation well into the 1960s.

THE UNANSWERABLE ARGUMENT.

Figure 3.2 A Typical Cartoon Depicting the Chinese in Vancouver, Early Twentieth Century

Source: K.L. Anderson, *Vancouver's Chinatown: Racial Discourse in Canada, 1875–1980* (Montreal and Kingston: McGill-Queen's University Press, 1991), 87, from the *Saturday Sunset*, 10 Aug. 1907.

THE CHANGING GEOGRAPHY OF IMMIGRANT SETTLEMENT: FROM CENTRAL CITY TO THE SUBURBS

Although immigrants to North American cities have settled in residential concentrations since the late 1800s, as described in Boxes 3.1 and 3.2, the focus of the chapter is on the contemporary period, beginning with the end of World War II. Because of the sea change in immigrant origins and settlement patterns in the mid-1960s we divide this part of the chapter into two sections, post-war European settlement (1945 to the 1960s) and the development of a more diverse suburban immigrant landscape (1970s to the present).

Post-War European Settlement, 1945–1960s

Between the end of World War II and the 1960s most immigrants to the US and Canada originated from Britain and continental European countries such as Germany, Greece, Hungary, Italy, the Netherlands, Poland, and Portugal. Restrictive legislation directed especially against Asians and national origin quotas curtailed the number of immigrants from particular source countries until the 1960s, when previous restrictions began to be lifted.

During this period, immigrant settlement, especially in the US, was thought to follow the spatial assimilation model (Park et al., 1925). The model defined the process whereby a group attains residential propinquity with members of the majority American- or Canadian-born population. In the United States and Canada, during the early post-war period, it generally involved the spatial proximity of European immigrant groups to the American- or Canadian-born white population. Shifts in neighbourhood ethnic composition were based on the principles of **invasion and succession**, an idea borrowed from plant ecology, whereby a newly arrived immigrant group invades an inner-city neighbourhood and ultimately succeeds the group living there as that group moves outward towards the suburbs.

The spatial assimilation model is illustrated schematically in Figure 3.3. Most newly arrived immigrants have limited economic resources and as a result settle in relatively inexpensive rental accommodation close to jobs in the inner city. There, immigrants from specific national or cultural origins tend to cluster spatially, forming enclaves such as the Jewish Quarter, Little Italy, Portugal Village, and Greektown. The development of these clusters is aided by chain migration whereby immigrants send information back home and are followed by others from the same city or town. Often the only housing option is a room or apartment in an overcrowded tenement or in a dwelling owned or rented by a member of the same ethnic group. If the enclave contains a substantial number of immigrants from the same cultural background, ethnic businesses and cultural and religious institutions may be established. Consequently, visible evidence of the group emerges on the urban landscape, and certain neighbourhoods come to be associated with specific ethnic groups. Neighbourhood labels, entrance gates, and ethnic retailing serve to identify particular groups with particular spaces, even if in reality these neighbourhoods are more multi-ethnic than homogeneous.

As their job prospects, incomes, and English-language abilities (French in Quebec) improve, immigrants (and/or their descendants) move outward to better housing, frequently as homeowners. Homeownership provides the newcomers with a sense of permanence in the new land and an investment opportunity with the possibility of future social, economic, and educational gain. At this point in the model, individuals may also lose some of their cultural identity and disperse spatially. Massey (1985), in particular, draws a direct link between acculturation (acquiring the language and values of the receiving society), social

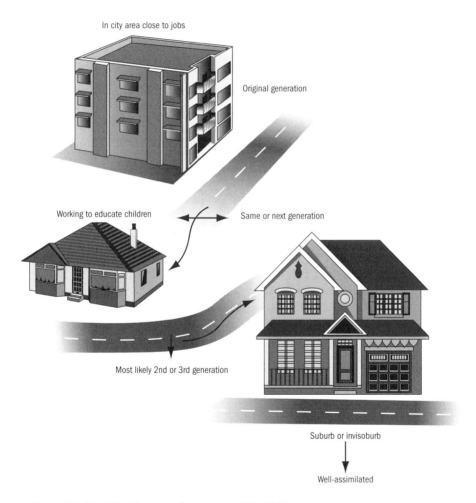

Figure 3.3 Post-War European Settlement, 1945–1960s

Source: From a sketch by E. Skop.

mobility (improved economic status), and spatial assimilation. Finally, by the third or fourth generation they continue moving outward to larger, more expensive suburban housing. At this stage the group is hypothesized to experience a further breakdown of cultural identity and become more dispersed. The educational attainment of each generation improves and incomes increase. Overall, through the process of suburbanization, the status of the group rises and eventually meets that of the American- or Canadian-born, white Anglo-Saxon majority. In the model, as a result of economic mobility and cultural assimilation, European ethnic neighbourhoods (and identities) weaken and eventually disappear into the suburbs.

European immigrant settlement in many large North American cities followed this ideal pattern, especially during the early post-World War II period. There were exceptions, however. Some ethnic groups did not disperse or dispersed slowly, regardless of their improved economic situation and movement to the suburbs. This result is perhaps more evident in Canadian cities. In Toronto and Montreal, for example, Italian immigrants and, to a lesser extent, Greek and Portuguese immigrants moved from central-city neighbourhoods in a sectoral fashion towards the outer suburbs where many remain spatially concentrated and have established a strong institutional, cultural, and commercial presence. At the same time, many non-European immigrants as well as their US- and Canadian-born descendants, especially those of Chinese background in both countries and of Mexican background in the US, remained spatially isolated and, in many cases, experienced extreme residential discrimination.

Due to the continued spatial concentration of some immigrant groups during this period, as well as the more recent emergence of radically different and more heterogeneous suburban immigrant geographies, the validity of the spatial assimilation model has been called into question. Researchers have suggested a substantial modification of the model, if not its elimination (Alba et al., 1999; Murdie and Teixeira, 2003; Skop and Li, 2003; Allen and Turner, 2005; Skop and Li, 2011). In the following discussion we consider these new immigrant geographies, especially as they have emerged in the suburbs of major cities in both countries.

The Development of a Suburban Immigrant Landscape, 1970s–Present

Beginning in the 1970s, both the US and Canada witnessed a decline in European immigration and increasingly large immigrant flows from Asia, the Caribbean, Latin America, and more recently, Africa and the Middle East. The new arrivals have been quite diverse, arriving with dissimilar racial and religious backgrounds, as well as various levels of educational attainment and job skills. A primary catalyst for this change was the introduction of immigration legislation in both countries in the 1960s that removed restrictions on race and national origins and placed more emphasis on education and employment skills, especially in Canada, and family reunification in the US (also discussed in the

Introduction to this volume). This process coincided with the recovery of Europe from the damaging effects of World War II and less incentive for Europeans to move to North America. Furthermore, the unprecedented growth of the global economy and the increased segmentation of the labour market have resulted in new linkages between North America and Africa, Asia, and Latin America, thereby creating a dramatic shift in the source countries of migrants.

During this period, the US received a sizable number of highly skilled and financially well-off immigrants as well as a substantial number of undocumented migrants. Along with the very large but primarily American-born black population in the US, the relatively larger number of low-skilled and undocumented workers coming to the US is perhaps the most important factor accounting for increasing differences in the composition of immigrants between the US and Canada. Although Canada has an expanding temporary foreign worker program, the impact, as yet, is not nearly on the same scale as that in the US.

Importantly, this shift in immigrant source countries as well as immigrant type has been accompanied by a significant change in the residential patterns of immigrants, with many cities witnessing the emergence of multi-ethnic suburban immigrant landscapes, as discussed in other chapters in this volume. This pattern results both from the relocation of central-city residents to suburban locales *and* from the settlement of most newly arrived immigrant groups directly in the suburbs (Murdie and Teixeira, 2003; Alba and Denton, 2004; Singer, 2008; Teixeira, 2007).

The reasons for the shift in immigrant settlement to the suburbs are complex (see Box 3.3). One factor is post-industrial restructuring and labour market segmentation, which has resulted in the decline of manufacturing jobs, a traditional source of employment for newly arrived immigrants, and also has created a shift in the location of manufacturing jobs and routine office and service functions to the suburbs. In addition, the intrinsic labour demands of the restructured economy require a flexible labour force to fill specific occupational voids. Thus, highly skilled migrants have been recruited to work in the burgeoning creative economy while less-educated, lower-skilled, and poorer migrants, especially in the US, have been enlisted to fill the lowest and least desirable jobs. Two outcomes have emerged from this economic restructuring. One is that many immigrants who first settled in central cities are moving to the suburbs to minimize their journey to work by following the jobs that have increasingly emerged outside of the central city; another is that newly arrived immigrants are bypassing the central city immediately to locate closer to suburban job opportunities.

At the same time, especially since 1970, urban geographies have become increasingly decentralized. Extensive suburbanization and dispersal have been the dominant spatial trends in both US and Canadian metropolitan areas over the past 40 years. This form of urban structure, with its automobile-oriented landscapes, its relatively open housing markets, and new suburban developments continually being built, provides a diversity of housing, including relatively new and expensive single-family houses on large lots and more

BOX 3.3	Twenty-First-Century Gateways: Regional and Local Trends

One feature of recent immigration to the US is that it has affected regions of the country that until a short time ago had little experience with post-1965 immigration. Small towns in Iowa, Kentucky, and western North Carolina find that they have significant numbers of Latinos in their communities, drawn there because of job opportunities in such industries as meat-packing and poultry processing, while towns in upstate New York and Maine have new populations of Somali refugees, resettled there with the assistance of refugee resettlement organizations. Despite this trend, the majority of post-1965 immigrants are still settling in the major metropolitan areas, including the established contemporary gateways of immigration: New York, Chicago, Los Angeles, and Miami.

There are, however, new medium-sized metropolitan immigrant gateways in the Southeast (Atlanta, Raleigh–Durham, and Charlotte), Southwest (Dallas and Fort Worth, Phoenix, and Las Vegas), Upper Midwest (Minneapolis–St Paul), and Northwest (Portland, Seattle), many of which have seen their immigrant populations triple and quadruple in size as a result of new immigration in the past 20 years. These twenty-first-century gateways are increasingly important destinations for new immigrants (Singer, 2008).

In addition to bolstering the populations in new and established gateways, immigrants are bypassing the city and settling directly in the suburbs in these places in large numbers; for the first time, more immigrants live in suburban areas than in cities. Demographer William Frey has pointed out that in 65 of the 102 largest metropolitan areas, minorities account for most of the suburban growth. More than half of Asian Americans in large metropolitan areas reside in the suburbs, as do half of Latinos (Frey, 2001). This means that suburbs are the new context for immigrant integration, a significant shift from post-war European patterns.

The majority of immigrants in Canada continue to settle in the established gateways of immigration: Toronto, Vancouver, and Montreal. In 2006, these three metropolitan areas received 69 per cent of the country's recent immigrants (Murdie, 2008). At the same time, medium-sized metropolitan immigrant gateways are holding their own as reception centres for new immigrants, and in some instances are becoming more important. Of the medium-sized metropolitan areas, Ottawa–Gatineau and Calgary have increased considerably in importance as gateways while Edmonton, Winnipeg, Hamilton, Kitchener, and Windsor are holding their own.

The suburbanization trend in immigrant settlement is equally present in Canada, if not more so. By 2006, almost all of Toronto's newly arrived immigrants were settling in the suburbs, and Vancouver's recent immigrant population was distinctly suburban. In Montreal, however, the vast majority of new immigrants remained centrally concentrated. Suburban enclaves pose particular challenges to service providers and municipal authorities, and highlight the importance of tracking these trends in promoting and encouraging successful integration.

affordable higher-density multiple-family housing in both newly developed and older suburbs. This variety of housing means that immigrants of varying socio-economic status can be accommodated in both older and newer suburbs, including previous arrivals who have achieved upward economic mobility and aspire to owner-occupied houses on large lots, recently arrived immigrants with substantial human and financial capital who can afford relatively expensive single-family housing on first arrival, and newly arrived immigrants and refugees of lower socio-economic status who have little alternative but to opt for higher-density apartment living or older single-family dwellings.

It is important to point out that there are contrasts between the US and Canada, however, particularly in the structure, form, and nature of cities and suburbs. Canadian suburbs tend to be more compact than their American counterparts and contain a higher proportion of relatively affordable high-rise rental housing. Compared to the US, Canada did not have a highly developed Interstate Highway system that in major US cities encouraged the rapid and extensive development of low-density suburbs as well as the decline and, in some cases, even the destruction of central-city neighbourhoods that formerly housed immigrant and minority populations.

Two major changes also were taking place in central cities that persuaded or forced immigrants who once lived there to leave for the suburbs, and these changes ruled out central cities as a housing option for most new immigrants. The first was the Administration resulting in expanded suburban housing opportunities for central city residents by insuring residential mortgage loans and the movement of blacks from the American South to the core areas of northern US cities, occurring at the same time that the Federal Housing Administration created policies that encouraged **white flight**; the result was that many whites, including European immigrants, relocated from central cities to the suburbs. In many US cities the outcome was widespread abandonment of central-city housing and depopulation of these areas, thus reducing their tax bases and ability to provide adequate services as well as the ability of property owners to obtain mortgage financing. In contrast, Canada has not experienced the same level of central-city abandonment and depopulation. Instead, Southern European immigrants used their skills in home renovation to fix up existing houses, often obtaining financing from well-established co-ethnics who preferred to invest in real estate rather than financial markets.

The second major change has been the **gentrification** of central-city areas, which has resulted in the decline of affordable rental housing and the displacement of working-class populations as more affluent groups move into professionally renovated housing. In the US, this trend has been characterized as a reversal of white flight. There are several potential implications for immigrants. Those who settled in central-city areas previously and achieved home ownership are able to sell at a considerable profit and relocate elsewhere, primarily in the suburbs. In contrast, central-city immigrant renters with relatively low incomes, many of whom are newcomers, have fewer options. For many of these people, the only possibility is an aging high-rise apartment in the inner suburbs further removed from efficient public transportation and the services on which they have come to rely. Gentrification also has affected the settlement options of new immigrants who cannot afford to or do not wish to settle in gentrified areas of the central city. Thus, the central city no longer plays the role it once did when the spatial assimilation model was first proposed.

A TYPOLOGY OF URBAN AND SUBURBAN ETHNIC SETTLEMENTS

Although the broad outline of immigrant settlement is similar in both countries, the extent and details vary from one urban area to another, partially based on immigrant composition and the historical patterns of immigrant settlement (Singer, 2008). In Figure 3.4 (substantially modified from Boal, 1976: 57) we summarize the diversity of contemporary immigrant settlement in North American cities by presenting a typology of urban and suburban ethnic settlements and providing examples of each. In this diagram several factors are shown to combine to identify the likelihood of spatial concentration: the type of initial resources that immigrants bring; the degree of similarity and difference between individual immigrants and the receiving society; the level of perceived and actual residential discrimination in the receiving society; and the level of solidarity and community in the receiving society. While this figure summarizes the potential experiences of immigrants in both societies, our discussion points out differences between the US and Canada where they exist. It must be emphasized, however, that in many instances the empirical evidence is thin or non-existent and often contradictory. It is therefore difficult to draw exact comparisons between the two countries.

No Spatial Concentration

If the immigrant group exhibits several of the following attributes—few initial resource differences from the receiving society, many similarities in individual traits, and low levels of discrimination—the group will probably exhibit no or very limited spatial concentration. Examples include well-established British and Western European groups such as those of Dutch, German, and Irish origin.

Low spatial concentration can also result from a group that exhibits high cultural retention, a process that Zelinsky and Lee (1998) refer to as 'heterolocalism' and Skop (2002) has characterized as an '**invisiburb**'. Examples include Asian Indians and Filipinos. In large part, the retention of group culture and identity without spatial concentration has been made possible by post-1960s improvements in transportation, especially high-speed expressways, and communication technologies, particularly the Internet (Skop and Adams, 2009).

Asian Indian immigrants are perhaps the most-researched example of a group that has settled directly in the suburbs and in certain US cities has developed a community based on social networks and organizations without the need for residential concentration. Price et al. (2005) note that, in Washington, DC, the

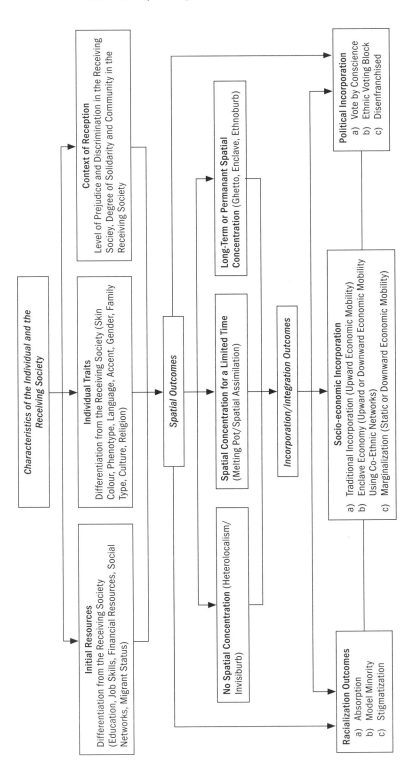

Figure 3.4 A Typology of Urban and Suburban Immigrant Settlements in US and Canadian Cities

Source: Drawing by R. Murdie and E. Skop.

group is fluent in English and well educated, with specialized skills, often in high-technology occupations. Consequently, most Asian Indians can choose where they wish to live and have no need to concentrate residentially. Skop and Altman (2006) further note that the lack of Indian residential clustering may be a deliberate strategy to avoid becoming racialized in a highly racialized society.

In contrast to findings from the US, Indo-Canadians (especially Punjabi Sikhs) are highly concentrated in suburban Toronto and Vancouver where they have developed a significant commercial and institutional structure (Hiebert, 1999; Murdie and Teixeira, 2003). This divergence from the more dispersed spatial patterns in several US cities may result from differences in occupational structure and varying attitudes in prejudice and discrimination. Indo-Canadians are not as strongly represented in professional occupations, probably due to a comparatively lower level of educational achievement and a lack of recognition of formal qualifications earned elsewhere. Also, while the spatial concentration by this group may be a strategy to avoid discrimination it may also be a desire to enhance their cultural identity, encouraged by Canada's pluralistic multiculturalism policy.

Spatial Concentration for a Limited Time

This is a classic representation of the spatial assimilation model identified in Figure 3.3 and discussed earlier. Many Southern and Eastern European immigrant groups (and their descendants) in the US and Canada have followed this ideal pattern, locating first in a spatially concentrated enclave in the central city and then settling in a more dispersed fashion in the suburbs (Peach, 2005). Empirical evidence in the contemporary period, however, is mixed. Italian immigrant neighbourhoods, for example, continue to exist in New York City and its suburbs, although they generally declined in total Italian-American population during the late twentieth century (Alba et al., 1997). In Montreal and Toronto, where Italian immigration is more recent than in the US, Italian neighbourhoods continue to flourish, both in the central city and the suburbs (Germain and Rose, 2000; Murdie and Teixeira, 2003); however, Italian populations in central-city neighbourhoods have declined as gentrification takes hold. Without a new stream of Italian immigrants, the ultimate viability of these neighbourhoods may be in doubt. Peach (2005) also notes that Italians are more highly segregated in Canada than in the US. This may be because large-scale Italian immigration to Canada is more recent, resulting in greater retention of cultural identity. It might also be evidence of the roles of multiculturalism and assimilation in prompting different spatial outcomes in divergent social contexts.

Long-Term or Permanent Spatial Concentration

Immigrant groups in long-term or permanent spatial concentrations show a considerable degree of distinctiveness from the receiving society. They often differ dramatically with respect to initial resources, although some groups catch up to and exceed the receiving population while remaining spatially concentrated. The group experiences varying degrees of prejudice and residential discrimination, and the levels vary depending on perceptions of difference by both the receiving community and the immigrants themselves. These concentrations are usually referred to as enclaves. The reasons why enclaves emerge are still debated; they include discriminatory behaviour and constraining forces within the receiving society, as well as the preference of individuals for living in ethnoracial neighbourhoods because of the supportive social systems and reciprocal relations that may exist there. As Marcuse (1997: 242) notes, although an immigrant enclave generally has a positive meaning for its residents it may share some of the negative characteristics of a ghetto, a term used in the current context primarily to capture the long-standing African-American experience in US cities, which emerged and continues largely because of the historic legacy of slavery, extreme prejudice, and residential discrimination.

The Jewish population in US and Canadian cities is the archetypical example of an enclave. Jews have achieved a high level of economic integration but have moved voluntarily in a highly segregated fashion from their original central-city locations to the suburbs, leaving little trace of their original enclaves. The Jewish population, however, is not homogeneous and groups distinguished by different levels of economic status and/or religious orthodoxy often live in spatially concentrated but separate enclaves in the suburbs.

Many other ethnic groups exhibit characteristics of an enclave. Italian Canadians in Toronto and Montreal have already been mentioned as a group that first settled in a spatially distinct area of the central city and later moved voluntarily to the suburbs, largely retaining a concentrated residential pattern and a strong complement of institutional and commercial facilities. In contrast to the Jewish population, however, residual concentrations of Italian settlement, often with a relatively high proportion of first-generation immigrants, remain in many central cities. Longer-established immigrant areas such as Chinatowns in North American central cities also function as traditional enclaves, although in the past they functioned more like isolated ghettos.

More recent groups who immigrated directly to the suburbs also have settled in enclaves and are often large enough to support an institutional and business structure. Although most of these groups can choose where to live in the suburbs some have more choice than others, depending largely on their financial resources. In Toronto, for example, there are two very different groups of suburban newcomers: (1) relatively low-income immigrants and refugees from Asian, African, and South American countries housed primarily in aging, rental high-rise buildings in the inner suburbs where they may experience discrimination by landlords; and (2) well-educated immigrants from countries such as China and India with considerable financial resources who can generally afford suburban home ownership upon first arrival (Murdie, 2008).

New suburban concentrations of Asian immigrant groups in large American metropolitan areas such as Los Angeles led Li (2009) to suggest a different model of ethnic settlement labelled **ethnoburb** (a multi-ethnic suburb). Ethnoburbs are characterized by a dynamic ethnic economy, strong links to the international economic system, and a multi-ethnic population whereby one group does not necessarily constitute a majority. Li argues that ethnoburbs differ from traditional enclaves in that they are larger in area and population and lower in ethnic population density. They are also more likely to contain persons of varying economic status and act as integrated residential and business centres with many more residents working nearby than in the typical enclave. The prototype for Li's ethnoburb is San Gabriel Valley, a rapidly growing Los Angeles suburb fuelled by waves of immigrants from Hong Kong, Indochina, mainland China, and Taiwan, many of whom have created a variety of businesses linked to the international economy. The ethnoburb model has since been tested in other Pacific Rim cities, notably the large areas of Asian settlement in Silicon Valley in California (Li and Park, 2006) and in suburban Toronto and Vancouver that exhibit many characteristics of ethnoburbs (Edgington et al., 2006; Lo, 2006).

THE SPATIAL ASSIMILATION MODEL, THE SPATIAL MOSAIC MODEL, AND THE INTEGRATION OF IMMIGRANTS

In the 1990s, Ceri Peach (1996) questioned whether there was such a thing as good or bad segregation and concluded that there are positive as well as negative outcomes from immigrant concentration. More than 10 years later, Smith and Ley (2008) revisited this idea by identifying the consequences for immigrants living in areas of concentrated poverty in Toronto and Vancouver. They argued that there are both 'places of stigma', areas viewed negatively by outsiders as a result of a high incidence of poverty, crime, and/or a racialized population, and 'neighbourhoods of hope', which also have a high incidence of poverty but are well served by public transportation and have good access to a wide range of local services, including assistance for recent immigrants. Essentially, they argue that where an immigrant lives influences how he/she will be incorporated into society, and that neighbourhood effects (transportation networks, public housing, and much needed services and employment) are important attributes distinguishing inclusion from exclusion.

At the bottom of Figure 3.4 we explore how immigrant settlement patterns (spatial outcomes) influence the immigrant integration process by outlining various dimensions of the process. Although any number of dimensions could be considered, we focus on how immigrants might become racialized, as well as how they may become socio-economically and politically incorporated into society. In each case, we point out three possible scenarios that might result from different spatial outcomes. In this way, we follow Portes and Zhou's (1993) idea of 'segmented assimilation' by identifying various integration experiences. The various scenarios in our diagram indicate that integration is not straightforward but rather a messy process with more than one result likely. Also, the lines and arrows flowing back to the 'spatial outcomes' cell suggest that the various integration scenarios can reinforce or weaken immigrant concentration.

At the heart of the US experience is the idea that residential change and social mobility are inextricably interwoven. The assumption is that the movement of immigrants away from concentrations indicates that social and structural barriers have weakened, thus suggesting eventual assimilation into the dominant social, economic, and political structures. Thus, as time passes and segregation lessens, race as a factor will disappear, immigrants will experience upward economic mobility, and individuals will vote along political lines perhaps determined by their personal world view and/or socio-economic status instead of as an ethnic voting bloc.

The assumption, however, that no spatial concentration, or spatial concentration for a limited time, is equivalent to assimilation no longer makes as much sense in the face of more complex immigrant geographies in the US. In reality, various integration outcomes are likely. For some immigrants, the straight-line progression from central-city enclave to suburban neighbourhood translates into privilege, power, and material advantages. But for those recent immigrants whose socio-economic status is already comparable to the middle class in the US, their direct settlement in the suburbs is hardly evidence of assimilation. For these newcomers, there is no steady progression through time and space to achieve their current status; instead, they arrive with those characteristics. In their critique of US research, Wright, Ellis, and Parks (2005: 113) warn against viewing the 'white, middle class, home-owning suburbanite as the single standard of cultural membership and belonging'.

The best example of this new process is the ethnoburb where 'ethnic suburbs offer minority people the opportunity to resist complete assimilation into the white cultural and social norms of the host society' (Li, 2009: 17). Here, immigrants may become upwardly mobile in terms of socio-economic status, but they also may turn to an ethnic voting bloc to achieve political goals. They also may be considered a **model minority**, which implies that they are still racialized by the dominant society and may believe they have attained middle-class status without realizing they are underemployed and overworked. The model minority idea also serves the majority population by rewarding certain immigrant behaviours and attitudes, while using the group as the standard that all other immigrant groups should emulate.

Lower-skilled immigrants who first locate in suburban areas also make the idea of moving outward and upward debatable. Indeed, the fact that poor immigrants are increasingly pushed to outlying suburban areas in the US provides further evidence against the notion that low levels of spatial concentration become the endpoint in the assimilation process. These immigrants often end up socially isolated and racially stigmatized, increasingly marginalized economically, and disenfranchised politically. This negative outcome is well illustrated in Hondagneu-Sotelo's (2007) *Doméstica*, which outlines the extreme isolation and anguish of immigrant domestic workers living in Los Angeles's suburbs. Of course, many of these domestic workers arrive in the US alone (e.g., live-in nannies and gardeners), which exacerbates the isolation felt in the suburbs. But those who immigrated with families often express the same feelings of separateness when asked about their experiences living in the suburbs.

Thus, even though more and more immigrants are spatially assimilated in the suburbs, they are not necessarily integrating into the normative, dominant structures of American-born, non-Latino, white America. Instead, some immigrants are selectively adopting receiving society traits while deliberately preserving community structures and values; at the same time, others are forced into a situation where the lack of a support system, especially for those located in older housing stock in the inner suburban ring, leaves many without adequate access to transportation, health care, legal services, and other social services.

The Canadian experience is driven by the idea that spatial concentration is potentially beneficial in achieving immigrant integration, partially because of the settlement support available from co-ethnics living in the same neighbourhood and extensive networks of community-based organizations. Ideally, over time, immigrants in these neighbourhoods achieve a considerable level of integration while retaining their unique traditions and engaging in a variety of transnational activities with their homeland. In reality, however, the **racialization**, socio-economic, and political outcomes are mixed.

The example of Vancouver's Hong Kong immigrants (Mitchell, 2004) illustrates how social networks created in concentrated communities can facilitate the adjustment process by helping immigrants find jobs and housing, and by providing access to social clubs and other systems of personal support. Ghosh (2007) also reveals how residential clustering of Bangladeshi immigrants in suburban Toronto and the social and symbolic ties they develop, as well as the institutionally complete neighbourhood they create, can have a positive effect and provide not only a means of survival, but also a way of acquiring a place to live and a job.

Although spatial concentrations can assist immigrant integration, the advantages of these neighbourhoods need to be tempered by their potential negative impacts. For example, the housing obtained may be of poor quality, acquisition of the official language may be hindered, jobs, especially in the ethnic economy, may not pay well, and information about better-paying jobs in the wider economy may not be available. Instead of 'neighbourhoods of hope' these areas may become 'places of stigma', making interaction with the wider society more difficult. For example, while Toronto's Bangladeshi immigrants have been successful in developing an institutionally complete and vibrant community, they remain stigmatized and economically marginalized, largely due to their recent arrival and non-recognition by Canadian employers and professional organizations of the education they received prior to their arrival in Canada (ibid.).They also are politically disenfranchised because many have not yet obtained the right to vote.

In addition, spatial concentration can have unintended consequences in regard to the integration of certain immigrants, especially racialized minorities who experience stigmatization and initial disadvantages in terms of social and economic capital. For example, Portuguese-speaking black African immigrants in Toronto tend to locate in or near Little Portugal, the historic area of Portuguese settlement in the inner city. This has obvious linguistic advantages for the newcomers, but the white Portuguese community is not always welcoming and does not play a particularly important role in the settlement of these newcomers. In part, this conflicted relationship derives from Portugal's long control over its African colonies (Teixeira, 2006).

What becomes quickly apparent from these and other case studies, both in the US and Canada, is that the integration process can take different forms, and the 'one pattern fits all' type of thinking needs to become 'this pattern fits this individual/group' and 'that pattern fits that individual/group'. Just as segmented assimilation allows for varied *socio-economic* outcomes where individuals/ groups become assimilated into fragmented, segregated, and unequal segments in society, research focused on the geography of immigrant settlement should allow for varied *spatial* outcomes where individuals/groups become incorporated into dispersed, disjointed, and segregated neighbourhoods in US and Canadian society (Skop and Li, 2005).

CONCLUSION

The immediate post-World War II period in both large US and Canadian cities was characterized by a major inflow of European immigrants who usually settled in enclaves in inner cities before moving outward to larger and more expensive housing in the suburbs, where it was anticipated that their ethnic concentrations would weaken and eventually disappear. To some extent, this happened in both countries, although immigrant groups in large Canadian cities tended to retain a higher level of spatial concentration following their move to the suburbs, reflecting, in part, the Canadian ideal of multiculturalism, which became a government policy initiative in 1971. In both countries the period beginning in the 1970s has been characterized by a continued relocation from the central city to the suburbs of previous inner-city immigrants, and the settlement of most newcomers directly in the suburbs, primarily in localized enclaves. Consequently, in the US, the spatial assimilation model has been increasingly questioned as a relevant description of the multi-ethnic suburban immigrant geography that emerged following 1970.

The likelihood of immigrant groups concentrating spatially is determined primarily by differences between the group and the receiving population in terms of initial resources, the degree of similarity in individual traits, and the level of prejudice and discrimination, as well as the degree of solidarity in the migrant community. The spatial outcomes (Figure 3.4) can be characterized in at least three ways: no spatial concentration (heterolocalism/invisiburb); spatial concentration for a limited time followed by spatial assimilation; and long-term concentration (ghetto, enclave, ethnoburb). All three outcomes can be found in US and Canadian cities. It is also likely that in the future this ethnic mosaic will become more diverse in terms of race, class, and national origins, just as the white suburb becomes less homogeneous, with implications for increased complexity in the system.

Integration into various sectors of society, but especially the acquisition of decent housing, language skills, additional education, and a good job, is crucial in the initial stages of settlement and may be assisted or impeded by the nature of ethnic concentrations. The outcomes of immigrant integration are variable in both countries—some groups achieve considerable success shortly after arrival in the new country and others do not. Settlement support from co-ethnics in the same neighbourhood and assistance from community-based organizations are especially important in speeding up integration. Even so, it is hard to make the case that one single undifferentiated process of integration is operative in either Canada or the US, and policy-makers must focus on various outcomes to ensure the most success.

QUESTIONS FOR CRITICAL THOUGHT

1. Why is suburbanization so important in understanding both old and new immigrant settlement patterns?

2. Is there such a thing as 'good' segregation? How does this differ from 'bad' segregation?

3. What are the implications of various spatial outcomes for second- and third-generation descendants of today's immigrants?

4. If you were a recently arrived immigrant, where would you like to live (which city and where within the city) and why would you like to live there?

5. As a newly arrived immigrant what kind of constraints do you think you would face in finding an appropriate place to live in your preferred city and/or neighbourhood?

SUGGESTED READINGS

1. Ghosh, S. 2007. 'Transnational Ties and Intra-Immigrant Group Settlement Experiences: A Case Study of Indian Bengalis and Bangladeshis in Toronto', *GeoJournal* 68: 223–42. Comparison of the housing experiences of two recently arrived South Asian groups with different transnational ties, illustrating how these differences affected their initial neighbourhood location and housing circumstances as well as their subsequent residential trajectories.

2. Murdie, R.A. 2008. *Diversity and Concentration in Canadian Immigration: Trends in Toronto, Montréal and Vancouver, 1971–2006.* Toronto: Research Bulletin 42, Centre for Urban and Community Studies, University of Toronto. Analysis of the increased ethnic diversity and spatial redistribution of immigrants in Canada's three major metropolitan areas over 35 years with particular focus on the increased suburban location of earlier arrivals and newcomers, especially in Toronto and Vancouver.

3. Singer, A. 2008. 'Twenty-First-Century Gateways: An Introduction', in A. Singer, S.W. Hardwick, and C.B. Brettell, eds, *Twenty-First-Century Gateways: Immigrant Incorporation in Suburban America*. Washington: Brookings Institution Press, 3–30. Introduction to a volume that documents the growth of foreign-born populations in a variety of twenty-first-century gateway cities, including Dallas, Phoenix, Atlanta, Washington, Sacramento, Minneapolis–St Paul, Portland, Austin, and Charlotte.

4. Skop, E., and W. Li. 2011. 'Urban Patterns and Ethnic Diversity', in J. Stoltman, ed., *21st Century Geography: A Reference Handbook*. New York: Sage. A valuable overview of urban ethnic geography, arguing that fundamental changes in urban structure and migrant characteristics require a reconsideration of the classic models of the socio-spatial behaviour of urban ethnic inhabitants.

REFERENCES

1. Alba, R., and N. Denton. 2004. 'Old and New Landscapes of Diversity: The Residential Patterns of Immigrant Minorities', in N. Foner and G.M. Fredrickson, eds, *Not Just Black and White: Historical and Contemporary Perspectives on Immigration, Race, and Ethnicity in the United States*. New York: Russell Sage Foundation, 237–61.

2. ———, J.R. Logan, and K. Crowder. 1997. 'White Ethnic Neighbourhoods and Assimilation: The Greater New York Region, 1980–1990', *Social Forces* 75, 3: 883–909.

3. ———, ———, B.J. Stults, G. Marzan, and W. Zhang. 1999. 'Immigrant Groups in the Suburbs: A Reexamination of Suburbanization and Spatial Assimilation', *American Sociological Review* 64 (June): 446–60.

4. Allen, J.P., and E. Turner. 2005. 'Ethnic Residential Concentrations in United States Metropolitan Areas', *Geographical Review* 95, 2: 267–85.

5. Anderson, K.L. 1991. *Vancouver's Chinatown: Racial Discourse in Canada, 1875–1980*. Montreal and Kingston: McGill-Queen's University Press.

6. Boal, F.W. 1976. 'Ethnic Residential Segregation', in D.T. Herbert and R.J. Johnston, eds, *Social Areas in Cities, Volume 1: Spatial Processes and Form*. London: Wiley, 41–79.

7. Edgington, D.W., M.A. Goldberg, and T.A. Hutton. 2006. 'Hong Kong Business, Money and Migration in Vancouver, Canada', in Li (2006: 155–83).

8. Frey, W.H. 2001. *Melting Pot Suburbs: A Census 2000 Study of Suburban Diversity*. Washington: Brookings Institution Press.

9. Germain, A., and D. Rose. 2000. *Montréal: The Quest for a Metropolis*. Chichester: Wiley.

10. Ghosh, S. 2007. 'Transnational Ties and Intra-immigrant Group Settlement Experiences: A Case Study of Indian Bengalis and Bangladeshis in Toronto', *GeoJournal* 68: 223–42.

11. Hiebert, D. 1999. 'Immigration and the Changing Social Geography of Greater Vancouver', *BC Studies* 121 (Spring): 35–82.

12. Hondagneu-Sotelo, P. 2007. *Doméstica: Immigrant Workers Cleaning and Caring in the Shadows of Affluence*. Berkeley: University of California Press.

13. Li, W., ed. 2006. *From Urban Enclave to Ethnic Suburb: New Asian Communities in the Pacific Rim Countries*. Honolulu: University of Hawai'i Press.

14. ———. 2009. *Ethnoburb: The New Ethnic Community in Urban America*. Honolulu: University of Hawai'i Press.

15. ——— and E. Park. 2006. 'Asian Americans in Silicon Valley: High Technology Industry Development and Community Transformation', in Li (2006: 119–33).

16. Lo, L. 2006. 'Suburban Housing and Indoor Shopping: The Production of the Contemporary Chinese Landscape in Toronto', in Li (2006: 134–54).

17. Marcuse, P. 1997. 'The Enclave, the Citadel, and the Ghetto: What Has Changed in the

Post-Fordist U.S. City', *Urban Affairs Review* 33, 2: 228–64.
18. Massey, D.S. 1985. 'Ethnic Residential Segregation: A Theoretical Synthesis and Empirical Review', *Sociology and Social Research* 69, 3: 315–50.
19. Mitchell, K. 2004. *Crossing the Neoliberal Line: Pacific Rim Migration and the Metropolis.* Philadelphia: Temple University Press.
20. Murdie, R.A. 2008. *Diversity and Concentration in Canadian Immigration: Trends in Toronto, Montréal and Vancouver, 1971–2006.* Toronto: Research Bulletin 42, Centre for Urban and Community Studies, University of Toronto.
21. —— and C. Teixeira. 2003. 'Towards a Comfortable Neighbourhood and Appropriate Housing', in P. Anisef and M. Lanphier, eds, *The World in a City.* Toronto: University of Toronto Press, 132–91.
22. Park, R.E., E. Burgess, and R.D. McKenzie. 1925. *The City.* Chicago: University of Chicago Press.
23. Peach, C. 1996. 'Good Segregation, Bad Segregation', *Planning Perspectives* 11, 3: 379–98.
24. ——. 2005. 'The Mosaic versus the Melting Pot: Canada and the USA', *Scottish Geographical Journal* 121, 1: 3–27.
25. Portes, A., and M. Zhou. 1993. 'The New Second Generation: Segmented Assimilation and Its Variants', *Annals, American Academy of Political and Social Science* 530 (Nov.): 74–96.
26. Price, M., I. Chang, S. Friedman, and A. Singer. 2005. 'The World Settles In: Washington, DC, as an Immigrant Gateway', *Urban Geography* 26, 1: 61–83.
27. Singer, A. 2008. 'Twenty-First-Century Gateways: An Introduction', in A. Singer, S.W. Hardwick, and C.B. Brettell, eds, *Twenty-First-Century Gateways: Immigrant Incorporation in Suburban America.* Washington: Brookings Institution Press, 3–30.
28. Skop, E. 2002. 'The Saffron Suburbs: Asian Indian Immigrant Community Formation in Metropolitan Phoenix', Ph.D. dissertation, Arizona State University.
29. —— and P.C. Adams. 2009. 'Creating and Inhabiting Virtual Places: Indian

Immigrants in Cyberspace', *National Identities* 11, 2: 127–47.
30. —— and C.E. Altman. 2006. 'The Invisible Immigrants: Asian Indian Settlement Patterns and Racial/Ethnic Identities', in J.W. Frazier and E. Tettey-Fio, eds, *Race, Ethnicity, and Place in a Changing America.* Binghamton, NY: Global Academic Publishing, 309–16.
31. —— and W. Li. 2003. 'From the Ghetto to the Invisiburb: Shifting Patterns of Immigrant Settlement in Contemporary America', in J.W. Frazier and F. Margai, eds, *Multicultural Geographies: Persistence and Change in U.S. Racial/Ethnic Patterns.* Binghamton, NY: Academic Publishing, 113–24.
32. ——and ——. 2005. 'Asians in America's Suburbs: Patterns and Consequences of Settlement', *Geographical Review* 95, 2: 167–88.
33. ——and ——. 2011. 'Urban Patterns and Ethnic Diversity', in J. Stoltman, ed., *21st Century Geography: A Reference Handbook.* New York: Sage.
34. Smith, H., and D. Ley. 2008. 'Even in Canada? The Multiscalar Construction and Experience of Concentrated Immigrant Poverty in Gateway Cities', *Annals, Association of American Geographers* 98, 3: 686–713.
35. Teixeira, C. 2006. 'Housing Experiences of Black Africans in Toronto's Rental Market: A Case Study of Angolan and Mozambican Immigrants', *Canadian Ethnic Studies* 38, 3: 58–86.
36. ——. 2007. 'Residential Experiences and the Culture of Suburbanization: A Case Study of Portuguese Homebuyers in Mississauga', *Housing Studies* 22, 4: 495–521.
37. Wright, R.A., M. Ellis, and V. Parks. 2005. 'Re-Placing Whiteness in Spatial Assimilation Research', *City and Community* 4, 2: 111–35.
38. Zelinski, W., and B.A. Lee. 1998. 'Heterolocalism: An Alternative Model of Sociospatial Behavior of Immigrant Ethnic Communities', *International Journal of Population Geography* 4, 1: 1–18.

CHAPTER 4

THE SPATIAL SEGREGATION AND SOCIO-ECONOMIC INEQUALITY OF IMMIGRANT GROUPS

Joe Darden and Eric Fong

INTRODUCTION

The study of residential segregation has often involved the examination of various racial and ethnic groups with different socio-economic characteristics within metropolitan areas in the same country (Massey and Denton, 1993). Such studies have employed spatial assimilation theory to explain why some groups are more residentially segregated than others. Spatial assimilation is the process whereby a group attains residential propinquity with members of the majority non-immigrant population. In the US and Canada, it has generally involved the spatial proximity of European immigrant groups to the non-immigrant white population.

Since Park's (1926) path-breaking work in Chicago, researchers have been linking residential segregation to socio-economic status (SES); however, we believe the theory, which was originally formulated to explain the pattern of residential segregation of white immigrant groups from Europe, is no longer adequate to explain the more recent immigration. For example, the spatial assimilation theory assumes that immigrants who arrive in the US or Canada are of low socio-economic status compared to the population born in the US or Canada. As a result, they settle in the least desirable neighbourhoods of the metropolitan area—often in the poorest areas of the central city. As the immigrant group acquires more social capital—i.e., education, a higher-status occupation, and higher income—its members move to more desirable residential locations, often in the suburbs among a higher percentage of the non-immigrant, high-status white population (Massey and Denton, 1993; South et al., 2005). According to the spatial assimilation theory, the only barrier immigrants need to overcome is socio-economic status or class.

Such spatial assimilation theory is limited because (1) most new immigrants (i.e., immigrants who have arrived since the 1960s) are not poor; (2) most do not settle in the poor sections of central cities; and (3) most are non-European (i.e., non-white). For these reasons we hypothesize that they do not follow the traditional path to residential integration or incorporation into the mainstream that the European, white, low-income immigrants followed. Thus we reject the spatial assimilation theory and apply instead the conceptual framework of differential incorporation.

Differential Incorporation

Differential incorporation describes a process in which the white majority differentially incorporates some non-white, non-European groups into the mainstream society to a greater extent than other groups (Henry, 1994). Incorporation is conceptualized on the basis of equal access to the rewards generated and distributed by the economic and political systems in Canada and the US (Breton et al., 1990).

Differential incorporation also is a two-way process. One aspect relates to the *internal characteristics* of the minority or immigrant group. The internal characteristics are the group's social capital as measured by education, skills, and experience. Each immigrant group possesses a quantity of social capital and arrives in a new country with that social capital. The other aspect of this process involves the *external forces* imposed on the minority or immigrant group by the white majority group, despite the internal characteristics of the minority or immigrant group. Racial discrimination is a major form of these external forces (Darden, 2004; Lieberson, 1980).

Since we are examining (1) different immigrant groups within the same country and (2) the same immigrant groups in two countries, we expect the incorporation process to reveal different 'place' results. Thus we use the concept of *differential incorporation*. The comparative approach will reveal the significance of *place*, as it takes into account that non-European immigrants may have been treated *differentially* by the host majority population in each country, and such differential treatment, which may involve degrees of racial and national origin discrimination, may be reflected in the non-European immigrant group's residential segregation and social and economic status (South et al., 2005).

Differential incorporation has important policy implications, because both Canada and the US use immigration policy to recruit immigrants, and the ability of immigrants to become easily incorporated based on their education skills and other characteristics is a factor in immigration policy (Darden, 2009).

Objectives

The objectives of this chapter are fourfold:

1. to determine the mean levels of residential segregation of selected non-European immigrant groups in the US and Canada;
2. to compare the mean levels of residential segregation of each country's non-European immigrant groups;
3. to compare the mean levels of residential segregation of the same groups in the United States and Canada; and
4. to compare each group's socio-economic status in both countries.

The results should reveal whether each group is (1) more or less residentially segregated in the US than in Canada, and (2) more or less socio-economically equal to the non-immigrant (predominantly white) population. The results have implications for public policy related to immigrant incorporation.

The Difficulty of Conducting Comparative Studies

Ideally, a comparative study of residential segregation and socio-economic inequality would use data enabling us to analyze the selected immigrant groups from the same country of origin at the time of their first entry into Canada and the US and to follow them over time to determine their patterns of incorporation in each country (Jasso et al., 2000; South and Crowder, 2005); however, such data are not readily available.

It is also important to account for time of arrival, since both socio-economic status and time of arrival are associated with a reduction in residential segregation as immigrants more readily move into neighbourhoods occupied by the majority group in the destination country (South and Crowder, 2005). Whether in Canada or the US, all things being equal, immigrants with a longer history of residence in a country are more likely to reside in more suburban locations as well (Dawkins, 2009).

Finally, it is important to compare immigrants (if possible) from a single country of origin. We have taken these points into consideration as we examine immigrants from non-European countries in Canada and the US.

DATA FOR THE US AND CANADA

Data for the United States were obtained from the US Bureau of the Census *Summary File 4* (2004) and *Profile of the Foreign-Born Population in the United States: 2000*. The Canadian analysis is based on Statistics Canada's 2001 Census Profile File and the Canadian Census Individual Public Use Microdata File.

The Study Areas: United States

For the US, this chapter examines 10 primary metropolitan statistical areas (PMSAs). The areas selected are the 10 areas with the highest percentages of foreign-born (i.e., immigrants). They also have the largest concentrations of non-European immigrant groups. The areas are New York City, Los Angeles, San Francisco, Miami, Chicago, Washington, DC, Houston, Dallas, Boston, and San Diego. Many of these areas are well-established gateways for the continuous entry of immigrants; however, Dallas and Washington, both with one million or more immigrants, emerged as immigrant gateways more recently (Singer, 2009). All have populations of more than 600,000 foreign-born (see US Bureau of the Census, 2001).

The population groups selected for the US consisted of six non-European immigrant groups: Iranians, Chinese, Filipinos, Caribbeans, Asian Indians, and Latinos. Most arrived in the US after Congress passed the 1965 Immigration and Nationality Act, which abolished the quota system that favoured European countries.

The Study Areas: Canada

The Canadian analysis includes the 10 largest census metropolitan areas (CMAs): Calgary, Edmonton, Hamilton, London, Kitchener, Montreal, Toronto, Ottawa–Gatineau, Vancouver, and Winnipeg. (We excluded Quebec City because of the very small visible minority population there.) These CMAs contain significant numbers of non-European immigrant groups and represent geographic areas in British Columbia, the Prairie provinces, Ontario, and Quebec. They have also served as gateways for immigrants, especially Toronto, Montreal, and Vancouver.

Included in the analysis are six non-European racial/ethnic groups. Most immigrant groups arrived in Canada after changes in immigration policies and practices after 1967 (Darden, 2004). Information about these groups has been drawn from two census questions regarding visible minority status and ethnic origins. Canada prefers the term 'visible minority' rather than 'racial minority', which is more generally used in the US. The six groups selected are South Asians (mostly Asian Indians), West Asians (mostly Iranians), Filipinos, Caribbeans, Chinese, and Latinos. The information at the census tract level on these six groups is drawn from the census question on visible minority status, except immigrant groups of Caribbean origin, which consist largely of blacks. The visible minority question does not include the Caribbean category. The Caribbean group information at the census tract level is obtained from the ethnic question.

Methods of Analysis

To determine the extent to which selected non-European groups share neighbourhood space with non-Hispanic whites we used the index of dissimilarity

(D). The index measures residential segregation, defined as the overall unevenness in the spatial distribution of two groups. Appendix 4.1 describes the computation of the index. Simple ratios were used to assess the extent of socio-economic inequality of each non-European group vis-à-vis the non-immigrant white population for both Canada and the US.

IMMIGRATION POLICY IN THE US AND CANADA

United States

Immigration policy in the US has a racist history. Few immigrants from non-European countries voluntarily entered the country before the 1960s. Early legislation, such as the 1924 National Origins Act and the Immigration and Nationalization Act of 1952, restricted the immigration of groups from non-European countries and strongly favoured immigrants from Europe (Jernegan, 2005). Only after the Immigration and Nationality Amendments Act of 1965 became law were restrictions on immigrants from non-European countries lifted (ibid.; Schafer, 1979).

These changes in immigration policy led to substantial increases in immigration from Latin America and Asia, which continue to characterize the immigration patterns today. In addition, the US has focused more recently on attracting highly skilled immigrants to fill technology jobs. In 2000, the US Congress passed the American Competitiveness in the Twenty-First Century Act, which increased the number of temporary work visas (H-1Bs) available each year (Jernegan, 2005). These changes increased the number of immigrants from Asia in particular.

Canada

Before the 1960s, one of the major criteria of Canadian immigration policies was the maintenance of the existing demographic composition (Reitz, 1999). A large proportion of immigrants to Canada were from European countries. Between 1955 and 1966, over 50 per cent of immigrants were British and Northern and Western European. Southern and Eastern Europeans together comprised 37 per cent (Kalbach, 1987).

In 1967, Canada introduced a new immigration policy to select immigrants based on a points system that emphasizes the potential adaptability of the newcomers. More points are given to those who have certain levels of education, skills, and language ability (Reitz, 1999). This alteration has drastically changed the sources of immigrants to Canada since the 1970s, and the number of immigrants from non-European countries has increased considerably (Hawkins, 1988; Citizenship and Immigration Canada, 2009).

We now turn to an assessment of each group's level of residential segregation and socio-economic inequality in comparative perspective.

THE UNITED STATES: RESIDENTIAL SEGREGATION AND SOCIO-ECONOMIC STATUS

Asian Indians

According to the 2000 census, there were 1,645,510 Asian Indians in the US (US Bureau of the Census, 2004). Indians are the third largest Asian immigrant group in the United States after the Chinese and Filipinos. The elimination of immigration quotas in 1965 increased Asian Indian immigration substantially (Rao, 2003). The PMSAs with the largest Indian populations are New York City, Chicago, and Washington.

Asian Indians were the least segregated from whites among the 10 PMSAs in this study. They also had the highest socio-economic status among the six groups (Tables 4.1–4.3). The mean index of dissimilarity was 52.0 (Table 4.1). Over all the PMSAs, Asian Indians were relatively evenly distributed, representing less than 1 per cent of the population. Residential segregation ranged from a high of 62 in Miami to a low of 48.7 in San Francisco.

Unlike the earlier immigrants from Europe who settled in central cities of metropolitan areas, especially in the poorer neighbourhoods (Dawkins, 2009), the settlement patterns of immigrants have shifted to the suburbs since the 1960s as a result, largely, of changes in immigration policies, the economic restructuring of the US in the eighties, and globalization and decentralization of employment opportunities (Frey, 2006; Singer, 2009). Increased population through immigration has led to the emergence of immigrant enclaves (Dawkins, 2009) as a new residential form in the suburbs (Li, 2009).

It appears that in order to take advantage of high-tech jobs, most Asian Indians settle in the suburbs instead of the central city (Frey, 2001; Mihalopoulos and Yates, 2001). Moreover, they tend to move directly to the suburbs upon arrival, bypassing the traditional location of immigrants, which has historically been the central city (Skop and Li, 2005). In 2000, nearly 60 per cent of Asian Indians had settled in the suburbs (Logan, 2001).

Filipinos

Filipinos, a second Asian immigrant group, consisted of a population of 1,864,120 in 2000 (US Bureau of the Census, 2004). In 1960, there were only 104,843 Filipino immigrants, constituting only 1 per cent of all immigrants in the US (Terrazas, 2008). The majority (51.9 per cent) entered the US before 1991; the largest concentration is in the Los Angeles area.

Among the six groups examined in this chapter, Filipinos ranked as the third least residentially segregated from whites and third in socio-economic status (Tables 4.1–4.3). The mean index of dissimilarity was 56.6 (Table 4.1). Their residential segregation ranged from a high of 66.8 in Boston and Dallas to a low of 46.6 in Washington.

Chinese

In 2000, there were 2,422,970 people of Chinese origin in the US (US Bureau of the Census, 2004); most are highly concentrated in New York, Los Angeles, and San Francisco. Among the six groups examined in this chapter, the Chinese ranked the second most residentially segregated from whites and fourth in socio-economic status (Tables 4.1–4.3). The mean index of dissimilarity for the 10 PMSAs was 60.2 (Table 4.1). The Chinese had the highest rate of home ownership (60 per cent) of any group in our study.

Iranians

Few studies have looked at the residential segregation of Iranians in the US. Iranian immigration to the US is very recent, and is related to the Islamic Revolution of 1978–9. Between 1980 and 1990, the number of Iranian immigrants increased by 74 per cent (Hakimzadeh and Dixon, 2006). Indeed, 154,857 Iranians were admitted to the US, a number more than seven times the number admitted to Canada (Hakimzadeh, 2006).This should not be surprising, however, given that the US has 10 times the total population of Canada. As of 2000, the US and Canada were the top two destination countries for Iranian immigrants. There were 291,040 in the United States and 75,115 in Canada (ibid.). The United States has the largest number of Iranians outside of Iran. Most of the Iranian immigrants lived in the PMSAs of California (Los Angeles and San Francisco). They also settled in Washington, New York City, Houston, and Dallas. In each of the PMSAs, they constituted less than 1 per cent of the total population.

Iranians have the highest level of residential segregation from whites. With a 67.5 mean index of dissimilarity based on the 10 PMSAs in this study, Iranians are the most residentially segregated immigrant group; however, Iranians have the second highest socio-economic status (Tables 4.1–4.3). They were most highly segregated in Boston (79.4) and least segregated in San Francisco (50.6).

Caribbean Immigrants

Caribbean immigrants have been coming to the US increasingly over the past 40 years (Gelatt and Dixon, 2006). Between 1990 and 2000, the number of Caribbean immigrants residing in the country increased 52.3 per cent, from 1.9 to 3 million (Brittingham and de la Cruz, 2004). Geographically, Caribbean immigrants are concentrated in the PMSAs of New York City, Miami, and Boston.

Caribbean immigrants are the fifth most residentially segregated from whites with a mean index of dissimilarity of 55.3 (Table 4.1). Also, Caribbean immigrants ranked fifth in socio-economic status (Tables 4.1–4.3). Caribbean immigrants and Latinos are the only immigrant groups with a socio-economic status total score below 5.00. (A total score of 5.00 means socio-economic equality with the white majority group.)

Latinos

Latino immigrants constitute the majority (53.3) per cent of the immigrants in the US (US Bureau of the Census, 2001). Of the PMSAs considered in this study, they are concentrated in Los Angeles, Houston, and San Antonio. They are also overwhelmingly from Mexico. Latinos were the fourth most residentially segregated group from non-Hispanic whites with a mean index of dissimilarity of 55.4 (Table 4.1). They ranked last, after Caribbean immigrants, in socio-economic status.

Summary

In sum, our analyses for the US reveal that the mean residential segregation level is high, i.e., above 50 per cent, for all non-European immigrant groups in the top 10 primary metropolitan areas with the largest (600,000 or more) foreign-born populations in the country. Residential segregation varied by group and primary metropolitan area. The mean level of residential segregation was highest for Iranians (67.5) and lowest for Asian Indians (52.0). Socio-economically, Asian Indians ranked first and Latinos ranked last.

CANADA: RESIDENTIAL SEGREGATION AND SOCIO-ECONOMIC STATUS

South Asians

In 2001, South Asians were the second largest visible minority group in Canada, with a population close to one million. Most are Asian Indians (Darden, 2004); however, South Asians include people of diverse ethnic backgrounds from a large geographic area, including Bangladeshis, Bengalis, Pakistanis, Punjabis, and Sri Lankans. Although South Asians have been in Canada for a long time, most of them arrived after Canada changed its racially restrictive immigration policy in the 1960s. According to the 2001 census, close to 70 per cent of South Asians are immigrants and arrived after 1981 (Tran et al., 2005). The majority emigrated from India (47 per cent), Sri Lanka (13 per cent), and Pakistan (12 per cent) (ibid.).

The majority of South Asians—61 per cent—live in Ontario and British Columbia. Like all other immigrant groups, South Asians cluster in the three major gateway immigrant CMAs, Toronto, Montreal, and Vancouver. Among those living in the three CMAs, about half of them live in Toronto (Lindsay, 2007b). Although they are clustered in a few major CMAs, the residential segregation levels of South Asians are moderate in all major areas, with an index of dissimilarity of 50.5. In addition, the proportions that have completed university and have administrative and managerial positions are moderate. South Asians rank fourth in overall socio-economic status.

Table 4.1   Mean Residential Segregation between Whites and Selected Non-European Immigrant Groups, US (2000) and Canada (2001)

Group	United States	Canada	Difference
Iranians (US)/West Asians (Canada)	67.5	58.7	-8.8
Chinese	60.2	49.1	-11.1
Filipinos	56.6	49.1	-4.7
Latin Americans	55.4	69.5	+14.1
Caribbeans	55.3	47.7	-7.6
Asian Indians (US)/ South Asians (Canada)	52.0	50.5	-1.5

Note: Mean residential segregation is measured using the index of dissimilarity by census tracts in 10 PMSAs in the United States and 10 CMAs in Canada. The means were calculated across the 10 CMAs and PMSAs to account for the different number of census tracts. The population of census tracts in the United States and Canada is comparable in size, ranging from 2,500–8,000, and averaging approximately 4,000 people.

West Asians

West Asians, who are mostly from Iran, are new in Canada. Statistics Canada also includes individuals of Afghan, Armenian, Israeli, Kurdish, and Turkish descent as being of West Asian ancestry. Most West Asians (especially Iranians) first arrived in Canada as visa students in the 1970s. The Iranian revolution in 1979 pushed many Iranians to immigrate to Canada in the 1980s. In 2001, there were about 205,000 living in Canada, a growth rate of about 45 per cent. Iranians comprised the largest share, 43 per cent of the total West Asian population (Lindsay, 2007c). As expected, most of them reside in the major immigrant gateway metropolitan areas (Darden, 2004). The majority are Muslims, who account for 53 per cent of the total West Asian population in Canada (Lindsay, 2007c). West Asians are the second most residentially segregated immigrant group among the six groups examined, with an index of dissimilarity of 58.7, and rank third in overall socio-economic status.

Chinese

The Chinese have a long history in Canada. The first major wave of Chinese arrived in the mid-nineteenth century in response to the discovery of gold on the Fraser and Thompson rivers in British Columbia. Later, migrant Chinese workers were brought to Canada by the Canadian Pacific Railway to work on the completion of the CPR line in the 1880s. The population of Chinese in Canada reached 16,000 in 1901, and increased to 35,000 in 1921 (Li, 1998). The passage of the **Chinese Immigration Act** in 1923 stopped further immigration, so the Chinese population in Canada remained at a similar level until the 1970s (ibid.).

Table 4.2   Mean Residential Segregation between Whites and Selected Non-European Immigrant Groups and Their Socio-economic Characteristics, US (2000) and Canada (2001)

Group	Mean Residential Segregation Index	Education	Occupation	Income ($)**	Home Ownership	Poverty Rate
Latin Americans: US	55.4	10.4	18.1	34,397	45.7	22.6
Latin Americans: Canada	69.5	14.4	14.0	16,089	38.6	29.4
Iranians: US	67.5	57.2	52.7	69,590	58.6	10.3
West Asians: Canada*	58.7	27.5	30.1	11,859	40.9	42.7
Chinese: US	60.2	48.0	52.3	60,058	60.9	13.5
Chinese: Canada	49.1	27.9	33.9	14,877	77.5	29.9
Filipinos: US	56.6	43.7	38.2	65,189	60.0	6.3
Filipinos: Canada	51.9	31.1	15.9	21,147	58.8	16.8
Caribbeans: US	55.3	19.3	28.6	43,650	36.4	15.8
Caribbeans: Canada	47.7	10.6	19.0	20,345	48.7	28.9
Asian Indians: US	52.0	63.8	59.9	70,708	46.9	9.8
South Asians: Canada***	50.5	23.3	19.8	12,617	44.3	39.3

Note: Mean residential segregation is measured by comparing non-Hispanic whites from each group based on census tracts in 10 US PMSAs and 10 Canadian CMAs. Education is measured based on the percentage of each group's population aged 25 and older in the US and 15 and older in Canada that has attained a bachelor's degree or higher level of education. Occupation is measured based on the percentage of each group's population aged 16 and older in the US and 15 and older in Canada that occupies a managerial or professional occupation. Home ownership is the percentage of each group's population in owner-occupied housing. The poverty rate is the percentage of each group's population living below the national poverty level.

*Iranians constitute the largest group of West Asians in Canada. Data are based on West Asians.

**American data are based on median household income; Canadian data are based on median individual income. Household income is not available for Canada.

***The majority of South Asians in Canada are of Indian origin. The remainder includes such groups as Pakistanis, Punjabis, and Sri Lankans.

Source: Computed by the authors from data obtained from Iceland et al. (2002); Rao (2003); Statistics Canada, (2006); US Bureau of the Census (2001, 2002, 2004).

In response to the change of immigration policy in 1967 and later in the 1970s, many Chinese who were students in Canada at that time took the opportunity to apply for immigrant status. Many Chinese from Hong Kong, especially the middle class with human capital resources, worried about the return of the island to mainland China in 1997 and moved to other countries (Fong et al., 2007). Canada was a popular destination. By 2001, the Chinese population in Canada exceeded one million, representing the largest visible minority group in the country. Although Chinese are located in many small and medium-size metropolitan areas in Canada, the majority of them settled in major gateway metropolitan areas. About 39 per cent of Chinese in Canada live in Toronto, 31 per cent in Vancouver, 5 per cent in Calgary, and 5 per cent in Montreal

Table 4.3   Mean Residential Segregation Rank and Socio-economic Status Rank for Selected Non-European Immigrant Groups, US (2000) and Canada (2001)

Group	Residential Segregation Rank	Education Ratio	Occupation Ratio	Income Ratio**	Home Ownership Ratio	Poverty Rate Ratio	Total SES Score	Overall SES Rank
Iranians: US	1	2.20	1.70	1.60	0.83	1.10	7.43	2
West Asians: Canada*	2	1.45	1.00	0.46	0.58	0.32	3.80	3
Chinese: US	2	1.80	1.70	1.40	0.87	0.82	6.59	4
Chinese: Canada	6	1.47	1.13	0.57	1.10	0.46	4.73	1
Filipinos: US	3	1.70	1.20	1.25	0.86	1.70	6.69	3
Filipinos: Canada	4	1.63	0.53	0.82	0.83	0.82	4.63	2
Latin Americans: US	1	0.40	0.58	0.83	0.65	0.49	2.95	6
Latin Americans: Canada	1	0.76	0.46	0.62	0.55	0.47	2.86	6
Caribbeans: US	5	0.75	0.92	0.82	0.52	0.70	3.71	5
Caribbeans: Canada	3	0.56	0.63	0.79	0.69	0.48	3.14	5
Asian Indians: US	6	2.50	1.90	1.70	0.67	1.10	7.87	1
South Asians: Canada***	5	1.22	0.66	0.49	0.63	0.35	3.35	4

Note: A ratio of 1.00 = equality for that variable with the non-immigrant non-Hispanic white population. A ratio of less than 1.00 = less than equality and a ratio greater than 1.00 = non-European immigrant group has a higher status than the non-immigrant non-Hispanic white population. The poverty rate ratio is interpreted differently from the other variables. The lower the ratio, the higher the degree of poverty between the non-European immigrant group and the non-immigrant non-Hispanic white population. A poverty rate ratio greater than 1.00 = non-immigrant non-Hispanic white population has a higher degree of poverty. A total SES score of 5.00 = total equality. For the residential segregation rank, the lower the score, the higher the degree of residential segregation from the non-Hispanic non-immigrant white population.

*Iranians constitute the largest group of West Asians in Canada. Data are based on West Asians.

**American data are based on median household income; Canadian data are based on median individual income. Household income is not available for Canada.

***The majority of South Asians in Canada are of Indian origin. The remainder includes such groups as Pakistanis, Punjabis, and Sri Lankans.

Sources: Computed by the authors from data obtained from Iceland et al. (2002); Rao (2003); Statistics Canada (2006); US Bureau of the Census (2001, 2002, 2004).

(Statistics Canada, 2003). The Chinese are the second least segregated among the six immigrant groups in our study, with an index of dissimilarity of 49.1, and ranked first in overall SES (Tables 4.1 and 4.2).

Filipinos

Filipinos are another recent immigrant group. Before 1961, only a very small number lived in Canada. The growth of the Filipino population was largely in response to labour demands in Canada in the 1970s, when many skilled and highly educated Filipinos immigrated. The age and gender composition of these immigrants showed a high proportion in the 20–39 age group, with females outnumbering males. This was due in part to Canada's recruitment of Filipina nurses, medical technicians, and caregivers. Over time, Filipinas became the most represented group of workers in the Live-in Care Giver Program (Darden and Kamel, 2004–5). It is not a surprise to find that about 70 per cent of Filipinos were immigrants (Statistics Canada, 2003). Their geographic distribution is similar to that of most recent immigrant groups. In 2001 a heavy concentration was in immigrant gateway cities such as Toronto (42 per cent) and Vancouver (17 per cent) (Darden, 2004). They are the third most residentially segregated group, with an index of 51.9, and they rank second in overall socio-economic status (Table 4.3).

Caribbean-Origin Groups

Caribbean immigration largely began in the 1960s. Based on the 2001 census, only about 16 per cent of those from the Caribbean came to Canada before 1961. Canadians of Caribbean origin comprise diverse groups. According to the 2001 census, 42 per cent of Caribbeans are of Jamaican origin, 16 per cent are Haitian, 10 per cent are Guyanese, and 10 per cent are from Trinidad and Tobago (Lindsay, 2007a).

The Canadians of Caribbean origin are heavily concentrated in Ontario (69 per cent) and Quebec (22 per cent) (ibid.). Together, they represent 91 per cent of the total Caribbean population in Canada, a concentration far higher than that of other new immigrant groups. Most Canadians of Caribbean origin live in Toronto and Montreal. They represent 6 per cent of the total population in Toronto and 3 per cent in Montreal (Statistics Canada, 2003).

Despite their recent arrival and heavy concentration in two metropolitan areas, the group ranks last among the six groups in regard to residential segregation, with a score of 47.7 on the index of dissimilarity. Although a lower percentage of Canadians of Caribbean origin have completed university, their average income is comparable to that of the other selected immigrant groups. Their poverty rate is also relatively lower than that of most other new immigrant groups. In overall SES, they rank fifth among the six groups (Table 4.3).

Latin Americans

Latin American immigration to Canada started after the 1960s when refugees escaping from oppressive dictatorships were provided the opportunity to settle in Canada. Most of Latin American immigration to Canada occurred between 1970 and 2001. Eighty per cent of the Latin Americans in Canada are immigrants and the overwhelming majority live in the gateway metropolitan areas of Toronto and Montreal (Statistics Canada, 2003). Latin Americans were the most residentially segregated group, with a mean index of dissimilarity of 69.5; also, Latin Americans ranked last in socio-economic status.

Summary

Most of the members of the six immigrant groups under study arrived in Canada in recent decades, and the populations of these groups have increased rapidly since the 1970s. Most of them have settled in a few major metropolitan areas. Residential segregation is highest among Latin Americans and lowest among Caribbean groups. At the same time, Chinese rank highest in overall socio-economic status and Latin Americans rank the lowest.

THE US AND CANADA COMPARED: RESIDENTIAL SEGREGATION AND SOCIO-ECONOMIC STATUS

In this final section, we assess the significance of 'place' on the level of residential segregation and socio-economic status of non-European groups by comparing the same groups in the United States and Canada with the non-immigrant white population.

Residential Segregation

As revealed in Table 4.1, all of the non-European groups except Latin Americans are more residentially segregated from whites in the US than in Canada. While the mean index of dissimilarity ranged from 52.0 to 67.5 in the 10 PMSAs in the United States, the mean indices in the 10 CMAs in Canada ranged from 47.7 to 69.5 (Table 4.1). Residential segregation is high (above 50 based on the index of dissimilarity) for all non-European groups in the US, but the indexes are generally at around 50 in Canada's CMAs.

Socio-economic Status

The non-European groups (except for Latin Americans and Caribbeans) are less residentially segregated from whites in Canada, yet, at the same time, they are closer in equality to whites in overall socio-economic status in the US. All US non-European groups in our study except Latin Americans had a high percentage of bachelor's degrees. The gap is very wide among Asian Indians, where 63.8 per

BOX 4.1   The Maintenance of Racial Inequality in Canada

Canada maintains racial inequality between whites and non-European immigrant groups primarily through discrimination in the labour market and secondarily in the housing market. Thus, non-European immigrants experience discrimination in employment generally more than in the US and are denied access to higher-status jobs. This discrimination impacts their overall earnings levels and their level of socio-economic status. Although the non-European immigrant groups experience employment discrimination differentially in general, all non-white, non-European immigrant groups in general do not have the same access to jobs as whites. As a result, many are not able to translate their level of education and skills into jobs and occupations that are equal to those of whites with the same skills and education. Thus, racial inequality is maintained. As a result of greater emphasis on the labour market than on the housing market, a higher level of socio-economic inequality occurs between whites and non European immigrants in Canada than in the US.

cent held a bachelor's degree or higher level of education in the US, compared to 23.3 per cent in Canada (Table 4.2).

The consistent pattern is found in the occupational status category. All of the non-European groups had achieved a higher occupational status in the US than in Canada. Occupational status was measured by the percentage of each group's population aged 16 and older that occupied a managerial or professional occupation. Among some groups, the gap was very wide. For example, the percentage of Filipinos in the US with professional and managerial positions was 38.2, compared to 15.9 in Canada, a ratio of 2.4 (Table 4.2).

Like occupation, all non-European groups had a higher income in the US than in Canada. Also, the income gap was quite wide, although data for the two countries are not directly comparable: US figures are for median household income whereas the Canadian numbers are for median individual income. Nonetheless, for example, those from the Philippines in the US had a median household income of $65,189, compared to $21,147 median individual income in Canada. Among the Chinese and Caribbeans, the rate of home ownership among the non-European groups was lower in the US. The rate of poverty among the non-European immigrant groups, however, was lower in the US than in Canada for all the groups (Table 4.2).

Finally, we compared socio-economic status scores of the non-European groups to the non-immigrant white population in each country (Table 4.3). A total SES score of 5 = total equality. All groups in our study had a higher SES score in the US than in Canada.

CONCLUSION

We have assessed the residential segregation and socio-economic status of selected non-European immigrant groups in two multicultural, predominantly white, advanced countries with a history of an ideology of white supremacy (see Darden, 2004, and Darden and Teixeira, 2009, for detailed discussion). White supremacy is an ideology held by many members of the white population, whether overt or assumed. It is the belief that in any relationship involving the white population and people of colour, the white population must have the superior position and competitive advantage (Darden, 2004). To provide those of Western and Northern European background with the competitive advantage in access to quality housing, jobs, and other amenities, members of this so-called white population have engaged in racial/ethnic discrimination in housing and employment in Canada and the US (ibid.; Darden and Teixeira, 2009). The evidence reveals that 'place' matters, as shown in the difference in the levels of residential segregation and socio-economic status. In general, there was a higher residential segregation level in the US. On the other hand, there was generally a higher level of socio-economic status achieved by the groups in the US, resulting in less overall socio-economic inequality.

Understanding Residential Segregation and Inequality

In this final section, we discuss the factors related to the similarities and differences in the patterns of residential segregation and socio-economic inequality among selected non-European groups in Canada and the US.

First, the data seem to indicate that the dominant populations in both countries have been strongly influenced by a common ideology of white supremacy. This ideology is differentially imposed based on skin colour and has influenced not only immigration policies but the differential incorporation process. The outcome is differential employment, housing, and residential segregation with the overall result of maintaining racial inequality. Institutional discrimination—defined as a process that indirectly excludes people of colour or provides an advantage to the white majority via policies, practices, and processes carried out by social institutions—is the primary mechanism by which racial inequality is maintained. Unlike overt individual racial discrimination, which is often rejected in society today, institutional discrimination is practised by 'established institutions' such as banks, brokers, builders, the real estate industry, and private and public institutions responsible for employment. The institutions do not use colour per se as the subordinating mechanism. Rather, they employ other mechanisms that indirectly result in discrimination, such as redlining by financial institutions of neighbourhoods where loans will not be made and accreditation and certification policies that refuse to recognize overseas credentials (Darden, 2004).

Institutional discrimination involving real estate agents, apartment managers, and financial institutions appears to be more evident in the housing industry in

BOX 4.2   Racial Discrimination in the United States

Research in the US using the audit method has documented that racial discrimination in housing, primarily in the form of racial steering, is still a problem in many metropolitan areas (Turner et al., 2002). Regardless of equal characteristics in terms of socio-economic status, whites are favoured over non-whites in the sale and rental of homes and apartments. Whites also receive more information assistance about financing and encouragement than comparable non-white home seekers. Thus, whites and non-whites are increasingly likely to be recommended and to ultimately purchase homes in different neighbourhoods, leading to a higher level of residential segregation than may be found in Canada. Also, in the mortgage industry, there is a differential rejection rate for conventional mortgages for whites and non-whites (Massey and Denton, 1993).

the US; however, a comparable audit method to measure housing discrimination in Canada has not been used (ibid.). Future comparative studies of discrimination in housing and employment in Canada and the US need to be conducted using the audit method. This method controls for all relevant characteristics of applicants except race or colour (Turner et al., 2002). In addition to institutional discrimination, the differences in metropolitan fragmentation and local vs regional political control may influence residential segregation levels.

Residential segregation is easier to maintain in the US through institutional means because in the US, most metropolitan areas are much more fragmented than in Canada. While several Canadian metropolitan areas are strongly affected by regional politics and demography, with a large share of power and influence at the federal and/or provincial level and few politically powerful municipalities, many members of the white majority, especially suburbanites in American metropolitan areas, favour local control and power by the individual municipality, resulting in metropolitan fragmentation and extremely uneven development between the city, where a disproportionate number of poor and racial minorities reside, and the suburbs where most affluent whites reside.

Given local control over land-use policies and zoning regulations, many suburban municipalities in the US have introduced policies and practices to indirectly exclude non-white population groups and poor immigrants from their communities (Walker, 2008). In Canada, where social housing is influenced by federal and provincial governments, the spatial distribution of non-white population groups is less concentrated in the central cities and is more dispersed throughout the census metropolitan areas, including the suburbs, thus providing more housing options that are affordable outside the central city.

Another factor that may help explain the differential patterns of residential segregation in Canada and the US is the relative size of the non-European (non-white) immigrant population, which poses a threat to white dominance. According to Blalock's theory of minority relations, the size of the non-European immigrant group relative to the white majority may influence the group's level of residential segregation and socio-economic status compared to the status of the majority group. More specifically, the relative size of the group produces a threat to the majority group in terms of *competition* for quality housing and neighbourhoods in the US and less in the area of employment. In Canada, on the other hand, the size of the non-European immigrant group presents more of a threat in the labour market than in the housing market. That is because the threat occurs first primarily in the competition for jobs, followed by the competition for housing and quality neighbourhoods (Blalock, 1967). While non-European immigrant groups are still perceived as a threat in the US, the severity of the threat related to employment has declined relative to the threat related to housing integration. Future research is needed to reveal whether Canada will follow a similar pattern with an increase in the threat over housing and neighbourhoods as the threat over employment competition declines.

The time of arrival may also have an influence on the differences in the area of socio-economic inequality between the non-European immigrant groups and the non-immigrant population. Usually, the shorter the period of time since a group's arrival in a country, the larger the socio-economic gap; however, since in both countries the periods of arrival for most members of the non-European immigrant groups did not differ significantly (1965–present in the US and 1967–present in Canada), the larger socio-economic gap between non-European groups and whites in Canada may be related to discriminatory practices in the employment area or weaknesses in anti-discrimination policies (Darden, 2009: 56–8). More comparative research using the audit method in Canada and the US is needed to determine the extent of discrimination in each country.

Based on prior research we hypothesize that Canadians are likely to impose and maintain racial inequality through discrimination in the employment sector and less likely to impose and maintain it through the housing sector. More than the US, Canada has for years engaged in a sophisticated form of institutionalized discrimination to deny non-European immigrant groups equal access to opportunities in the workplace through systemic discrimination, a covert type of discrimination not done by individuals but by policies and practices implemented by government and private institutions that impact visible minorities differently in hiring and promotions in the employment sector. It is the most prevalent type of discrimination in Canada (Galabuzi, 2006).

Although racial discrimination may be imposed differentially among non-white, non-European immigrant groups, there is general consistency in that access to the housing and labour market between all white European groups and all non-white non-European groups will be different, providing an advantage to the former groups and a disadvantage to the latter. The result is differential levels of residential segregation but a general level for all non-white immigrant groups in the US that is higher on average than the levels for the same groups in

Canada. In general, non-white immigrant groups in the US are not able to translate their socio-economic status (income, education, and occupation) into equal-quality neighbourhoods due to racial discrimination in the housing market.

The Process in the United States

In the US, the ideology of white supremacy, with its outcome of racial inequality, is implemented through politics, power, and social action, which are translated into explicit programs and discriminatory behaviours that operate among a large segment of members of the dominant group in both the private and public sectors. Those actors most responsible for carrying out the vision and values of the white dominant group are real estate brokers, apartment managers, builders, mortgage lenders, and various elected officials of some suburban municipalities.

Local municipalities in many American metropolitan areas engage in zoning restrictions, land-use policies, and selected service provisions designed to exclude non-white immigrants from residence within them. Such actions are facilitated more in the US than in Canada due to the fragmented metropolitan structure in the US that gives land-use and other decisions to local municipalities. Since the racial composition of municipalities influences some residents' choice of municipality as a place to live, non-white immigrants are often excluded from some suburban municipalities (Feagin, 1999; Dawkins, 2009). These actions influence the residential location of non-European immigrants in the suburbs, resulting in higher levels of racial residential segregation and neighbourhood socio-economic inequality in general in comparison to Canadian metropolitan areas.

The Process in Canada

In Canada, the ideology of white supremacy, with its primary focus on the labour market, maintains racial inequality through policies, power, and social action that transfer the responsibility for certain employment policies and credentialling to certain professional associations. Many of these associations have an inherent conflict of interest in the area of employment preferences. Research has shown for years that, due to Canada's credentials barrier, non-white, non-European immigrants, although trained and experienced, have been prevented from putting their knowledge and experience to use in Canada's labour market (Darden, 2009).

Since the labour market discrimination process is a primary focus in Canada and a secondary focus in the US, non-white immigrant groups in Canada have a higher degree of socio-economic inequality to whites in Canada than in the US. In sum, differential incorporation is found in both the US and Canada. We have argued that such processes and the resultant patterns of residential segregation and socio-economic inequality by race/ethnicity are deeply rooted in the ideology of white supremacy that exists in both countries.

APPENDIX 4.1: COMPUTATION
OF THE INDEX OF DISSIMILARITY

The index of dissimilarity can be stated mathematically as:

$$D = 100 \left(1/2 \sum_{i=1}^{k} x_i - y_i \right)$$

where:

x_i = the percentage of a PMSA or CMA's non-Hispanic white population living in a given census tract (neighbourhood)

y_i = the percentage of a PMSA or CMA's non-European group population living in the same census tract (neighbourhood)

k = the number of census tracts

D = the index of dissimilarity. It is equal to one-half the sum of the absolute differences (positive and negative) between the percentage distributions of the non-European group population and non-Hispanic white population in the PMSA or CMA (Darden and Tabachneck, 1980; Iceland et al., 2002).

The index value may range from '0', indicating no residential segregation, to '100', indicating complete residential segregation. The higher the index, the greater is the degree of residential segregation. Conceptually, the index of dissimilarity measures the minimum percentage of a group's population that would have to change residence for each neighbourhood (census tract) to have the same percentage of that group as the PMSA or CMA area as a whole.

QUESTIONS FOR CRITICAL THOUGHT ───────────────────────

1. What do the conclusions in this chapter reveal about the differences in the mean level of residential segregation in the United States and Canada?

2. What do the conclusions in this chapter reveal about the differences in the extent of socio-economic inequality between non-European immigrants and the non-immigrant populations in Canada and the United States?

3. What does the evidence in this chapter demonstrate as to why the Asian immigrant groups should be analyzed separately rather than as a single group?

SUGGESTED READINGS ───────────────────────────────

1. Darden, Joe T. 2003. 'Residential Segregation: The Causes and Social and Economic Consequences', in C. Stokes and T. Melendez, eds, *Racial*

Liberalism and the Politics of Urban America. East Lansing: Michigan State University Press. This chapter examines three theories of black residential segregation: socio-economic status, neighbourhood preference, and racial discrimination in housing.

2. ——— and C. Teixeira. 2009. 'The African Diaspora in Canada', in J. Frazier, J.T. Darden, and N. Henry, eds, *The African Diaspora in the United States and Canada at the Dawn of the 21st Century.* Binghamton, NY: Global Academic Publishing. This chapter examines the African Diaspora in Canada from the conceptual framework of differential incorporation. It measures the extent to which people of African descent have been incorporated into the mainstream of Canadian society.

3. Fong, Eric, Wenhong Chen, and Chiu Luk. 2007. 'A Comparison of Ethnic Businesses in Suburbs and the City', *City and Community* 6: 119–36. This study examines the differences in the emergence of ethnic businesses in the city compared to the suburbs.

4. South, Scott, Kyle Crowder, and E. Chavez. 2005. 'Migration and Spatial Assimilation among US Latinos: Classical versus Segmented Trajectories', *Demography* 42: 497–521. This study explores the spatial assimilation process that Latin Americans have used to move from the city to the suburbs in the United States.

REFERENCES

1. Blalock, H. 1967. *Towards a Theory of Minority Group Relations.* New York: Capricorn Books.
2. Breton, R., W. Isajiw, W. Kalbach, and J. Reitz. 1990. *Ethnic Identity and Equality: Varieties of Experiences in a Canadian City.* Toronto: University of Toronto Press.
3. Brittingham, Angela, and Patricia de la Cruz. 2004. *Ancestry: 2000.* Washington: US Bureau of the Census, Department of Commerce.
4. Citizenship and Immigration Canada. 2009. *Facts and Figures 2008: Immigration Overview: Permanent and Temporary Residents.* Ottawa: Citizenship and Immigration Canada.
5. Darden, Joe T. 2004. *The Significance of White Supremacy in the Canadian Metropolis of Toronto.* Lewiston, NY: Edwin Mellen Press.
6. ———. 2009. 'The African Diaspora: Historical and Contemporary Immigration and Employment Practices in Toronto', in J. Frazier, J.T. Darden, and N. Henry, eds, *The African Diaspora in the United States and Canada at the Dawn of the 21st Century.* Binghamton, NY: Global Academic Publishing.

7. ——— and S. Kamel. 2004–5. 'Filipinos in Toronto: Residential Segregation and Neighbourhood Socioeconomic Inequality', *Amerasia Journal* 30, 3: 25–38.
8. ——— and A. Tabachneck. 1980. 'Algorithm 8: Graphic and Mathematical Descriptions of Inequality, Dissimilarity, Segregation or Concentration', *Environment and Planning A* 12: 227–34.
9. ——— and C. Teixeira. 2009. 'The African Diaspora in Canada', in J. Frazier, J.T. Darden, and N. Henry, eds, *The African Diaspora in the United States and Canada at the Dawn of the 21st Century.* Binghamton, NY: Global Academic Publishing.
10. Dawkins, Casey. 2009. 'Exploring Recent Trends in Immigrant Suburbanization', *Cityscape: A Journal of Policy Development and Research* 11, 3: 81–97.
11. Feagin, J. 1999. 'Excluding Blacks and Others from Housing: The Foundation of White Racism', *Cityscape: A Journal of Policy Development and Research* 4, 3: 79–91.
12. Fong, Eric, Wenhong Chen, and Chiu Luk. 2007. 'A Comparison of Ethnic Businesses in Suburbs and the City', *City and Community* 6, 2:119–36.

13. Frey, William H. 2001. *Melting Pot Suburbs: A Census 2000 Study of Suburban Diversity.* Washington: Brookings Institution, 2001.

14. ———. 2006. *Diversity Spreads Out: Metropolitan Shifts in Hispanic, Asian, and Black Populations since 2000.* Washington: Brookings Institution Living Cities Census Series.

15. Galabuzi, Grace Edwards. 2006. *Canada's Economic Apartheid: The Social Exclusion of Racialized Groups in the New Century.* Toronto: Canadian Scholars' Press.

16. Gelatt, J., and D. Dixon. 2006. 'Detailed Characteristics of the Caribbean-born in the United States', in *Migration Information Source.* Washington: Migration Policy Institute.

17. Hakimzadeh, Shirin. 2006. 'Iran: A Vast Diaspora Abroad and Millions of Refugees at Home', in *Migration Information Source.* Washington: Migration Policy Institute.

18. ——— and David Dixon. 2006. 'Spotlight on the Iranian Foreign-born', in *Migration Information Source.* Washington: Migration Policy Institute.

19. Hawkins, Freda. 1988. *Canada and Immigration: Public Policy and Public Concern.* Montreal and Kingston: McGill-Queen's University Press.

20. Henry, Frances. 1994. *Caribbean Diaspora in Toronto.* Toronto: University of Toronto Press.

21. Iceland, John, Daniel Weinberg, and Erica Steinmetz. 2002. *Racial and Ethnic Segregation in the United States: 1980–2000* Washington: US Bureau of the Census.

22. Jasso, G., Douglas S. Massey, M.R. Rosenzweig, and J.P. Smith. 2000. 'The New Immigrant Survey Pilot (NIS-P): Overview and New Findings about US Legal Immigrants at Admission', *Demography* 37: 127–38.

23. Jernegan, Kevin. 2005. 'A New Century: Immigration and the US', in *Migration Information Source.* Washington: Migration Policy Institute.

24. Kalbach, Warren. 1987. 'Growth and Distribution of Canada's Ethnic Populations, 1871–1981', in L. Driedger, ed., *Ethnic Canada.* Toronto: Copp Clark Pitman.

25. Li, Peter. 1998. *Chinese in Canada.* Toronto: Oxford University Press.

26. ———. 2005. 'The Rise and Fall of Chinese Immigration to Canada: Newcomers from

Hong Kong Special Administrative Region of China and Mainland China, 1980–2000', *International Migration* 43: 9–32.

27. Li, Wei. 2009. *Ethnoburb: The New Ethnic Community in Urban America.* Honolulu: University of Hawai'i Press.

28. Lieberson, S. 1980. *A Piece of the Pie: Black and White Immigrants since 1880.* Berkeley: University of California Press.

29. Lindsay, Colin. 2007a. *The Caribbean Community in Canada, 2001.* Profiles of Ethnic Communities in Canada. Ottawa: Statistics Canada.

30. ———. 2007b. *The South Asian Community in Canada, 2001.* Profiles of Ethnic Communities in Canada. Ottawa: Statistics Canada.

31. ———. 2007c. *The West Asian Community in Canada, 2001.* Profiles of Ethnic Communities in Canada. Ottawa: Statistics Canada.

32. Logan, John R. 2001. *The Ethnic Enclaves in America's Suburbs.* Albany, NY: Lewis Mumford Center for Comparative Urban and Regional Research.

33. Massey, Douglas, and N. Denton. 1993. *American Apartheid: Segregation and the Making of the Underclass.* Cambridge, Mass.: Harvard University Press.

34. Mihalopoulos, Dan, and J. Yates. 2001. 'Indian Immigrants Flock to Suburbs to Fill High Tech Jobs', *Chicago Tribune*, 29 May.

35. Park, Robert E. 1926. 'The Urban Community as a Spatial Pattern and a Moral Order', in E.W. Burgess, ed., *The Urban Community.* Chicago: University of Chicago Press.

36. Rao, K.V. 2003. 'Indian Americans'. At: www.asian-nation.org/indian.shtml. (26 July 2009)

37. Reitz, Jeffrey G. 1999. *Warmth of the Welcome: The Social Causes of Economic Success for Immigrants in Different Nations and Cities.* Boulder, Colo.: Westview Press.

38. Schafer, R. 1979. *Racial and Ethnic Groups.* Boston: Little, Brown.

39. Singer, Audrey. 2009. *The New Geography of United States Immigration.* Washington: Brookings Institution.

40. Skop, E., and W. Li. 2005. 'Asians in America's Suburbs: Patterns and Consequences of Settlement', *Geographical Review* 95, 2: 167–88.

41. South, Scott, and Kyle Crowder. 2005. 'Geographic Mobility and Spatial

Assimilation among US Latino Immigrants', *International Migration Review* 39, 3: 577–607.

42. ——, ——, and E. Chavez. 2005. 'Migration and Spatial Assimilation among US Latinos: Classical versus Segmented Trajectories', *Demography* 42: 497–521.

43. Statistics Canada. 2003. 'Selected Demographic and Cultural Characteristics (102), Visible Minority Groups (15), Age Groups (6) and Sex (3) for Population (97f0010xcb01044), 2001', Special Interest Tables. At: www.statcan.gc.ca. (June 2008)

44. ——. 2006. 2001 Public Use Microdata Files (PUMF). Ottawa: Minister of Industry.

45. Terrazas, A. 2008. 'Filipino Immigrants in the United States', in *Migration Information Source*. Washington: Migration Policy Institute.

46. Tran, Kelly, Jennifer Kaddatz, and Paul Allard. 2005. 'South Asians in Canada: Unity through Diversity', *Canadian Social Trends* 78 (Fall).

47. Turner, M., S. Ross, G. Galster, and J. Yinger. 2002. *Discrimination in Metropolitan Housing Markets: National Results from Phase 1 HDS, 2000*. Washington: Urban Institute.

48. US Bureau of the Census. 2001. *Profile of the Foreign-born Population in the United States: 2000*. Washington: US Department of Commerce.

49. ——. 2002. *2000 Census Summary File 3 (SF3)*, State of Michigan.

50. ——. 2004. *2000 Census Summary File 4 (SF4)*. Washington: US Government Printing Office.

51. Walker, Kyle. 2008. *Immigration, Suburbia, and the Politics of Population in US Metropolitan Areas*. Minnesota Population Center Working Paper No. 2008–05. Minneapolis: University of Minnesota.

CHAPTER 5
IMMIGRANTS, REFUGEES, AND HOUSING

Thomas Carter and Domenic Vitiello

INTRODUCTION

When immigrants and refugees arrive in a new country, some live temporarily with family and friends or in transitional housing provided by settlement organizations, but finding their own housing is a high priority. Often the experience is not easy. Sometimes it takes many years to achieve the housing circumstances they desire. Good housing, however, is crucial to successful resettlement and an important facilitator of integration into a new society (Chera, 2004). Adequate, **affordable housing** becomes the stable base from which immigrants access other formal and informal supports and services, strengthen family, and build social networks. Without such housing, newcomers may have no security of tenure, compromised health, educational, and employment outcomes, and an impaired social and family life (Danso and Grant, 2000). Good housing provides an environment that enables refugees and immigrants to rebuild their personal and cultural identity and facilitates the building of a new 'home' and community (Ready, 2006).

Drawing on literature from Canada and the United States, this chapter discusses the housing experiences and trajectories of newcomers. It examines how the increasing diversity and number of arrivals affects housing demand and market dynamics as well as the challenges and opportunities the market presents for these newcomers.

The diversity of newcomers and of housing markets across North America results in such varied patterns that there is no such thing as a 'typical' immigrant housing experience. Increasing diversity changes housing preferences, attitudes, and requirements, which filter through the market. The chapter will point out that these changes are occurring in settings as diverse as central-city Chinatowns and small agricultural processing centres. The effects are also visible in the residential landscapes of migrant-sending communities, as people in towns from Mexico to the Philippines witness their neighbours building new homes with modern amenities financed by relatives in the United States or Canada. The bifurcation of immigrants' educational and work experiences is reflected in housing demand and market diversity. Low-wage earners and less-educated immigrants often face affordability problems in the rental sector, while professional-class immigrants raise demand in the higher-priced ownership sector. Diversity also translates into different tenure ratios, different designs and sizes, and a changing metropolitan geography of immigrant housing. The discussion that follows explores these dynamics.

Other questions discussed include: How do the housing circumstances of newcomers compare to those of the non-immigrant population? Do newcomers experience different housing circumstances and trajectories depending on their ethnocultural group? What are the effects of other variables such as labour force integration and educational levels? How do patterns vary between cities and regions? Important indicators of newcomers' housing experiences to be examined include housing affordability, condition, household crowding, rates of ownership, discrimination, and landlord–tenant relationships, among others. Changes in labour force characteristics are also critical for understanding housing experiences and outcomes. Housing analysts and policy-makers will find the information in this chapter useful in structuring more appropriate housing responses that contribute to successful settlement and integration.

THE GEOGRAPHY OF IMMIGRANT HOUSING DEMAND

Immigration is a key driver of housing demand. Since 1995, the foreign-born population has accounted for one-third of household growth in the United States and close to two-thirds in Canada (Statistics Canada, 2005; US Bureau of the Census, 2008). Demand has increased, or at least remained at a higher level than would have been the case without immigrant demand, especially given declining rates of economic and population growth and an aging demographic among the non-immigrant populations in both countries. The changing geography of immigrant settlement has dispersed this demand more widely across the entire urban hierarchy in both countries and generated demand in both large and small cities and rural areas (Carter et al., 2009).

Outside the major metropolitan centres an influx of immigrants and temporary foreign workers to smaller places, particularly those with agricultural plants or resource-based industries, has transformed some small centres overnight. Larger immigrant families and their limited incomes put pressure on the demand for rental housing, which is generally in limited supply in such places. Demand increases in surrounding communities as families commute from other residential locations when the supply of rental housing is exhausted in the production centre. California's central valley cities and towns, with a tremendous number of Mexican farm workers, and Brooks, Alberta, with its meat-packing

BOX 5.1   The Impact of Immigration on Small Centres

Steinbach, Manitoba, has been transformed by immigration. With a 2006 population of 11,066 Steinbach and area has witnessed an influx of more than 5,000 immigrants since 2000. Steinbach's population increased 20 per cent between 2001 and 2006. Smaller centres near Steinbach grew even faster. Immigrants are arriving to address labour shortages in agriculture, trucking, and various industries—notably food processing, furniture, and farm machinery. The impact on housing has been significant. Demand has spawned construction of some affordable housing for purchasers, such as the row housing in Steinbach shown in Figure 5.1. Rental accommodation, however, is extremely scarce. Vacancy rates are near zero and rents have been increasing at double the rate of inflation for the past five years. Renters in the first year or two after arrival experience major housing affordability and suitability problems, particularly because many are large families requiring three or more bedrooms—a product that is almost non-existent in the rental sector.

Figure 5.1 Affordable Immigrant Housing Changes the Architectural Landscape of Small Centres: Row Houses in Steinbach, Manitoba. Photo by Thomas Carter.

plant employing over 2,000 Sudanese workers, are two cases in point (Farm Foundation, 2005; Broadway, 2007). Trailer housing is common and particularly suited to temporary foreign workers (Bruce and Carter, 2003). Similar housing pressures also characterize some tourist and retirement centres such as Kelowna, British Columbia, where immigrants often live in crowded circumstances in basement apartments (Teixeira, 2009).

Traditionally, housing demand generated by new arrivals focused initially on the rental sector. Few new immigrants land with sufficient funds to purchase a home. The dominance of immigrants in rental housing remains significant in some areas. In Florida, immigrants accounted for 60 per cent of the increase in rental-occupied housing during the 1990s and all of the net increase in California, New York, and Illinois (Myers and Liu, 2005). Refugees and **unauthorized immigrants** remain in rental housing for longer periods of time, often their entire lives. Carter and Osborne (2009) found more than 85 per cent of refugees settling in Winnipeg remained in the rental sector after three years, and work by Murdie (2008) and Hiebert and Mendez (2008) reveals almost negligible rates of ownership among refugee households in Toronto and Vancouver four years after their arrival.

Recent immigrants, however, also are increasing demand in the home-buying sector. Some of the growing numbers of professional workers and entrepreneurs, with greater purchasing power and often more assets, are in a position to purchase on arrival. Others spend less time in the rental sector. Among recent immigrants in Toronto, 32 per cent are owners (Preston et al., 2006) and over 40 per cent of recent immigrants in Vancouver own their own homes (Hiebert et al., 2006). In New York City, 34 per cent of immigrants live in owner-occupied housing (Fiscal Policy Institute, 2007). These centres are high-priced markets. Carter (2009) found that in Manitoba, a more modestly priced market, 76 per cent of skilled and business-class immigrants were homeowners five years after arrival. The Manitoba average for all households is 69 per cent. Immigrants represent a strong counterbalancing force in the demand for ownership housing as the aging baby boomers pass their peak home-buying age and move to other housing options (Myers, 2007; National Association of Realtors, 2002). The greater purchasing power of many immigrants today has profound implications for the housing market.

MARKET TRENDS CREATE CHALLENGES AND OPPORTUNITIES

Recent immigrants have come to North America during a time of significant changes in the housing market. In the ownership sector these changes include: price increases in most major metropolitan centres; low mortgage interest rates; greater flexibility in mortgage lending; and lower down payment requirements. These trends have expanded access to credit, permitting lower-income households to buy homes, although price escalation in some major centres has negated such advantages (Joint Centre for Housing Studies, 2005; CMHC, 2009). Skilled, entrepreneur, and investor immigrants, with their higher incomes and assets, were able to take advantage of these factors to move into ownership earlier than newcomers in the past.

Rents have increased in recent years, although at rates that approximate increases in the consumer price index in both Canada and the United States (Joint Centre for Housing Studies, 2005; Hiebert and Mendez, 2008), in contrast to the inflationary increases in the ownership sector. However, as Goodman (2001) points out, for the lower 20 per cent of the income profile, rent increases have outpaced income growth for some time, creating hardships for lower-income immigrants and refugees.

Rental supply has been tight in Canada during the past two decades, with vacancy rates in most major metropolitan areas in Canada below 3 per cent, and 1 per cent or lower in several cities (CMHC, 2009). Rent increases have been significant in centres such as Vancouver, Toronto, and Calgary. In the United States, although there has been significant variation depending on the region and the city, rates in most areas climbed from just over 7 per cent to close to 10 per cent between 1995 and 2007 (Joint Centre for Housing Studies, 2008). With recent foreclosures, rates are starting to fall as people move back into the rental sector (ibid.). The lowest vacancy rates are in the West and Northeast and in gateway cities such as San Francisco, Los Angeles, and New York. The most recent survey reveals a city-wide vacancy rate of 2.9 per cent in New York and approximately 5 per cent in Los Angeles and San Francisco (ibid.).

Immigrants and refugees who rent also have faced challenges because the affordable rental stock is shrinking in both countries (Carter, 2008; Joint Centre for Housing Studies, 2008) due to demolition of older but cheaper units and 'condominiumization' of older and better-quality units. This process hits low-income immigrants and refugees hardest because it reduces the stock they can best afford. New York City alone lost 250,000 affordable units in the period 2002–5 (Pratt Centre, 2008). This loss has been exacerbated by limited construction of new rental units in both countries. Almost all recent additions in the private market are at the upper end of the rental range. In addition, construction of publicly subsidized rental stock has been meagre and the public portfolios in both countries are declining (CHRA, 2008). In the United States, there has been a net reduction of more than 177,000 public housing apartments since 1995 (Sard et al., 2008). Rezoning and conversion of single-room occupancy housing in central cities have erased housing stock for many single migrant men, notably in Chinatowns. In many US and Canadian cities, gentrification of older neighbourhoods also has reduced housing choices for working-class immigrants and refugees.

In summary, a bifurcated pattern emerges. Market trends in the rental sector have increased the barriers for low-income newcomers to access affordable housing. Skilled immigrants with greater purchasing power, however, moved into the inflationary ownership market, taking on high mortgages but also profiting from inflationary prices. It is too early to determine accurately the effects of the recent 'housing bubble' and ensuing recession on immigrants. Housing price inflation certainly made home ownership more difficult for newcomers in general and particularly for those in major cities where inflation was greatest. It is also certain that many immigrants have faced a housing crisis with loss of jobs and income during the recent recession. The recession combined with the sub-prime mortgage crisis led to mortgage defaults, foreclosures, and loss of homes for some. The early evidence illustrates, however, that the rate of home ownership among immigrants in the United States declined at a slower pace in 2008 than it did for the US-born (Kochhar et al., 2009). Perhaps this is because fewer immigrant households, with their lower incomes, had the very high mortgages that were in greatest jeopardy during the recent market and lending crisis. Immigrants also make higher down payments. Some immigrants, as well as many non-immigrants, have been forced back into the rental market with the loss of their homes, leading to lower vacancy rates, which increase the barriers for low-income immigrant renters. The effects of the housing bubble and recession were much less significant in Canada. Canadian house prices appreciated but not to the same extent as in the United States, there was no meltdown in the mortgage market, and the ensuing recession was considerably more modest with fewer people losing their jobs.

THE HOUSING EXPERIENCES OF NEWCOMERS

Overall, the housing experiences of immigrants and refugees are similar in Canada and the United States, varying more by ethnic group and other characteristics than between the two countries. In both countries housing outcomes for immigrants converge with those of non-immigrants over a period of 20 years. Ownership rates and dwelling sizes increase, and a larger proportion live in detached houses. Like non-immigrants, however, bifurcated labour market experiences sustain bifurcated housing trajectories.

Class of Entry and Social Profile

Immigrants' and refugees' ability to access affordable and adequate housing varies according to their class of entry and social profile (Hiebert and Mendez, 2008). Generally, immigrants are more vulnerable in the housing market, whether owners or tenants, than are non-immigrants. Renter immigrants are more disadvantaged than owner immigrants; recent immigrants are more vulnerable than less recent arrivals; refugees and unauthorized immigrants are more vulnerable than immigrants; and visible minorities are more disadvantaged than immigrants of European origin. For many new immigrants, however, housing choice has expanded significantly. Higher levels of skill and education among many immigrants have brought higher incomes and increased purchasing power and flexibility in the housing market, reflected most noticeably in patterns of home ownership for skilled and business-class immigrants.

Gender and age can also make a difference. Women, particularly those on social assistance, who are single parents, or who have large families, often face serious barriers in accessing housing. If they are black, their problems are multiplied. Vulnerability to discrimination in housing based on gender is certainly exacerbated when combined with race, source of income, and family size (Murdie, 1995). Pruegger and Tanasescu (2007) found that immigrant women with children were common clients of emergency shelters in Calgary because they were often turned away by private landlords. Elderly immigrants are a growing proportion among new arrivals. Often they are brought in by sons and daughters already in the country. They are generally in their sixties or seventies and have a high prevalence of low incomes (Dempsey, 2007). Often the elderly live in crowded circumstances with little privacy in their children's homes, where they may be pressed into service as babysitters. Alternatively, some may live in poor-quality apartments where they become isolated with little chance for socialization, particularly if their language skills are limited. Many feel bewildered and depressed, having been taken from 'home and community' and ending up in a foreign environment (Shavelson, 2008).

High Levels of Core Need among Immigrants and Refugees

In Canada, immigrants experience higher levels of core need, that is, the inability to afford adequate, suitable shelter. Thirty-six per cent of immigrant renters are in core need versus 20 per cent of non-immigrants. For owners the respective figures are 18 and 11 per cent. Recent immigrants experience much higher levels of core need. Thirty-two per cent of immigrant renters arriving prior to 1981 were in core need compared to 44 per cent of those arriving between 2001 and 2006. For immigrant owners the respective figures are 13 and 35 per cent (CMHC, 2009). Refugees face the most difficult housing challenges. Virtually all refugees, who make up just over 10 per cent of the foreign-born in the United States and Canada, initially constitute core need households (Hiebert and Mendez, 2008; Murdie, 2008; Carter et al., 2009). In the US, housing bureaucracies and scholars measure need in different ways, with attention to income and local housing conditions. Acute housing needs are most prevalent among visible minority groups from Africa, the Caribbean, and Latin America, especially in New York, Atlanta, and other 'hot' markets (Adams and Osho, 2008; Pratt Center, 2008).

Housing Affordability

Housing affordability ranks among the most pervasive and persistent of issues for immigrants and refugees. In the Longitudinal Survey of Immigrants to Canada (LSIC), six months after arrival nearly three-quarters of households spent 30 per cent or more of their income on housing. After two years this figure had dropped to approximately 50 per cent, and after four years to less than 40 per cent (Hiebert and Mendez, 2008). Affordability problems mean households are less likely to experience progressive housing careers and positive integration experiences. They are 'stuck' in the rental market. Their circumstances often become worse, leaving them vulnerable to homelessness, particularly for the large undocumented population in the United States (Goodman, 2001; Joint Centre for Housing Studies, 2008). High-priced markets also present challenges. In New York 57 per cent of foreign-born renters pay more than 30 per cent of their income in rent (Pratt Center, 2008). More affordable housing in Philadelphia, Hazelton, and other old industrial cities has helped drive secondary migration from New York.

Discrimination

When attempting to access housing, immigrants and refugees, particularly those who are 'visible minorities', often face discrimination (Li, 2003). This is generally most common in the private rental market, where most new arrivals access housing. Renters are more likely to experience discrimination than are prospective homebuyers, particularly black and Latino renters. Although social housing may not be entirely faultless in regard to discriminatory practices, the mandate and regulations under which social housing operates reduce the likelihood that recent arrivals will face the same discriminatory practices they face in the private market.

Discrimination may be based on race, religion, immigrant status, or other factors, but it is often camouflaged under some other pretext (Ghosh, 2006). Some landlords also may resort to cultural and racial discrimination, refusing to accept tenants or ignoring their rights, and their influence in local politics can encourage public-sector discrimination. In addition, some landlords may fail to maintain units occupied by immigrants, charge extra fees for services that should be covered by rent, and harass immigrant renters by forbidding visitors. A lack of knowledge about their rights and responsibilities often helps make new arrivals victims of unscrupulous landlords (Mattu, 2002). Undocumented immigrants live in some of the most vulnerable housing situations, with little viable legal recourse in the face of landlord abuse or neglect.

Municipalities engage in discriminatory practices by failing to provide landlord–tenant information in multiple languages, selectively enforcing standards, enforcing overly restrictive requirements regarding the number of people who can live in an apartment, conducting spot inspections at unreasonable hours, condemning property with little notice, and not providing administrative or judicial recourse for tenants (Schechter, 2007; Roth, 2009). In the United States some local governments have passed restrictive housing codes aimed at preventing undocumented immigrants from renting in their towns. Others have targeted overcrowding by using building codes and maintenance and occupancy bylaws to discourage larger families from renting. Municipalities regularly discourage the development of new rental housing with three or more bedrooms for fear the units will attract low-income immigrant families (Harwood, 2009; Myers et al., 1996).

The discriminatory practices of rental and real estate agents also can present formidable barriers. Some agents avoid doing business with particular immigrant groups, steer them to neighbourhoods that are less safe, and provide a lower level of service (Teixeira, 1995; Novac and Darden, 2002). The National Housing

Discrimination Study found that in Chicago, Latinos are 15 per cent less likely to be offered rental incentives, while in home buying they are shown fewer units and are often actively discouraged from moving forward on a purchase. Overall, at least one in three immigrant renters or purchasers in the United States faces discrimination (HUD, 2005).

Not all relationships between landlords and tenants are adversarial. Sinclaire (2002) found in a study in Toronto that 86 per cent of tenants had good relationships with landlords even though 76 per cent were unaware of their rights. The experience was particularly positive with landlords of the same background. Carter and Osborne (2009) found a high level of satisfaction with landlords among refugees in Winnipeg, who noted that landlords often helped them find other services.

Public or Social Housing

Despite persistent myths of newcomers occupying disproportionate amounts of social housing in both the United States and Canada, the vast majority of immigrants and refugees find housing in the private market. In the US, 6–7 per cent of immigrants receive housing assistance, slightly more than the American-born (Basolo and Nguyen, 2009). The situation in Canada can be different, especially for refugees. Carter et al. (2009) found that after two years approximately 40 per cent of refugees arriving in Winnipeg were able to access social housing. Murdie (1994) found similar patterns in Toronto.

Social housing presents a more affordable option for recent arrivals with limited incomes. In the Winnipeg study, refugees in social housing saved over $200 on monthly housing payments compared to those in private-sector housing (Carter and Osborne, 2009). Immigrant and refugee access to social housing may facilitate the resettlement process, reducing the transition time and the long-term costs to society in areas such as health, education, and social assistance. Murdie (1994) points out, however, that a concentration of visible minority groups in social housing may lead to stigmatization and continued marginalization.

Homelessness among Immigrants and Refugees

Homelessness is not as common among immigrants and refugees as it is among some groups within the non-immigrant population, notably Aboriginal people in Canada and African Americans and American military veterans in the United States. Increasing numbers of refugees and a smaller proportion of immigrants require temporary shelter and other housing services. Some immigrants and refugees in Toronto, particularly refugee claimants, are homeless and others are at risk of becoming homeless (Access Alliance, 2003). In Vancouver, Hiebert et al. (2005) found that relative and absolute homelessness is less than would be expected among newcomers given their income levels and other challenges they face in the market. Social networks help mitigate the worst forms of homelessness. Still, many live in high-risk housing circumstances and can be considered among the **hidden homeless**. Hidden homelessness is pervasive, although not well documented, particularly among African immigrants in older urban and suburban neighbourhoods of New York, Philadelphia, and Atlanta (Vitiello, 2009). In the United States, most recent reporting and debate on immigrant homelessness have focused on squatter camps of undocumented Mexicans, many of whom are day labourers.

HOME OWNERSHIP EXPERIENCES AND TRAJECTORIES

Newcomers have a strong desire to own their own homes, making home ownership an indicator of integration into the destination society. In a study undertaken by Fannie Mae (Federal National Mortgage Association) in the United States, more than 70 per cent of immigrants identified buying a home as a 'milestone' in life. Most immigrants view home ownership as an investment, providing autonomy and security against job loss or financial hardship. Ownership also contributes to a sense of permanency, of belonging and commitment to community, and serves as a base for further economic and social advancement (National Association of Realtors, 2002).

High immigration levels have spurred home-buying demand over the past decade. Although the rate of home ownership among the foreign-born (48.8 per cent in 2000) in the US lags behind the national average (68 per cent), immigrants represent 14 per cent of new homebuyers (Drew, 2002). Despite the arrival of more highly educated newcomers with well-paying jobs who take shorter paths to home ownership, the rate of ownership for immigrants has been declining since the 1970s in both the United States and Canada (Myers and Liu, 2005; Hiebert and Mendez, 2008). It takes them longer to achieve the rates of ownership of US- and Canadian-born residents. In the United States approximately 18 per cent of immigrants who have been in the country less than five years are owners. This rises to 50 per cent for those resident 15 to 20 years. Rates reach nearly 68 per cent after 20 years, equal to the national ownership rate (Joint Centre for Housing Studies, 2008). The picture in Canada is similar, although immigrants move more quickly into the ownership market. Home ownership rates increase with elapsed time since arrival from about 20 per cent six months after arrival to more than half after four years, but still below the national average of 68 per cent (Hiebert and Mendez, 2008).

Home ownership rates also vary by ethnicity and class. Immigrants from Europe and Asia exhibit much higher rates of home ownership than newcomers from other parts of the world (Myers and Liu, 2005; Hiebert and Mendez, 2008). With more arrivals from Africa and Latin America with lower incomes there is less potential for these groups to become owners (Herbert and Belsky, 2006).

Despite low interest rates and flexible mortgages in recent years, research from the Joint Centre for Housing Studies (2005) found that first-time foreign-born homebuyers in the US still opted to make larger down payments compared to US-born homeowners. Thus, foreign-born households tend to allocate a greater share of their income to housing compared to US-born households. Close to 40 per cent of first-time foreign-born buyers claimed that they allocated 30 per cent or more of their income towards their mortgage payments compared to just 28 per cent among the US-born. The sub-prime market and non-traditional mortgage products increased access to capital, but they also increased the risks for already vulnerable immigrants, particularly among lower-income Latinos. Although immigrant households in Canada have also benefited from lower interest rates, more flexible mortgage financing, and lower down payment requirements, there has been less risk to immigrant home purchasers in Canada because of more careful lending guidelines.

Immigrant homebuyers often cluster at smaller geographic levels, concentrating in neighbourhoods where others of their ethnocultural group reside (Li and Skop, 2007). Clusters of particular racial and ethnic groups have developed in many suburbs, forcing communities unaccustomed to immigration to grapple with the housing (and other) challenges of new neighbours (McDonald, 2004).

OVERCROWDING AND IMMIGRANT HOUSING ISSUES

The incidence of crowding among newcomers has been growing since the 1980s in the United States, although it varies by centre and ethnocultural group (Myers et al., 1996). The household size of new arrivals is, on average, larger than households of the US-born population, with larger families and a higher proportion of extended families, multiple families, and composite households sharing a single dwelling (Butterfield, 2008). Although some ethnocultural groups may have less demanding residential space preferences and some traditionally reside with extended family, the larger household size does lead to higher proportions of crowded households. In Canada there are few data by ethnocultural group, but information from the LSIC survey indicates that about one-quarter of households were crowded six months after they arrived. This proportion fell to 15 per cent after four years (Hiebert and Mendez, 2008), compared to less than 3 per cent within the Canadian population (Statistics Canada, 2008). Carter et al. (2009) found that almost 50 per cent of refugee households were crowded in their first year in Winnipeg and after three years this figure had fallen to 31 per cent.

Many households trade space for lower housing costs if they are renters, or share space and pool limited resources to become owners (Rosenbaum and Friedman, 2007). Economizing on housing in the United States or Canada is sometimes also part of a transnational housing strategy in which immigrants save money to invest in new construction in their hometowns, building family homes to which they typically expect to return (Levitt, 2001; Smith, 2006).

Immigrants' larger households increase the demand for larger units. The ownership sector has been more responsive to this demand, as purchasers have more control over what is purchased, especially if they are having a new home built. The rental sector, despite the demand, tends to build for the typical US- or Canadian-born rental household requiring one or two bedrooms. Larger immigrant households have a difficult time finding three-bedroom rental units, even if they can afford them, leaving them with little choice but to live in crowded circumstances.

In the United States, regulating crowded housing has recently been one way for local governments to discourage the settlement of undocumented immigrants. Fear of health and safety risks associated with people living in crawl spaces and basements can be genuine. This concern sometimes also intersects with less well-founded anxieties about the economic and cultural impacts of immigration, which in turn collide with mundane yet contentious issues of parking, noise, and unfamiliar smells from the neighbours' kitchens. Local authorities have narrowed the legal definition of family housing, instituted tougher penalties for violations, and stepped up enforcement against both landlords and tenants. The pioneering Illegal Immigration Relief Act of Hazleton, Pennsylvania, and its many copycats stipulate that landlords are responsible for verifying the legal status of tenants, and some towns require all renters to register at city hall. A smaller number of local governments have established affirmative policies protecting all immigrants' access to housing, regardless of status (Harwood, 2009; NILC, 2008).

IMMIGRANTS ARE CHANGING
THE FACE OF HOUSING AND CITIES

Immigration is the driving force behind many changes in housing and neighbourhoods. Immigrants moving into older neighbourhoods can create a 'demographic dynamism'. The energy and money they invest change housing and retail markets in these areas. Conversely, when North American cities have lacked sufficient immigration to replace central-city residents departing for the suburbs, inner-city economies have eroded and neighbourhoods have become blighted (Myers, 1999). In most North American cities, the recent arrival of immigrants to populate these areas has complemented the major demographic shifts of US- and Canadian-born residents, and their residential and commercial investments support urban revitalization and gentrification.

While immigrants have always settled in North America's inner cities, they are now spreading their impacts in the housing market across metropolitan areas, particularly to suburban communities. In places with large concentrations of high-wage jobs, such as the 'boomburgs' of Boston and California's Silicon Valley, immigrants have accounted for a large proportion of recent home and condominium purchases, as much as 56 per cent in Massachusetts between 2000 and 2004, even though the foreign-born make up just 15 per cent of the state's population (Blanton, 2005). Ethnically oriented shopping centres have become common in suburbs of major North American cities, supported by residential concentrations of what scholars call 'ethnoburbs' (Fong et al., 2008; Li and Skop, 2007).

Finally, immigrants have helped change the physical and institutional landscapes of housing and neighbourhoods in North America. Immigrant housing organizations have sometimes employed architectural design as a strategy for cultural preservation and, in a sense, to mark their territory. This has been most evident in Chinatowns and, in the United States, Puerto Rican and Chicano barrios. It is also evident in the suburbs of Vancouver and Toronto, the 'monster' homes for extended families in Vancouver, for example. Undocumented immigrants, by contrast, often try not to draw attention to their homes for fear of attracting immigration authorities, although they play vital roles in building and maintaining North America's residential environments through their work in construction, landscaping, and housekeeping, helping to keep housing and related services affordable for others even while they help sustain housing markets through their own demand for shelter.

IMMIGRANTS REQUIRE DIVERSE HOUSING STRATEGIES

Addressing the housing problems of people with different socio-economic characteristics, financial resources, and cultural backgrounds is one of the most important challenges facing immigrant-receiving societies. Diverse housing strategies are employed, mostly in the private market and at the level of individual families, but also through civil society organizations and government initiatives.

The Role of Community-Based Non-Profit Organizations

Community development corporations, ethnocultural organizations, refugee resettlement agencies, and tenants' rights organizations have played important roles in providing and preserving immigrant housing in North American cities. Non-profit organizations develop, manage, and provide educational workshops on housing in Chinatowns from Philadelphia to Vancouver, and from the Sudanese community of Minneapolis to the Mexican neighbourhoods of Dallas. Funding sources for third-sector housing include philanthropy, government, and fundraising campaigns; however, housing production, management, and counselling have not been a primary strategy for most immigrant service providers. They are typically underfunded and lack staff with formal expertise in housing (Carter, 2009). Nonetheless, their front-line workers often assist newcomers with housing on an informal basis (Theodore and Martin, 2007).

BOX 5.2   Civil Society and Housing in Philadelphia

Civil society organizations play diverse roles in mediating the housing experiences of immigrants and refugees. The recent history of the involvement of newcomer and receiving-community institutions in housing assistance, advocacy, production, and management in Philadelphia illustrates how these roles vary among ethnic groups with different settlement and housing experiences.

Due to deindustrialization and the loss of one-third of the city's population between the early 1950s and the 2000s, until recently Philadelphia has been a region of low immigration compared to the major gateways; however, with growth of the service economy since the 1990s, the city has become a 're-emerging gateway'. Neighbourhood decline, suburbanization, and recent growth all have shaped newcomers' housing experiences. Migrant communities have advocated and developed affordable housing in response to urban renewal, gentrification, and the devolution of housing production to community development corporations (CDCs) since the 1970s.

The Philadelphia Chinatown Development Corporation was one of the city's first CDCs, established in 1969 out of the community's fight against a highway that resembled other 'highway revolts' in US Chinatowns. Since the 1990s, however, the organization has struggled to acquire property amid a condominium boom that has also converted many formerly affordable apartments and workplaces of recent Chinese immigrants who have moved to working-class suburbs. Puerto Rican and pan-Latino CDCs have built more housing in the still depressed land market of North Philadelphia, and their counselling programs on financial literacy and home ownership have served many Latin American migrants.

Community organizations often impact recent immigrant communities' housing experiences outside of formal housing programs. African and Caribbean mutual aid and faith organizations accommodate people in apartments above their offices. The location decisions in regard to Korean churches and Sikh temples have reshaped housing networks and played key roles in their constituencies' suburbanization. African-American-led community development corporations have hired community organizers to help prevent foreclosure among African and Caribbean neighbours.

Refugee resettlement in Philadelphia, as in most US and Canadian cities, has transformed old settlement houses that had worked with earlier generations of immigrants. Federal funding in both countries has supported transitional housing, but this funding has focused more on health, education, language, and employment than on long-term housing strategies. The typical approach of resettlement agencies has been to scramble to find housing in the rental market, generally in neighbourhoods where case workers believe refugees will find transit, services, and sometimes co-ethnics to help ease their transition. In the 1980s, landlord abuse sparked tenant activism among recent Southeast Asian immigrants who won important battles in West Philadelphia, but many of these people moved to new neighbourhoods and regions partly due to poor, sometimes violent, housing and neighbourhood experiences. Today, Burmese and Bhutanese refugees are resettled in Vietnamese and Cambodian South Philadelphia, and Iraqis are in the city's main working-class Palestinian neighbourhood. One resettlement agency seeks housing among church congregations in rural York and Lancaster counties, while others continue to practise urban resettlement and argue it helps revitalize city neighbourhoods.

The Role of Social Support Networks

Many new arrivals settle in receiving communities with help from co-ethnics who support them in accessing housing, work, and other services. This is arguably the most pervasive 'housing assistance' for new arrivals and can include: staying with relatives or friends; help in finding an apartment and paying the rent; lending pools fostering ownership; interpreting leases; protecting tenant rights; and providing furnishings. Although this assistance can be one of the most generous human acts, it does not always encourage integration. In some instances the ties that bind newcomers to their housing supporters are exploitative, as in the case of people indentured to pay off their passage. Evidence in both countries, however, suggests the social support network is all that keeps many newcomers from being homeless.

Housing Assistance from Employers

In metropolitan centres across North America employers aid immigrants with housing. Big corporations often provide housing search assistance and housing allowances as part of a broader 'relocation package'. For skilled economic immigrants arriving to address labour shortages, housing assistance generally is contractual, within a limited time frame, and can vary depending on the importance of the skills the immigrant brings to the workplace (Carter, 2009). Similar assistance is also available to many temporary foreign workers who provide labour in resource communities, the meat-processing industry, and farm labour sector. Often, however, the accommodation provided to temporary foreign workers is sparse and better suited to single individuals than families, and when the workers are undocumented it can be substandard and abusive. Abusive situations also arise when immigrant workers reside with their employers, for example, in Canada's temporary foreign worker program for child and elder care.

Government Housing Policy and Programs

Governments can, and do, make a difference in housing policies and programs that facilitate the settlement and integration of refugees and immigrants. Studies suggest household affordability improves significantly when newcomers are able to access government-sponsored public or social housing. Additions to the stock of publicly subsidized housing or shelter subsidies to allow newcomer households to access better-quality private-sector rental units help facilitate resettlement and integration. Access to social housing provides more opportunities to locate larger three- and four-bedroom units not available in the private market. If the need for larger units is to be addressed it probably has to be a public initiative; however, governments in both Canada and the United States have not funded much new affordable housing stock to meet this need for many years (Carter, 2008).

Governments also fund transitional housing units—a place to stay until newcomers get settled and learn more about the housing process and housing market. Those who live in transitional units speak positively about the circumstances. Such units are generally managed by community or ethnocultural housing organizations.

Immigrants and refugees need information and assistance in finding adequate, affordable housing in good neighbourhoods. Housing and home ownership counselling is more common in the US than in Canada, although sometimes information in both places is neither current nor well tailored to newcomers' needs and preferences (Carter et al., 2009). Newcomers also need better education on tenant–landlord rights and responsibilities. Conversely, the property management sector requires better education and awareness of the cultural characteristics and housing needs of diverse immigrant households. Resources are required to support delivery of all of these services, mainly through community-based organizations.

A proactive immigration policy by governments to address labour force shortages has not been matched by a proactive housing policy to improve the housing circumstances of those who arrive. Improved housing clearly matters, but this is not a stand-alone solution to the problems faced by newcomers—it is only part of a complex set of initiatives required to improve their life chances in a new country.

Transnational Housing Strategies

It is important to recognize that many immigrants' housing strategies operate over a larger geography—transnationally. Undocumented Mexican, Central American, and Brazilian immigrants in the United States often view themselves as temporary migrants. Migration and the endurance of uncomfortable, often crowded housing are typically part of a family strategy commonly aimed towards the construction or expansion of a family residence in their country of origin. This housing strategy is partly a result of weak mortgage markets in sending countries and of barriers to credit, banking, and overall housing choice for people without legal status (Durand and Massey, 2004; Smith, 2006).

Immigrants at the other end of the legal and economic spectrum also pursue transnational housing strategies, usually with fewer constraints. For example, immigrants from Hong Kong and China have invested their own and relatives' money in housing in New York and Vancouver seeking secure outlets for capital (Kwong, 1996). Some have returned from Canada to Hong Kong and China to earn higher salaries, yet their retired parents and grandparents remain in Canada, maintaining family homes, as someday their children plan to return to share Canada's greater health and social benefits (Kobayashi et al., 1998).

In sum, immigrants, refugees, and receiving community organizations and governments pursue a great variety of housing strategies, reflecting the great diversity of migrants and the places they settle. Although many secure housing that supports positive integration trajectories, adequate, affordable housing remains a dream as opposed to a reality for many newcomers.

CONCLUSION

Few immigrants and refugees experience the quality of housing of the majority population for many years after arrival. Their rates of ownership are lower, and more live in crowded circumstances and have affordability problems. Most experience progressive housing careers, but evidence suggests their housing circumstances are not improving as fast as they were a generation ago. This slower progress is related to a combination of the characteristics of new arrivals, changing circumstances in the housing market, and poorer labour force integration. Nonetheless, today many immigrants arrive with greater earning potential and considerable assets. Some move into good housing immediately, sometimes as homeowners.

Housing circumstances vary by class of entry and ethnocultural characteristics. Skilled workers ultimately have the highest propensity to purchase a home and are less likely to lose their home. Refugees, particularly refugee claimants, and undocumented immigrants have the lowest rates of ownership and experience the highest levels of affordability problems and crowded circumstances. European immigrants have more success in accessing quality, affordable housing, although they do not necessarily exhibit higher rates of ownership. At the other end of the housing continuum are visible minorities who are most likely to experience the worst housing conditions. Significant differences among the visible minorities are manifested in the marketplace. Those from South and Southeast Asia have high rates of ownership and good housing circumstances. Arab, African, and Latin American newcomers report high levels of housing difficulties and much lower rates of ownership.

The housing experiences of immigrants and refugees illustrate many similarities between Canada and the United States. This is not surprising. The housing process, systems, and market circumstances are very similar in the two countries. The source countries of immigrants and refugees to these two countries are also similar, as are the characteristics of new arrivals. Two broader differences between the two countries are worth noting. The first is the large undocumented population in the United States, which faces additional barriers in accessing adequate housing. The second has more to do with the non-immigrant population, which in most parts of the urban United States has a fundamentally different experience of diversity. While visible minorities in Canada share some of the same discrimination that darker-skinned immigrants and non-immigrants experience in the United States, little in Canada's history of housing and urban development quite compares to the US legacy of African-American housing and segregation. This difference has profoundly affected immigrant housing and settlement experiences as well as the context of housing policy in the US, which has generally been viewed 'in black and white'.

Looking to the future, immigration will continue to drive much of the housing demand in both countries. Some suggest immigration will fix housing policy as it generates long-term demand. Others identify a need for housing policy to facilitate more successful resettlement and integration, suggesting there is much work to be done in structuring better housing policy to support newcomers to North America.

QUESTIONS FOR CRITICAL THOUGHT

1. How do the housing market dynamics of different cities and neighbourhoods affect the housing choices available to immigrants and refugees (e.g., in expensive New York City vs more affordable Montreal)?

2. Identify and describe five 'best practices' that could be used as a basis for improving housing policy for newcomers.

3. How has the arrival of different ethnocultural groups affected the style and design of housing?

SUGGESTED READINGS

1. Adams, M., and G. Osho. 2008. 'Migration, Immigration and the Politics of Space: Immigration and Local Housing Issues in the United States', *Research Journal of International Studies* 8: 5–12. This article surveys the impacts of immigration on local housing markets in the US, including immigrants' experiences and receiving community responses.

2. Harwood, S., ed. 2009. 'Planning in the Face of Anti-Immigrant Sentiment', *Progressive Planning* (Winter): 8–26. This collection of articles surveys restrictive measures against immigrants and some affirmative approaches taken by local and state governments in the US to regulate immigrant housing and related activities.

3. Hiebert. D., and P. Mendez. 2008. *Settling In: Newcomers in the Canadian Housing Market, 2001–2005*. Ottawa: CMHC. The Longitudinal Survey of Immigrants to Canada (LSIC) is used to investigate the participation of immigrants in Canada's housing market during the first four years of the settlement process, beginning in 2000–1. The analysis focuses on the changing rate of home ownership, crowding, and affordability.

4. Myers, D., and C. Liu. 2005. 'The Emerging Dominance of Immigrants in the US Housing Market, 1970–2000', *Urban Policy and Research* 23, 3: 347–65. The authors summarize and analyze the explosive growth of immigrant housing demand in the US and in five individual states with the largest immigrant populations—California, New York, Florida, Texas, and Illinois.

REFERENCES

1. Access Alliance. 2003. *Best Practices for Working with Homeless Immigrants and Refugees*. Toronto: Access Alliance Multicultural Community Health Centre.

2. Adams, M., and G. Osho. 2008. 'Migration, Immigration and the Politics of Space: Immigration and Local Housing Issues in the United States', *Research Journal of International Studies* 8: 5–12.

3. Basolo, V., and M. Nguyen. 2009.

'Immigrants' Housing Search and Neighborhood Conditions: A Comparative Analysis of Housing Choice Voucher Holders', *Cityscape* 11, 3: 99–126.

4. Blanton, K. 2005. 'Immigrants Fuel Housing Boom: State Attracting Wave of Affluent, Educated Workers', *Boston Globe*, 6 Oct. At: www.boston.com/business/globe/articles/2005/10/06/immigrants_fuel_housing_boom/.

5. Bruce, D., and T. Carter. 2003. *The Housing Needs of Low Income People Living in Rural Areas*. Ottawa: CMHC.

6. Broadway, M. 2007. *Meatpacking, Refugees and the Transformation of Brooks, Alberta*. Edmonton: Department of Rural Economy, University of Alberta.

7. Butterfield, E. 2008. 'Immigrants Maintain Larger Households', *Builder Magazine*: 5–6.

8. Canadian Housing and Renewal Association (CHRA). 2008. *Creating a Modern Housing Policy: A Legacy for Tomorrow's Leaders*. Ottawa: CHRA.

9. Canada Mortgage and Housing Corporation (CMHC). 2009a. *Canadian Housing Observer*. Ottawa: CMHC.

10. Carter, T. 2008. *Canadian Policy Paper: Canadian Policy Update*, prepared for the Tri-County Conference. Ottawa: CHRA.

11. ———. 2009. *An Evaluation of the Manitoba Provincial Nominee Program*. Winnipeg: Manitoba Department of Labour and Immigration.

12. ——— and J. Osborne. 2009. 'Housing and Neighbourhood Challenges of Refugee Re-settlement in Declining Inner City Neighbourhoods: A Winnipeg Case Study', *Journal of Immigrant and Refugee Studies* 7, 3: 308–27.

13. ———, C. Polevychok, and J. Osborne. 2009. 'The Roles of Housing and Neighbourhood in the Re-settlement Process: A Case Study of Refugee Households in Winnipeg', *Canadian Geographer* 53, 3: 305–22.

14. Chera, S. 2004. *The Making of 'Home': Housing as a Vehicle for Building Community among Newcomers to Edmonton*. Edmonton: Edmonton Mennonite Centre for Newcomers.

15. Danso, R.K., and M. Grant. 2000. 'Access to Housing as an Adaptive Strategy for Immigrant Groups: Africans in Calgary', *Canadian Ethnic Studies* 20, 3: 19–43.

16. Dempsey, C. 2007. *Elderly Immigrants in Canada: Income Sources and Self-Sufficiency*. Ottawa: Citizenship and Immigration Canada.

17. Drew, B.R. 2002. *New Americans, New Homeowners: The Role and Relevance of Foreign-born First-time Homebuyers in the US Housing Market*. Cambridge, Mass.: Joint Center for Housing Studies, Harvard University.

18. Durand, J., and D. Massey. 2004. *Crossing the Border: Research for the Mexican Migration Report*. Urbana: University of Illinois Press.

19. Farm Foundation. 2005. *Immigrants Change the Face of Rural America*. Oak Brook, Ill.: Farm Foundation.

20. Fiscal Policy Institute. 2007. *Working for a Better Life: A Profile of Immigrants in the New York State Economy*. New York: Fiscal Policy Institute.

21. Fong, E., et al. 2008. 'Suburban Residential Clustering of Racial and Ethnic Groups', *Journal of Population Studies* 35: 37–74.

22. Ghosh, S. 2006. 'We are Not All the Same: The Different Migration, Settlement Patterns and Housing Trajectories of Indian Bengalis and Bangladeshis in Toronto', Ph.D. thesis, York University.

23. Goodman, J. 2001. *Housing Affordability in the United States: Trends, Interpretations, and Outlook*. Washington: Millennium Housing Commission.

24. Harwood, S. 2009. 'Immigration and Racialized Regulation: Planning in the Face of Anti-Immigrant Sentiment', *Progressive Planning* (Winter): 8–9.

25. Herbert, C., and E. Belsky. 2006. *The Homeownership Experience of Low-Income and Minority Families: A Synthesis and Review of the Literature*. Washington: HUD.

26. Hiebert, D., S. D'Addario, K. Sherrell, and S. Chan. 2005. *The Profile of Absolute and Relative Homelessness among Immigrants, Refugees, and Refugee Claimants in the GVRD*. Vancouver: MOSAIC.

27. ——— and P. Mendez. 2008. *Settling In: Newcomers in the Canadian Housing Market, 2001–2005*. Ottawa: CMHC.

28. ———, ———, and E. Wyly. 2006. *The Housing Situation and Needs of Recent Immigrants in the Vancouver CMA*. Ottawa: CMHC.

29. Housing and Urban Development (HUD). 2005. *Discrimination in Metropolitan Housing Markets: National Results from Phase 1, 2, and*

3 of the Housing and Discrimination Study. Washington: HUD.

30. Joint Center for Housing Studies. 2005. The State of The Nation's Housing: 2005. Cambridge, Mass.: Harvard University.

31. ———. 2008. The State of The Nation's Housing: 2008. Cambridge, Mass.: Harvard University.

32. Kobayashi, A.L., E. Moore, and M. Rosenberg. 1998. Healthy Immigrant Children: A Demographic and Geographical Analysis. Ottawa: Human Resources and Social Development Canada, Applied Research Branch, Strategic Policy.

33. Kochhar, R., A. Gonzalez-Barrera, and D. Dockterman. 2009. Through Boom and Bust: Minorities, Immigrants and Homeownership. Washington: Pew Hispanic Center.

34. Kwong, P. 1996. The New Chinatown. New York: Hill and Wang.

35. Levitt, P. 2001. The Transnational Villagers. Berkeley: University of California Press.

36. Li, P. 2003. Destination Canada: Immigration Debates and Issues. Toronto: Oxford University Press.

37. Li, W., and E. Skop. 2007. 'Enclaves, Ethnoburbs, and New Patterns of Settlement among Asian Immigrants', in M. Zhou and J. Gatewood, eds, Contemporary Asian Americans: A Multi-Disciplinary Reader, 2nd edn. New York: New York University Press, 222–36.

38. McDonald, J. 2004. 'Ethnic Clustering and the Locational Choice of Immigrants to Canada', Canadian Journal of Urban Research 13, 1: 85–101.

39. Mattu, P. 2002. A Survey of the Extent of Substandard Housing Problems Faced by Immigrants and Refugees in the Lower Mainland of British Columbia. Vancouver: MOSAIC.

40. Murdie, R. 1994. 'Blacks in Near Ghettos? Black Visible Minority Population in Metropolitan Toronto Housing Authority Public Housing Units', Housing Studies 9, 4: 435–57.

41. ———. 1995. 'Housing Issues Facing Immigrants and Refugees in Greater Toronto: Initial Findings from the Jamaican, Polish and Somali Communities', paper presented to Habitat II Research Conference, Ankara, Turkey, Nov.

42. ———. 2008. 'Pathways to Housing: The Experience of Sponsored Refugees and Refugee Claimants in Accessing Permanent Housing in Toronto', Journal of International Migration and Integration 9, 3: 81–101.

43. Myers, D. 1999. 'Demographic Dynamism and Metropolitan Change: Comparison of Los Angeles, New York, Chicago, and Washington, D.C.', Housing Policy Debate 10, 4: 919–54.

44. ———. 2007. Immigrants and Boomers: Forging a New Social Contract for the Future of America. New York: Russell Sage Foundation.

45. ———, W. Baer, and S. Choi. 1996. 'The Changing Problem of Overcrowded Housing', Journal of the American Planning Association 62, 1: 66–84.

46. ——— and C. Liu. 2005. 'The Emerging Dominance of Immigrants in the US Housing Market, 1970–2000', Urban Policy and Research 23, 3: 347–65.

47. National Association of Realtors. 2002. Housing Opportunities in the Foreign-Born Market. A Report from the Research Group of the National Association of Realtors.

48. National Immigration Law Center (NILC). 2008. Laws, Resolutions and Policies Instituted across the US Limiting Enforcement of Immigration Laws by State and Local Authorities. At: www.nilc.org/immlawpolicy/LocalLaw/locallaw-limiting-tbl-2008-12-03.pdf. (3 Jan. 2009)

49. Novac, S., and J. Darden. 2002. Housing Discrimination in Canada: The State of the Knowledge. Ottawa: CMHC.

50. Pratt Centre for Community Development. 2008. Confronting the Housing Squeeze: Challenges Facing Immigrant Tenants and What New York Can Do. New York: Pratt Centre for Community Development.

51. Preston, V., R. Murdie, and A. Murnaghan. 2006. The Housing Situation and Needs of Recent Immigrants in the Toronto CMA. Ottawa: CMHC.

52. Pruegger, Valerie, and Alina Tanasescu. 2007. Immigrants and Refugees: Homelessness & Affordable Housing. Calgary: Housing Issues of Immigrants and Refugees in Calgary, Sept.

53. Ready, T. 2006. Hispanic Housing in the United States. South Bend, Ind.: Institute for Latin Studies, University of Notre Dame.

54. Rosenbaum, E., and S. Friedman. 2007. The Housing Divide: How Generations of Immigrants Fare in New York's Housing Market. New York: New York University Press.

Carter/Vitiello: Immigrants, Refugees, and Housing

55. Roth, B. 2009. 'Housing Overcrowding in the Suburbs: The Politics of Space and the Social Exclusion of Immigrants', *Progressive Planning* (Winter): 18–20.

56. Sard, B., L. Straub, and W. Fischer. 2008. *Preliminary Facts on Public Housing.* Washington: Center on Budget and Policy, HUD.

57. Schechter, G. 2007. *Fair Housing and Immigrants in the Northern Suburbs: Issues and Challenges.* Chicago: Interfaith Housing Centre of the Northern Suburbs.

58. Shavelson, Lonny. 2008. *Elderly Immigrants Flow into California.* Los Angeles: National Public Radio (documentary program).

59. Sinclaire, B. 2002. *Housing New Canadians: Bibliography.* Toronto: City of Toronto Housing Department.

60. Smith, R. 2006. *Mexican New York: Transnational Lives of New Immigrants.* Berkeley: University of California Press.

61. Statistics Canada. 2005. *Population Projections for Canada, Provinces and Territories.* Ottawa: Statistics Canada.

62. ———. 2008. *Population Projections for Canada, Provinces and Territories 2005–2031.* Ottawa: Statistics Canada.

63. Teixeira, C. 1995. 'Ethnicity, Housing Search, and the Role of the Real Estate Agent: A Study of Portuguese and Non-Portuguese Real Estate Agents in Toronto', *Professional Geographer* 47, 2: 176–83.

64. ———. 2009. 'New Immigrant Settlement in a Mid-Sized City: A Case Study of Housing Barriers and Coping Strategies in Kelowna, British Columbia', *Canadian Geographer* 53, 3: 323–39.

65. Theodore, N., and N. Martin. 2007. 'Migrant Civil Society: New Voices in the Struggle over Community Development', *Journal of Urban Affairs* 29, 3: 269–87.

66. US Bureau of the Census. 2008. *The Net International Migration Component.* Washington: Population Division, US Bureau of the Census.

67. Vitiello, D. 2009. 'The Migrant Metropolis and American Planning', *Journal of the American Planning Association* 75, 2: 245–55.

CHAPTER 6

ECONOMIC EXPERIENCES OF IMMIGRANTS

Lucia Lo and Wei Li

INTRODUCTION

Immigration has always been one of the biggest public policy issues in Canada and the US (Green and Green, 1999; Borjas, 2000). Public concerns usually centre on the economic dimension. Whether immigration is good for the country and whether migration is good for the individual are hot topics for debate. There are three major arguments for admitting immigrants, especially from a governmental perspective: (1) immigration increases economic growth; (2) it fills labour and skills shortages; and (3) it generates fiscal benefits. Public opinion is more mixed, ranging from the negative stance that immigrants take away jobs from the locals, depress wages, and drain public coffers, to the more positive view that immigrants do the dirty jobs locals are unwilling to perform, expand the consumer and human capital bases of the country, and promote international trade.

Concerns have been raised lately, especially in Canada, about the economic well-being of the current generation of immigrants. Ample evidence indicates that immigrants who arrived after the 1980 are not doing as well as their predecessors in terms of their labour market experiences, raising concerns about

the appropriateness of immigration policies and Canadian claims of accommo-
dating diversity. We ask:

- Are immigrants able to practise in the fields in which they studied or
  worked in their home countries before emigration?
- Do they earn as much as the American/Canadian-born?
- If less, how long does it take them to catch up with the American/
  Canadian-born population?
- Are immigrants generally at a disadvantage in the labour market compared
  to American/Canadian-born citizens with the same level of education,
  skills, or experience? Why?

This chapter focuses on the economic experiences of immigrants in Canadian
and US cities. First, we examine the theoretical context underpinning immigrant
economic experiences and their impact on the economy. Next, we compare
the two countries' social, economic, political, and institutional structures and
identify the differential contexts under which immigrants arrive and settle. The
following two sections focus on immigrants' labour market experiences and
outcomes, and where possible examine differences due to country of origin,
gender, ethnoracial origins, and period of immigration. The last section looks
at the immigrant profiles of selected large and medium-sized cities in the two
countries to determine how place affects immigrants' economic experiences,
while the conclusion points to the need for further, more detailed research.

THEORIES ON ECONOMIC EXPERIENCES

How immigrants fare in the labour arenas of their destination societies has
been the subject of substantial theoretical and empirical work (Abbott and
Beach, 1993; Baker and Benjamin, 1994; Bloom et al., 1995; Borjas, 1990, 2000;
Chiswick, 1978; LaLonde and Topel, 1992; Picot and Sweetman, 2005; Raijman
and Tienda, 1999). The dominant approaches focus on the human capital and
status attainment traditions in economics and sociology, and the structural
conditions facing immigrants in the destination society.

Classical **assimilation theory** argues that immigrants are at a disadvantage
and receive considerably lower economic rewards than American/Canadian-
born workers of comparable skills because of their unfamiliarity with the
new labour market, limited access to information and social ties, inadequate
command of English (or French in Quebec), non-transferable occupational
skills, or discrimination; however, over time, most experience upward mobility
and close the earnings gap (Borjas, 1990, 2000; Chiswick, 1978; LaLonde and
Topel, 1992; Lofstrom, 2002). Nonetheless, no consensus exists on the size of the
so-called entry effect and the amount of time it takes to close the gap (Abbott
and Beach, 1993; Baker and Benjamin, 1994; Bloom et al., 1995).

Sometimes immigrants can surpass the American/Canadian-born, as was the
case with those who arrived in the US in the 1950s and 1960s. The theory of

immigrant **self-selection** argues that economic immigrants represent the more ambitious, motivated, risk taking, and able elements in their origin countries and these traits underlie their success (Chiswick, 1978).

Yet not all able and highly skilled immigrants are equally successful in establishing themselves in their new countries. Transferability of skills and human capital differ across countries and occupational markets. Some occupations (e.g., scientists, craftspeople) may be highly transferable while others (e.g., lawyers, accountants, doctors, teachers) are country-specific and require knowledge of laws, licensing permits, or language proficiency. Certain occupations may be in great demand while others may be a liability in a saturated market. The occupational market in which immigrant workers operate may thus affect their economic opportunities in the destination society's labour market (Raijman and Semyonov, 1995).

Recent research also suggests that immigrants' economic well-being depends on such factors as a country's immigration policies, citizenship laws, labour market opportunities, and welfare institutions (Lewin-Epstein et al., 2003). Financial or other resources provided by a country and the granting of citizenship and voting rights within a reasonable period of time facilitate incorporation. Meanwhile, whether immigrants are able to convert their human capital into economic resources depends greatly on the **context of reception** or the social climate towards immigrants (Portes and Rumbaut, 1990). Immigrants arriving during periods of mass migration and economic decline or facing hostility from the dominant group are more likely to have difficulty finding employment that matches their qualifications (Henry, 1994; Raijman and Semyonov, 1995).

THE EMPIRICAL CONTEXT

We hypothesize that the history of immigration to Canada and the US accounts for many of the differences in well-being experienced by different immigrant groups. Variations between Canada and the US can be explained by their admission and integration policies, the types of immigrants they recruit, as well as the size of the economy and the business environment. Immigration history, geopolitics, and admission and integration policies to some extent determine immigrant characteristics. The size of the economy and the business environment impose different opportunities and challenges. The country's prevailing social climate, its racial structure before large-scale migration, variations in group size, and the degree of **institutional completeness** among immigrant groups can mean different opportunity sets for immigrants.

Role of Immigration Policies

Historically, American and Canadian policies favoured immigrants from Europe, and both countries enacted discriminatory laws against certain groups (see Chapter 2). The 1965 **Immigration and Nationality Act** in the US and the 1967 Immigration Act in Canada, which abolished race-based immigration quotas, marked a fundamental change in admission policies; however, they also established the divergent paths along which immigrants must now travel and contributed to the different socio-economic and demographic characteristics of immigrants in the two countries. Those admitted to Canada and the US since 1967 can be categorized as economic, family, or humanitarian immigrants. Canadian immigration policy values human capital (measured by age, education, and English/French proficiency) and Canadian experiences and connections. Until the Canadian Experience Class was introduced in 2008 to recruit temporary foreign workers and international students graduating from Canadian universities, the points system had been the basis for recruiting highly skilled migrants. Conversely, the immigration admission quota system in the US focuses on family reunification. Up to 80 per cent of the country's annual immigration quota was designated for this class until 1990, when the annual quota for employment-based immigrants was tripled and an H-1B non-immigrant visa was created to recruit college-educated international migrants. The ratios between economic and family class immigrants in the two countries are almost mirror images of each other: around 2004–5, the ratio between economic and family class immigrants is 58:25 in Canada and 22:58 in the US (Li, 2009).

Permanent immigrants, however, do not represent the complete picture of international migration dynamics in the US. The two million temporary migrants who arrived in 2005 (many highly skilled) almost doubled the number of permanent, skilled migrants in the country. The US reliance on temporary skilled migrants has created many legal and logistical issues for the migrants themselves (see Box 6.1). In contrast, temporary migrants count for a very small percentage of skilled migrants in Canada.

Differential Human Capital Levels

Such policy differences result in noticeable differences in the immigrant profiles of the two countries (Table 6.1).[1] Immigrants admitted to Canada have a higher overall educational attainment than the Canadian-born. While the proportion of immigrant degree holders is similar in both countries, the proportion with no high school certificate in the US is much higher than in Canada. Overall, educational levels among the US foreign-born are more polarized than among Canadian immigrants. The proportion of foreign-born at either end of the education spectrum remains stable over time, with over half having at most a high school certificate and about a quarter at least a bachelor's degree. The Canadian points system has achieved its goal of admitting immigrants based on their human capital: 46 per cent of the pre-1991 arrivals had attained no higher than a high school certificate, but this proportion dropped to 36 per cent for those arriving between 2001 and 2006. The proportion with degrees, on the other hand, doubled from 20 per cent before 1991 to 42 percent in the 2001–6 period.[2]

In terms of official-language capability, immigrants to Canada again fare better overall than those in the US (Table 6.2). In 2006, 6.4 per cent of all immigrants

BOX 6.1   A US Story

Set up by the 1990 Immigration Act, the H-1B is a non-immigrant visa program designed for temporary workers employed in 'specialty occupations' requiring specialized knowledge and at least a bachelor's degree or its equivalent. It is a de facto immigration program. Sponsored by American employers, H-1B visas have an initial three-year term, renewable for another three years. During this period, holders are eligible to apply for permanent residency status. They can bring their immediate families when entering the US; however, regardless of their qualifications, their spouses are not permitted to seek employment in the US until permanent resident status is granted. The annual H-1B visa quota has varied over the years. It is currently set at 85,000. Universities and non-governmental organizations (NGOs) are exempt from the quota.

The H-1B program has invoked the hottest debate among American policy-makers and the public related to highly skilled international migration. Proponents, especially American high-tech firms, citing the inadequacy of US-trained American scientists, stress that the H-1B program is one of the key means for attracting the best and brightest to the US. Reducing the number will yield more return migration of talents and increase competition against the US. Opponents, citing abuses, disagree that the program attracts the best and brightest, and condemn it as undermining American high-tech employees.

Both sides of the debate fail to consider what is happening in the lives of H-1B visa holders and their spouses. One issue is disenfranchisement. During the prolonged wait for visa application/renewal and legal status adjustment, H-1B visa holders are no one's political constituency despite their high level of human and financial capital and political activism. Another issue is brain 'waste', as indicated in the following story.

Sanjay Mavinkurve, a 29-year-old Google engineer born in India who helped create the foundation for Facebook while studying at Harvard, and his wife, 28-year-old Indian-born Samvita Padukone, who studied engineering and finance in Singapore, have been living in a modest apartment in Toronto. Although he has studied and worked in the US for most of his adult life, Sanjay commutes regularly between the Google headquarters in California and Google's Toronto sales office, where he is a solo engineer among marketers. As an H-1B visa holder, Sanjay has no problem staying and working in California, but Samvita cannot legally work in the US. Theirs is not a unique case. A number of Sanjay's Google colleagues and many others in the American high-tech industry reside in Canada for the same reason. Alberta actually targets American H-1B visa holders and lures them to work and live there (Marlow, 2009). Microsoft Inc., citing visa problems, opened a facility across the border in Richmond, BC (Yoo, 2007). These cases, reflecting the gendered brain waste among highly skilled female spouses of the largely male-dominant H-1B visa holders, represent a societal issue that has yet to be fully explored, let alone addressed.

Table 6.1   Educational Levels by Immigration Status

Canada	Canadian-Born	Immigrants	Period of Immigration		
			Before 1991	1991–2000	2001–6
Total population 15 years and over by highest certificate, diploma, or degree	19,592,380	5,841,245	3,408,415	1,546,035	886,795
Percentage in immigrant cohorts	n.a.	100%	58.3%	26.5%	15.2%
No certificate, diploma, or degree	24.6%	21.3%	23.7%	19.0%	16.4%
High school certificate or equivalent	26.4%	22.7%	22.5%	25.1%	19.3%
Apprenticeship or trades certificate or diploma	11.4%	9.1%	11.5%	6.4%	4.5%
College, non-university certificate, or diploma, or university certificate, or diploma below bachelor's level	21.8%	21.5%	22.7%	21.0%	17.9%
Bachelor's degree; degree in medicine, dentistry, veterinary medicine, or optometry; or higher	15.7%	25.5%	20.8%	28.4%	41.8%

United States	US-Born	Foreign-Born	Before 1990	1990s	Since 2000
Total population 15 years and over by highest education attainment	201,108,615	36,849,574	17,910,197	10,905,296	8,034,081
Percentage in foreign-born cohorts	n.a.	100%	48.6%	29.6%	21.8%
No high school diploma	18.3%	32.9%	30.0%	35.3%	35.8%
High school graduate or GED	29.9%	23.9%	23.4%	24.5%	24.1%
Some college, or associated degree	28.7%	18.9%	21.5%	17.2%	15.3%
Bachelor's degree, professional degree, or higher	23.0%	24.5%	25.1%	22.9%	24.8%

Table 6.2   Official-Language Capability by Immigration Status

			Period of Immigration		
Canada	Canadian-Born	Immigrants	Before 1991	1991–2000	2001–6
Total population	24,788,725	6,186,950	3,408,415	2,513,175	1,109,985
% speaking neither English nor French	0.4%	6.4%	4.8%	7.5%	9.3%
United States	US-Born	Foreign-Born	Before 1990	1990s	Since 2000
Total population age 5+	239,300,295	38,962,768	17,910,197	29,643,000	9,319,768
% not speaking English	0.1%	10.9%	6.6%	8.2%	19.2%

in Canada and 10.9 per cent of those in the US did not speak English (or French) at all. The cohort effect is also stronger in the US. Among the most recent, 9.3 per cent in Canada and 19.2 per cent in the US are non-official-language speakers, reflecting the more polarized education levels of the US foreign-born. The ability to speak an official language is not only an indication of human capital, but often a prerequisite for better-paid jobs and social mobility.

The Economic Environment

Canada and the US differ on various economic fronts. The US economy, with a GDP of US$14.3 trillion in 2009, is 10 times as big as the Canadian economy, and so is its total population (IMF, 2010). Although the per capita GDP is similar in the US and Canada, a larger economy provides more jobs and business opportunities.

The two countries also share similar industrial structures. Five sectors— manufacturing; wholesale and retail trades; professional, scientific, technical, and education services; health care and social services; and public administration and company management—account for almost two-thirds of the labour force (Table 6.3). Canada, however, still has a stronger primary resource sector, whereas quaternary services are much more prominent in the US.

The US's much larger, more advanced economy, its more variegated economic structure, and its lower percentage of foreign-born population likely provide more economic opportunities and better economic outcomes for immigrants. Its bifurcated industrial sectors offer ample jobs not only to highly skilled immigrants, but also to those less educated but willing to accept 'dirty, difficult, and demeaning' jobs. Whether the Canadian economy can absorb and make maximum use of the drastically increased human capital level among its most recent immigrants remains questionable. Empirical evidence outlined in the following section suggests otherwise.

It is generally believed that the US, which has fewer business regulations, offers an environment more conducive to business development. Only recently did we recognize that a more laissez-faire policy is a double-edged sword. On the one hand, a relatively hands-off regulatory regime provides more opportunities for immigrants. This is evident even in highly professional, well-regulated economic sectors such as the financial industry (Li and Lo, 2008; Dymski et al., 2010). At the same time, Canada's more stringent regulatory rules with regard to regulated professions (engineering, medicine, nursing, and teaching) mean that foreign credentials and work experiences are often not recognized and recertification is a painful, if not impossible, process (Zietsma, 2010). Overall, the US appears to offer greater opportunities for income growth, productivity, innovation, risk-taking, and investment than Canada.

LABOUR MARKET EXPERIENCES

Canada and the US attract migrants with different degrees of human capital. A migrant's human capital and labour market experience do not always match perfectly, however. This section examines migrant workforce activities at the national level, which is important given that 95 per cent of immigrants/foreign-born concentrate in cities (Chui et al., 2007; Singer, 2008).

Labour Force Participation

Among American/Canadian-born and for both men and women, Canada has higher **labour force participation rates** than the US. The opposite, however, is true of immigrants. This is likely due to two reasons. First, the large number of undocumented migrants in the US and a generally lower education attainment among US foreign-born make it more necessary for immigrants to work. Second, immigrants in Canada find it increasingly difficult to find gainful employment, especially in the early years of their settlement, given that foreign credentials are often not recognized by employers. Immigrants' job prospects are often further hampered by a lack of English/French proficiency. Overall, male immigrant labour force participation is higher in the US than in Canada (78.6 vs 69.2 per cent), whereas the rates are similar for female immigrants (around 55 per cent). In both countries, immigrants from the developing countries have higher than average labour force participation rates.

Proportionally, there are more full-time American-and foreign-born workers in the US than in Canada. This is especially so among recent immigrants: 75 per cent in the US are working full-time compared to 39 per cent in Canada. Also, in contrast to Canada, more migrants from the developing countries are working full-time in the US.

Female participation in the workforce contributes to household income, and has implications for gender and family relations. In both countries, however, more female immigrants experience unemployment than do their male counterparts (14.7 per cent among the most recent cohort in Canada; 10.6 per cent in the US). Unemployment rates among Canadian immigrants (twice as high as that of the Canadian-born) are alarmingly high, especially among those from the Caribbean, Africa, and Central America.

Table 6.3   Industrial Participation among Different Population Groups, Canada and the US

Canada	Total	Canadian-Born	Foreign-Born*	Period of Immigration		
				Before 1991	1991-2000	2001-6
Total labour force age 15+ years by industry (North American Industry Classification System 2002)	16,861,180	13,194,625	3,666,550	1,922,275	1,059,675	565,095
Agriculture, forestry, fishing and hunting; mining and oil and gas extraction	4.5%	5.1%	2.3%	2.5%	1.8%	2.2%
Manufacturing	11.9%	10.9%	15.4%	14.9%	16.4%	16.3%
Construction	6.3%	6.7%	5.1%	5.8%	4.1%	4.4%
Accommodation and food services	6.7%	6.4%	7.6%	6.1%	9.1%	9.9%
Transportation and warehousing; utilities	5.7%	5.7%	5.5%	5.7%	5.9%	4.4%
Wholesale and retail trade	15.8%	16.0%	15.0%	13.9%	16.6%	16.6%
Information and cultural industries; arts, entertainment, and recreation	4.5%	4.7%	3.9%	3.9%	4.0%	3.8%
Public administration; management of companies and enterprises	15.1%	15.4%	14.0%	14.2%	13.0%	13.9%
Finance and insurance; real estate and rental and leasing	5.9%	5.6%	7.0%	7.7%	6.8%	5.5%
Health care and social assistance	10.2%	10.3%	9.9%	10.8%	9.0%	8.0%
Professional/scientific/technical/educational services	13.5%	13.2%	14.3%	14.4%	13.2%	15.0%

United States	Total	US-Born	Foreign-Born	Before 1990	1990s	Since 2000
Total labour force age 16+ by industry (North American Industry Classification System 1997)	177,825,457	149,843,058	27,982,399	13,399,607	8,572,595	6,010,197
Agriculture, forestry, fishing and hunting; mining and oil and gas extraction	1.9%	1.8%	2.5%	2.0%	2.5%	3.5%
Manufacturing	11.3%	11.0%	13.4%	13.9%	13.6%	11.8%
Construction	7.8%	7.3%	10.5%	7.8%	11.5%	15.1%
Accommodation and food services	7.5%	7.0%	10.2%	7.6%	11.7%	13.6%
Transportation and warehousing; utilities	4.8%	5.0%	4.3%	5.3%	4.0%	2.8%
Wholesale and retail trade	15.3%	15.6%	13.5%	13.7%	14.0%	12.5%
Information and cultural industries; arts, entertainment, and recreation	4.7%	4.9%	3.4%	3.7%	3.4%	3.0%
Public administration; management of companies and enterprises**	9.3%	9.4%	8.6%	9.4%	7.9%	7.5%
Finance and insurance; real estate and rental and leasing	6.8%	6.9%	5.6%	6.9%	5.1%	3.5%
Health care and social assistance	11.8%	12.0%	10.9%	12.7%	10.2%	7.9%
Professional/scientific/technical/educational services	18.8%	19.1%	17.1%	17.0%	16.1%	18.8%

*Less than 3.4 per cent of Canadian foreign-born citizens are non-permanent residents.

**The 1997 NAICS does not have a separate category for 'Management of companies and enterprises'. This category includes the residual not accounted for by all other categories.

Immigrant Entrepreneurship

Immigrant entrepreneurship or immigrant businesses are seen as the backbone of many immigrant neighbourhoods and a means of upward mobility. Given the employment situation for immigrants in Canada and the US, it is not surprising that more Canadian immigrants have established their own businesses. Indeed, the rate of self-employment is higher in Canada. The generally higher educational attainment of Canadian immigrants may account for this resourcefulness. The difference in self-employment rates is notably larger between their immigrant populations (14.3 per cent in Canada and 11.1 per cent in the US) than between their total populations (11.2 per cent in Canada and 10.0 per cent in the US). Among all groups in the two countries, male immigrants in Canada have the highest self-employment rate (17.9 per cent).

Not all immigrant groups are equally entrepreneurial. In the US, immigrants from West Asia, the Middle East, and Europe are more entrepreneurial, at around 20 per cent; this percentage is double that for immigrants from Africa, Central America, and Southeast Asia, many of whom have less human capital or arrived as refugees. In Canada, immigrants from Western Europe are the most entrepreneurial (25 per cent), followed by West Asians and Middle Easterners, the other Europeans, and East Asians (16 to 18 per cent). In both countries, immigrants from Southeast Asia, the Caribbean, and Central America are less likely to be self-employed. Immigrant small businesses are often associated with unpaid family labour, mostly from women. The percentage of immigrant women engaged in unpaid work outside the domestic sphere is about 0.5 per cent in both countries.

Labour Segmentation and Industry of Employment

Industrial classification systems commonly divide industries into four sectors: the primary sector (agriculture and raw materials), the secondary sector (manufactured goods), the tertiary sector (services), and the quaternary sector (research and development). Generally, primary and secondary industries are more labour-intensive. Tertiary industries consist of both low-end (e.g., accommodation and food) and high-end services (e.g., education and health care). Participants in the primary, secondary, and lower tertiary sectors are often less educated and skilled. Jobs associated with these sectors are considered less prestigious.

Labour market segmentation is evident in both countries. Compared to the American/Canadian-born, immigrants' industrial participation differs in the two countries, being more bifurcated in Canada. Table 6.3 shows proportionally more Canadian immigrants than Canadian-born working in lower-end manufacturing and accommodation and food services as well as in the high-end professional, scientific, and educational services. US foreign-born, relative to the American-born, however, concentrate more in lower-end manufacturing, construction, and accommodation and food services. The immigrant/foreign-born population in Canada are more likely to be employed in low- to mid-prestige sectors such as manufacturing, wholesale/retail trade, and public administration

and management of companies whereas those in the US are more likely to participate in high-prestige professional, scientific, and education services sectors.

This trend echoes the immigrant admission policy and integration practices
of each country. A skilled labour recruitment program in the US has led to
increased immigrant participation in advanced services, whereas the temporary
foreign worker program in Canada attracts workers from countries like Mexico to
work on farms in Ontario and British Columbia, and in the oil fields in Alberta.
More importantly, despite increasing educational attainment, a large proportion
of recent immigrants to Canada work in the less prestigious secondary and lower
tertiary sectors. One can argue that this pattern is due to newer immigrants
taking time to find jobs and adjusting to a new environment; however, the US
situation implies this is not the only explanation. The recent cohorts of the US
foreign-born, while less educated than immigrants in Canada, are less engaged
in the secondary sector compared to the earlier cohorts and their Canadian
counterparts. The comparison among different cohorts in the US perhaps reflects
more on structural changes to the US economy, but the comparison to the same
cohort of Canadian immigrants is telling. In fact, a vast literature discusses the
problems of foreign credential devaluation and discrimination in Canada as
key reasons for the gloomy employment picture among its newer immigrants
(Alboim et al., 2005; Aydemir and Skuterud, 2005; Conference Board of Canada,
2001; Ferrer and Riddell, 2008; Hum and Simpson, 1999; Li, 2001; Picot and
Sweetman, 2005).

Immigrant men and women participate in different industries. The pattern
among men is quite similar in the two countries. Compared to the American/
Canadian-born, immigrant men are over-represented in both the low-prestige
and high-prestige sectors, and under-represented in public administration,
which often requires citizenship status. Immigrant women's industrial participation differs slightly between the two countries in comparison to the American/
Canadian-born. While in both countries they are more likely to work in manufacturing as well as in professional, scientific, and technical services, and less likely
to be in retail trade, educational services, and public administration, Canadian
immigrant women are more likely to be employed in administrative support,
wholesale, finance, and insurance and their US counterparts in accommodation and food services; the difference generally reflecting the women's countries
of origin and education level. Many women immigrants from Asia to Canada,
for example, are highly educated and prefer pink-collar jobs to blue-collar jobs,
whereas Mexican and South American female immigrants to the US, who are less
educated, often can only find work in the hospitality industry.

There is also a division of labour among immigrant groups. In Canada, education, arts, entertainment, and recreation, and public administration appear to
be the domain of Western and Northern Europeans, while Italians dominate the
construction sector and immigrants from South and Southeast Asia and from
Central and South America predominate in manufacturing. In the US, higher
proportions of Central American immigrants (dominated by Mexicans) work in
the secondary sector and immigrants from South Asia, West Asia, and the Middle

East are proportionately more involved in retail trade. Such ethnic niching has been widely discussed in the literature (Light and Gold, 2000; Waldinger, 1994; Wilson, 2003). One argument is that group affiliation provides access to information and social networks linked to particular sectors of the labour market.Of course, some groups are better niched than others because of their size and/or history in the two countries.

Occupational Concentration

Another way of examining economic experiences is through occupational status. Table 6.4 compares the occupational structures of the two countries. The first few occupations in the list are often seen as superior to the rest in terms of both status and rewards. In general, the US foreign-born's occupational structure is more varied than its Canadian counterpart. Compared to the American-born, US foreign-born are over-represented in five of the 10 occupation groups, ranging from highly skilled (natural and applied sciences) to semi-skilled (sales and service, trades, transport, and equipment operation) and low-skilled jobs (primary industries, processing, manufacturing, and utilities). Canadian immigrants are proportionately more engaged than the Canadian-born only in natural and applied sciences, processing, manufacturing, and utilities. In addition, immigrants in both countries are less likely to work in social sciences, education, government service, and religion. The trend for highly skilled and highly educated immigrants to work in natural and applied sciences instead of social science, education, government service, and religion occupations has been similarly observed in international student enrolment figures.

A major difference between Canada and the US is that immigrants to Canada are less likely to work in trades, transport, equipment operation, and primary industry, whereas US immigrants are more averse to management, business, finance, and administration. This being said, immigrant jobs are bifurcated. If we compare participation in what are perceived as 'good' jobs versus 'bad' jobs, a much higher proportion of Canadian immigrants participate in management; business, finance, and administration; natural and applied sciences; health; and social science, education, religion, and government services (50 per cent vs 38 per cent in the US) and a higher proportion of US foreign-born work in sales and services; trades, transport, and equipment operation; jobs unique to primary industry; and jobs unique to processing, manufacturing, and utilities (61 per cent vs 48 per cent in Canada). These differences indicate a generally more educated immigrant population in Canada.

In addition, a number of differences occur between established and recent immigrants in the two countries. In Canada, recent immigrants increasingly orient towards natural and applied sciences, sales and services, and processing, manufacturing, and utilities; those in the US orient towards sales and service and trades, transport, and equipment operation. Established immigrants (those who immigrated before 1990) in both countries, on the other hand, are more likely to work in management, business, finance, and administration, and health. Established immigrants in Canada are also fairly well represented in trades, transport, and equipment operation.

Occupation is gendered. Men—both immigrants and non-immigrants—are more likely to work in management; natural and applied sciences; trades, transport, and equipment operation; and primary and secondary industries. Women are more often found in business, finance, and administration; health; social science, education, and government services; sales and services, and, to a lesser extent, in art, culture, recreation, and sport. When we compare immigrants to the American/Canadian-born along gender lines, important variations are observed. Immigrant women are two to three times more likely than American/Canadian-born women to work in business, finance, and administration; health; and social sciences, education, and government services. The data, however, do not allow us to discern if immigrant women are overwhelmingly concentrated in pink-collar jobs or support roles within these occupations. Further comparisons of skill levels required for these jobs would be useful.

Immigrant men in Canada are generally doing better compared to their counterparts in the US. As Table 6.4 illustrates, immigrant men in the US are almost three times as likely as American-born men to be working in the primary industries whereas Canadian immigrant men are one-third as likely as their Canadian-born counterparts to be employed in the primary sector. Among the male-dominated occupations, immigrant men in the US are less likely to be in management occupations, and immigrant men in Canada are under-represented in primary industries, as well as in trade, transport, and equipment operation. This difference reflects in part the difference in human capital between Canadian immigrants and the US foreign-born, and in part the origin of the US foreign-born. The last column in Table 6.4 shows a spatial division of work in Canada: East Asians and East Europeans are more likely to be employed in natural and applied sciences than either the total population or the immigrant population; Northern and Western Europeans in social sciences, education, and government services; and South and Southeast Asians and Central and South Americans in processing, manufacturing, and utilities. The latter illustrates how differences in language ability, training orientation, and education attainment shape immigrants' occupational status in the labour market.

Canadian immigrants enjoy higher job prestige overall than their US counterparts. They are more drawn to natural sciences than social sciences and incline towards white-collar rather than blue-collar jobs. But the increasing concentration of newer immigrants in processing, manufacturing, and utilities jobs is an alarming sign; many are degree-holders whose credentials are considered inferior to those granted in Canada or other Western countries.

LABOUR MARKET OUTCOMES

As a result of the greater challenges faced by immigrants in the Canadian labour market, unemployment is a more serious problem among Canadian immigrants than among the US foreign-born. The immigrant unemployment rate in Canada

Table 6.4   Occupational Concentration of Immigrants Relative to Non-Immigrants and by Place of Birth*

Occupational Groups	Canada		US		Domination in Canada by Place of Birth
	Men	Women	Men	Women	
Management**	1.1	0.6	0.7	0.9	E. and W. Asia, N. and W. Europe
Business, finance, and administrative	1.1	2.7	0.7	3.4	Africa, N. Europe, Caribbean, S. America
Natural and applied sciences**	1.5	0.2	1.2	0.3	E. Asia, E. Europe
Health	1.5	3.4	1.2	3.7	Africa, Caribbean, SE. Asia
Social science, education, government service, and religion	1.0	2.6	0.7	3.5	N. and W. Europe, Africa
Art, culture, recreation, and sport	0.9	1.6	0.7	1.6	N. and W. Europe
Sales and service	1.0	1.5	1.0	1.2	E., SE., and W. Asia, S. Europe, C. America
Trades, transport, and equipment operation**	0.8	0.1	1.1	0.1	S. Asia, E. and S. Europe, Central America
Primary industry**	0.4	0.8	2.8	0.1	W. Europe, Central America
Processing, manufacturing, and utilities**	1.4	0.3	1.3	0.4	S. and SE. Asia, C. and S. America

*The table reports concentration indices. A concentration index measures the proportion of immigrants in a particular occupation group relative to the proportion of non-immigrants working in the same occupational group. An index equal to 1 indicates similar representation; an index greater than 1 indicates over-representation or concentration, and vice versa.

**Male-dominated occupations.

(7 per cent) is 75 per cent higher than in the US (4 per cent). In both countries, the unemployment rate among immigrant women is two percentage points higher than among their male counterparts. The most vulnerable groups are from West Asia/the Middle East and Africa in Canada, and from the Caribbean and Central and South America in the US.

On the whole, immigrants earn less than the American/Canadian-born. Immigrants in Canada also make less than the foreign-born in the US. The average employment income for immigrants is $35,876 in Canada and $40,244 in the US (both adjusted to the 2006 Canadian dollar). Immigrant men earn more than immigrant women: 50 per cent more in Canada and 40 per cent more in the US. Earning potential also varies among immigrant groups. Those who came to the US from the Caribbean and Central and South America (the least educated groups) are earning less than the average US foreign-born. Added to this list in Canada are immigrants from everywhere else except Europe and Africa.

Central Americans, West Asians and Middle Easterners, South Asians, and East Asians, in that order, fare the worst in dollar terms; again, these are some of the most highly educated immigrant groups. Earning potential also varies within origin groups. West Asians and Middle Easterners, for example, are high-income earners in the US, but are almost at the bottom of immigrant earning potentials in Canada. In 2006, they made 170 per cent of the average foreign-born income in the US but 83 percent of the average immigrant income in Canada.

Given these labour market outcomes, higher levels of poverty among immigrants in Canada should come as no surprise. Poverty, here defined by Statistics Canada's low-income cut-offs (LICOs), refers to the income thresholds—which vary by family and community size—below which families devote a larger share of income to the basic necessities of food, clothing, and shelter than does the average family of similar size (Statistics Canada, 2006: 7). Almost one in five immigrants in Canada is living in a low-income family, a figure twice as high as the number among the Canadian-born. The situation is especially dire among West Asians and Middle Easterners, East Asians, and Africans; one in three lives in poverty. It also appears that only those immigrant groups from Northern and Western Europe live above the low-income cut-offs.

The poverty measure in the US differs from Canada's LICO in that it does not vary by community size and is solely determined by food consumption. Poverty status is determined by an individual's family income relative to the appropriate poverty threshold, which is based on the Economy Food Plan designed to address the dietary needs of families on an austere budget (US Census Bureau, 2009). In the US, three in 20 foreign-born, mostly from Central America, Africa, the Caribbean, West Asia, and the Middle East, experience poverty.

In both countries, predictably, more women immigrants than men live in poverty. The gender difference is smaller in Canada (18.6 per cent for men; 19.9 per cent for women) than in the US (13.7 per cent for men; 16.8 per cent for women).

There is no question that immigrants, generally, are doing worse than the American/Canadian-born. Even those who successfully enter the labour market face an 'entry effect' (Li, 2003). The earnings disadvantage gets bigger with each successive cohort of immigrants (Abbott and Beach, 1993; Borjas, 1995). In Canada, for example, those who arrived between 2001 and 2006 earn only 78 per cent of those who landed five years earlier and there are estimates that it now takes 15 years or more to catch up to the American/Canadian-born (Baker and Benjamin, 1994; Borjas, 1995; Bloom et al., 1995). The picture is even gloomier for less-skilled, less-educated immigrants, such as those from Mexico and Central America to the US, who may never catch up to the American/Canadian-born (Borjas, 1995).

So why do highly educated, highly skilled immigrants in Canada fare so dismally in the labour market? One explanation is that Canadian-born workers' educational attainments have improved substantially in the last several decades, which, when coupled with changes in the labour market favouring those trained for a knowledge-based economy, means that immigrant workers face increased

Table 6.5   Characteristics of Selected Large and Medium-Sized Cities

Canada (rank in size)	Toronto (1)	Montreal (2)	Vancouver (3)	Calgary (5)	Winnipeg (8)	Halifax (13)	Sudbury (24)	Fredericton (42)
Total population in millions	5.1	3.5	2.1	1.1	0.69	0.37	0.16	0.09
% immigrants	45.7	20.6	39.6	23.6	17.7	7.4	6.7	7
% recent immigrants	8.8	4.6	7.2	5.4	3.5	1.4	.4	1.5
% immigrants not speaking English	7.6	5.3	11.1	6.5	4.3	1.7	2.5	2.0
Immigrant unemployment rate	7.1	11.1	6.3	4.2	4.9	6.5	6.6	7.4
Immigrant self-employment rate	12.9	13.7	15.5	13.1	9	17.8	17.2	12.3
Industries in which immigrants are over-represented	MAN ACC	MAN ACC ADM	AFF MAN ACC	MAN ACC	MAN HEA	MGT EDU PRO	EDU MAN REA PRO	AER EDU ACC PRO
Occupations in which immigrants are over-represented	PMU SCI TTE	PMU SCI	PMU SCI	PMU SCI	PMU HEA	SCI SEG HEA ACR	SCI SEG HEA	SEG ACR HEA
Immigrant average employment income as a percentage of non-immigrant income	77	82	81	83	95	112	131	103

US (rank in size)	New York (1)	Los Angeles (2)	Miami (7)	Seattle (15)	Charlotte (37)	Grand Rapids (66)	Reno (123)	Waco (220)
Total population in millions	18.8	12.9	5.4	3.3	1.6	0.77	0.40	0.19
% immigrants	30.8	35.4	39.7	17.1	9.9	7.5	16.3	7.5
% recent immigrants	6.2	6.5	9.5	4.8	3.6	2.1	4.4	1.2
% immigrants not speaking English	8.9	14.6	17.2	5.4	10.2	10.5	8.3	17.5
Immigrant unemployment rate	6.8	5.5	5.3	5.2	6.3	7.4	2.8	6.5
Immigrant self-employment rate	11.3	14.2	16.8	9.4	8.5	6.2	6.4	7.1
Industries in which immigrants are over-represented	ACC CON MAN OTH	AFF MAN CON ACC OTH	OTH CON MAN WHO	ACC MAN	CON ACC	AFF MAN PRO	AFF CON ACC AER MAN	CON AFF MAN ACC OTH
Occupations in which immigrants are over-represented	PMU TTE HEA	PRI PMU TTE	PMU TTE PRI	PMU PRI SCI	PRI TTE PMU SCI	PRI PMU	PRI PMU TTE	PRI TTE PMU
Immigrant average employment income as a percentage of non-immigrant income	71	73	82	87	79	76	81	81

Note: *Industry abbreviations*: ACC = accommodation and food; ADM = administration support; AER = arts, entertainment, and recreation; AFF = agriculture; CON = construction; EDU = education; HEA = health care and social services; MAN = manufacturing; MGR = management of companies; OTH = other services; PRO = professional, scientific, and technical; REA = real estate; WHO—wholesale trade. *Occupation abbreviations*: ACR = arts, culture, and recreation; HEA = health; PMU = processing, manufacturing, and utilities; PRI = related to primary industries; SCI = natural and applied sciences; SEG = social sciences, education, and government; TTE = trades, transport, and equipment operation.

competition from the Canadian-born labour force (Reitz, 2002). Another more disturbing possibility is systemic discrimination by employers against immigrants who, since the 1980s, have come overwhelmingly from non-traditional source countries, and who tend to be non-white and whose mother tongue is not English (Reitz et al., 2009).

URBAN VARIATIONS

Unlike those who arrived years ago looking for good farmland to till, today's immigrants flock to the largest Canadian and American cities, where they typically make up 30–45 per cent of the total population, the proportion being higher in Canada than in the US. For example, in 2006, immigrants accounted for almost 46 per cent and 40 per cent of the Toronto and Vancouver respective populations, yet only 31 per cent and 35 per cent of the New York and Los Angeles populations (Chui et al., 2007). While the proportion of immigrants in small cities is small or negligible, it varies widely among medium-sized cities, ranging from a low of 5 per cent to a high of 20 per cent. How immigrants fare in diverse city sizes is an interesting topic for further research. Within the constraints set by available (i.e., free) data, we examined eight metropolitan areas in each country, all within the top 25 per cent or so of their respective urban hierarchies (Table 6.5). The Canadian metropolitan areas range in size from just over 85,000 to 5.1 million people whereas the American metropolises range from just under 200,000 to 18.8 million.

While the rate of immigrant self-employment is higher in Canada, it is not uniform across all city sizes. One important distinction is that larger cities in the US, as represented by Miami and Los Angeles, experience higher immigrant self-employment rates, whereas in Canada the rate is much higher in some medium-sized cities such as Halifax and Sudbury.

In the top three Canadian cities—Toronto, Montreal, and Vancouver—immigrants are concentrated primarily in the manufacturing, accommodation, and food industries. Going down the urban hierarchy, over-representation of immigrants is observed in professional, scientific, technical, and educational services, health care, and social assistance, with or without simultaneous over-representation in manufacturing and/or accommodation and food services. This is an interesting difference. The industrial concentration of immigrants in the US is also remarkably different from that in Canada. In our selected sample of US metropolitan areas, immigrants are always over-represented in manufacturing, construction, accommodation and food; other services; and in some cases agriculture, forestry, and fishing. Many immigrants in large cities are engaged in wholesale trade, and many in medium-sized cities, like those in Canada, are in professional, scientific, and technical services.

In terms of occupation, immigrants in all of the American cities and in the larger Canadian cities are over-represented in processing, manufacturing, and utilities—occupations commonly classified as blue-collar and low-skilled. This concentration is complemented by a focus on trade, transport, and equipment operation and on primary industries in most American cities, but not in any of the Canadian cities in the sample. On the other end of the skill and job prestige spectrum, we see a concentration in natural and applied sciences and related occupations in larger Canadian cities, and in health, social science, education, and government services in medium-sized Canadian cities.

These differences most likely reflect the different niches/functions of the various American and Canadian cities as well as their position within each country. It does appear, however, that immigrants in medium-sized Canadian cities are doing better: they hold better, more skilled jobs than the Canadian-born in the same cities and than their immigrant peers in the larger Canadian cities. Medium-sized cities in Canada house fewer immigrants, and a larger percentage are able to speak English. There are two possible explanations for this: (1) these immigrants come from less diverse origins and are more likely to be of European background; (2) less-educated immigrants prefer not to settle in mid-sized cities because of fewer opportunities for employment in manufacturing or services. In fact, the average employment income of immigrants relative to the Canadian-born increases as we move down the urban hierarchy. For example, in 2006, the average employment income of immigrants in Toronto was about 77 per cent of that for Canadian-born residents, a percentage that rises to 95 per cent in Winnipeg and over 100 per cent in Halifax and Sudbury. On the other hand, the percentage in American cities is consistently below 85 per cent.

CONCLUSION

Immigration policies in Canada and the US began to diverge after the immigration reforms of the 1960s, the former emphasizing human capital and the latter family reunification. They recently converged again in their attempt to recruit temporary workers. Such policy differences result in immigrant profiles being similar and yet different. In the last two decades, immigrants to Canada generally have higher education and official-language capability than those to the US. These differences, coupled with different economic structures, business environments, and regulatory regimes, have contributed to distinctive immigrant experiences in the two countries.

The foreign-born in the US are more likely to participate in the workforce and to work full-time, and they earn more than immigrants in Canada. Immigrants to Canada are more likely to operate their own business. While proportionally more Canadian immigrants are employed in the mid-range industrial sectors, the US foreign-born are more likely to be employed in both the lower and upper ends of the industrial continuum. In Canada, only a small proportion of immigrants work in the traditionally well-paid and secure professional, scientific, technical, educational, government, and business service sectors. Despite challenges in the labour market, the overall higher educational achievement of Canadian immigrants means they are more likely to be in high-status white-collar jobs than their US counterparts.

The economic well-being of immigrants is a complex story. Important differences exist along gender and race/ethnicity lines (Pendakur and Pendakur, 2004; Wright and Ellis, 2002). Country of origin, time of arrival, and place of settlement also matter (Ellis et al., 2007; Newbold and Foulkes, 2004). In both countries, immigrant men and women often work in different industries. Immigrant women, however, are more likely than immigrant men to suffer from precarious and poorly paid employment.

Ethnoracial background affects immigrants' economic well-being. Due to the different opportunities and challenges faced in each country of settlement, immigrants coming to Canada and the US from the same country of origin may experience different economic outcomes. In both countries, West Asians and the Middle Easterners have one of the highest participation rates in wholesale and retail trade. On the other hand, East Asians and East Europeans in Canada but South Asians and West and North Europeans in the US that have the highest representation in professional, scientific, technical, and education services. Immigrants of colour are also often over-represented in the least desirable jobs. Their reliance on manufacturing and hospitality jobs renders them especially vulnerable in economic downturns. Generally, immigrants from Europe or migrating between the two countries fare better than those from other parts of the world. Immigrants from Central and South America, the Caribbean, and Africa are at the greatest disadvantage, facing more widespread unemployment and poverty.

Ethnoracial inequality extends to period of immigration. Recent immigrants suffer more. Increasing concentration of newer and highly educated immigrants in processing and manufacturing jobs is an alarming sign in Canada. Important differences also exist between large metropolitan areas, often traditional immigrant gateways, and medium-sized urban areas, the emerging gateways. Immigrants settling in smaller cities, especially in Canada, appear to do better.

Our attempt to provide a broad outline of the variations that exist between the two countries is framed by what we conceptualized earlier in the chapter. Immigrant admission and integration policies, the types of immigrants each country recruits, the size of each country's economy, and its business environment all play a role in the differences uncovered. To understand more fully these differences and similarities, future research needs to focus on more sophisticated analyses where age, education, gender, and ethnicity can be controlled simultaneously. Were data available in both countries, it would be ideal to add immigration class as a variable of interest. Because large-scale data sets such as the census can only describe aggregate trends, understanding the social and economic processes contributing to immigrants' differential experiences requires comparative qualitative research that may reveal the factors contributing to immigrants' economic success or struggle. In particular, detailed investigations in cities of different sizes and functions may shed light on the role space and place play in the economic well-being of immigrants.

Recent immigrants' disappointing experiences in Canada suggest that it may be useful to review the assistance and programs provided to immigrants during

BOX 6.2   A Canadian Story

Charles So, a successful businessman with a college diploma and 16 years of senior management experience in the Hong Kong tourist industry, arrived in Toronto with his family in 1995. He has not been able to maximize his education/experience in Canada but considers himself luckier than most because he brought enough money to buy a home, and he was able to use connections to help land his first job in Toronto, albeit only at $30,000 a year. Later, he traded his white-collar job for a unionized factory job with better pay and benefits. In 2007, the company closed, and he was out of work. The union helped him retrain as a supply chain and logistics manager, but he was unable to find work in his new field after sending out 140 resumes. Eventually, he turned to friends for a significantly lower-paying job in a small manufacturing company. 'After 13 years, I am almost back to zero . . . starting all over again', said So.

Yisola Taiwo, from Nigeria, helped design the athletes' village for Abuja's All-Africa Games. Three years after moving to Canada in 2007 with a pregnant wife and big dreams, he has yet to land his first architecture job. Having worked for no pay at a Toronto architecture firm, he finally secured a two-month contract working with architectural drawings to design building security systems—not a bad gig, but something still unrelated to his field.

These stories are heard across Canada. Immigrants arriving since the 1980s are better educated than ever, but find it harder to get jobs in their fields. Compared to the Canadian-born, immigrant men made 85 cents on the dollar in 1980 and 63 cents in 2010. They also feel downturns the most. In the last recession, recent immigrants saw a 5.7 per cent drop in employment levels, compared with a 3 per cent drop for established immigrants and a 1.6 per cent drop for Canadian-born workers. Integrating into Canadian society has never been easy. As immigrants increasingly settle in the suburbs, as ethnic enclaves become more discrete and more isolated from transit and social services (Lo et al., 2007; Lo et al., 2009), the integration process becomes harder and more complex, and the catch-up time has lengthened significantly. This issue goes beyond equity and equal opportunity. Economists estimate that the Toronto region is losing as much as $2.25 billion annually because people are unable to get jobs commensurate with their training and qualifications or because they are underpaid. As Toronto Board of Trade president Carol Wilding says, the gap 'is hurting Toronto's economy, its livability, and its competitiveness on the world stage.' Charles So says, 'We came here to help build this country. But we are not given a chance to contribute. Employers have to change. Government retraining programs have to put in more effort to help people get jobs in their areas of education.'

United Way president Frances Lankin notes that a simple solution is to create links between small- and medium-sized businesses and new immigrants who don't have the extensive social networks required to navigate Canada's corporate world. Suggestions are plenty. Progress is slow. Yisola Taiwo is still struggling to get his architect's licence, a process that only recently has become easier.

Sources: Monsebraaten (2008); Paperny (2010).

their initial settlement. To unravel the paradox between immigrant admission policy practices and the reality they face upon arrival (see Box 6.2), we need to determine if current settlement programs in Canada prepare recent immigrants for a contemporary job market in which greater educational, skill, and language requirements are the norm (Reitz, 1998) and to what extent hiring practices are discriminatory. Recently introduced specialized language training, bridging, and mentoring programs designed for physicians, nurses, engineers, and other professionals need to be evaluated for effectiveness. It also would be useful to compare the mandates and measures for assessing foreign credentials adopted by professional regulatory bodies in Canada and the US. Canada might learn a valuable lesson from the US experience.

Finally, this chapter has focused only on the economic experiences of immigrants in the labour market. It does not examine their economic impacts. A systematic comparative approach to any single attribute, such as immigrants' contribution to establishing and maintaining international business links or creating jobs in the domestic economy, would be a valuable starting point.

QUESTIONS FOR CRITICAL THOUGHT

1. What are the major similarities and differences of immigrant economic experiences in general and by major origins between Canada and the US?

2. How do you explain such similarities and differences?

3. What are the major similarities and differences of immigrant economic experiences in urban areas of different sizes between Canada and the US?

SUGGESTED READINGS

1. Aydemir, A., and G. Borjas.2006. 'A Comparative Analysis of the Labour Market Impact of International Migration: Canada, Mexico, and the United States', National Bureau of Economic Research Working Paper No. 12327. This comparative analysis examines the impact of immigration on wage structure in Canada, Mexico, and the US.

2. Li, P.S. 2003. *Destination Canada: Immigration Debates and Issues*. Toronto: Oxford University Press. This book theoretically and empirically addresses the historical, social, demographic, and economic merits of Canada's immigration policies.

3. Peri, G. 2010. *The Impacts of Immigrants in Recession and Economic Expansion*. Washington: Migration Policy Institute. This report analyzes the short-run and long-run impacts of immigration on the US labour market during both economic upturns and downturns.

4. Reitz, J.G. 2007. 'Immigrant Employment Success in Canada, Part I: Individual and Contextual Causes', and 'Part II: Understanding the Decline', *Journal of International Migration and Integration* 8, 1: 11–62. Reitz

applies an analytical framework that considers individual characteristics of immigrants as well as the context of the labour market to understand recent trends in immigrant employment success in Canada.

NOTES

We are grateful to the US National Science Foundation for grant SES0852424, which partially funded the research that led to this chapter. We also thank Wan Yu of Arizona State University and Anne Marie Murnaghan of York University for their assistance in data compilation and analysis in the preparation of this chapter.

1. All US census data presented in this chapter are statistics on all 'foreign-born population'; Canadian census data, unless otherwise noted, are specifically for landed immigrants. Immigrants and foreign-born are not identical populations. In most common understanding, immigrants are those who gain permanent resident status in a country whereas foreign-born include workers and students who are given temporary resident status. Ideally, we should compare all foreign-born or all permanent residents in the two countries. The latter is impossible because the US census does not report details on permanent residents alone. In the case of Canada, ample data exist on all aspects of immigrants but not on non-permanent residents, who account for only 3.4 per cent of the Canadian foreign-born. Therefore, the educational attainment levels and socio-economic characteristics for the US foreign-born are somewhat 'inflated' by the inclusion of temporary skilled migrants and international students/scholars, and need to be used with caution when directly matched to the Canadian data.
2. Unless otherwise noted, all data used in this chapter, including Tables 6.1–6.5, are from the 2006 Canadian Census and the 2005–7 American Community Survey (ACS) in the US.

REFERENCES

1. Abbott, M.G., and C.M. Beach. 1993. 'Immigrant Earnings Differentials and Birth-year Effects for Men in Canada: Post-war–1972', *Canadian Journal of Economics* 26, 3: 505–24.
2. Alboim, N., R. Finnie, and R. Meng. 2005. 'The Discounting of Immigrants' Skills in Canada: Evidence and Policy Recommendations', *IRPP Choices* 11, 2: 1–26. At: www.irpp.org.
3. Aydemir, A., and M. Skuterud. 2005. 'Explaining the Deteriorating Entry Earnings of Canada's Immigrant Cohorts, 1966–2000', *Canadian Journal of Economics* 38: 641–71.
4. Baker, M., and D. Benjamin. 1994. 'The Performance of Immigrants in the Canadian Labour Market', *Journal of Labour Economics* 12, 3: 369–405.
5. Bloom, D.E., G. Grenier, and M. Gunderson. 1995. 'The Changing Labour Market Position of Canadian Immigrants', *Canadian Journal of Economics* 28, 4b: 987–1005.
6. Borjas, G.J. 1990. *Friends or Strangers: The Impact of Immigrants on the US Economy.* New York: Basic Books.
7. ———. 1995. 'Assimilation, Changes in Cohort Quality Revisited: What Happened to Immigrant Earnings in the 1980s', *Journal of Labour Economics* 13, 2: 201–45.
8. ———. 2000. 'Economic Research and the Debate over Immigration Policy', in A.V. Deardorff and R.M. Stern, eds, *Social Dimensions of U.S. Trade Policy.* Ann Arbor: University of Michigan Press, 65–82.
9. Chiswick, B. 1978. 'The Effect of Americanization on the Earnings of Foreign-born Men', *Journal of Political Economy* 86, 5: 897–921.
10. Chui, T., K. Tran, and H. Maheux. 2007. *Immigration in Canada: A Portrait of the*

Foreign-born Population, 2006 Census. Ottawa: Statistics Canada.

11. Conference Board of Canada. 2001. *Brain Gain: The Economic Benefits of Recognizing Learning and Learning Credentials in Canada.* Ottawa: Conference Board of Canada.

12. Dymski, G., W. Li, C. Aldana, and H. Hyo Ahn. 2010. 'Ethnobanking in the United States: From Antidiscrimination Vehicles to Transnational Entities', *International Journal of Business & Globalisation* 4, 2: 163–91.

13. Ellis, M., R. Wright, and V. Parks. 2007. 'Geography and the Immigrant Division of Labour', *Economic Geography* 83, 3: 255–81.

14. Ferrer, A.M., and W.C. Riddell. 2008. 'Education, Credentials and Immigrant Earnings', *Canadian Journal of Economics* 41, 1: 186–216.

15. Green, A.G., and D.A. Green. 1999. 'The Economic Goals of Canada's Immigration Policy: Past and Present', *Canadian Public Policy* 25, 4: 425–51.

16. Henry, F. 1994. *The Caribbean Diaspora in Toronto: Learning to Live with Racism.* Toronto: University of Toronto Press.

17. Hum, D., and W. Simpson. 1999. 'Wage Opportunities for Visible Minorities in Canada', *Canadian Public Policy* 25, 3: 379–94.

18. International Monetary Fund (IMF). 2010. At: www.imf.org/external/pubs/ft/weo. (6 June 2010)

19. LaLonde, R.J., and R.H. Topel. 1992. 'The Assimilation of Immigrants in the U.S. Labour Market', in G.J. Borjas and R.B. Freeman, eds, *Immigration and the Workforce: Economic Consequences for the United States and Source Areas.* Cambridge, Mass.: National Bureau of Economic Research, 67–92.

20. Lewin-Epstein, N., M. Semyonov, I. Kogan, and R.A. Wanner. 2003. 'Institutional Structure and Immigrant Integration: A Comparative Study of Immigrants' Labour Market Attainment in Canada and Israel', *International Migration Review* 37, 2: 389–420.

21. Li, P.S. 2001. 'The Market Worth of Immigrants' Educational Credentials', *Canadian Public Policy* 27, 1: 23–38.

22. ———. 2003. 'Initial Earnings and Catch-up Capacity of Immigrants', *Canadian Public Policy* 29, 3: 319–37.

23. Li, W. 2009. 'Changing Immigration, Settlement and Identities in the Pacific

Rim', *New Zealand Population Review* 33, 1: 69–93.

24. ——— and L. Lo. 2008. 'Ethnic Financial Sectors: The US and Canada Compared', *Migracijske i etničketeme* 24, 4: 301–22.

25. Light, I., and S.J. Gold. 2000. *Ethnic Economies.* New York: Academic Press.

26. Lo, L., P. Anisef, R. Basu, V. Preston, and S. Wang. 2009. 'Infrastructure in York Region: A GIS Analysis of Human Services', research report submitted to Infrastructure Canada. At: www.yorku.ca/yisp.

27. ———, L. Wang, S. Wang, and Y. Yuan. 2007. *Immigrant Settlement Services in the Toronto CMA: A GIS-Assisted Analysis of Demand and Supply.* Working Paper 59. Toronto: CERIS–Ontario Metropolis Centre.

28. Lofstrom, M. 2002. 'Labour Market Assimilation and the Self-employment Decision of Immigrant Entrepreneurs', *Journal of Population Economics* 15, 1: 83–114.

29. Marlow, I. 2009. 'U.S. Red Tape Forces Gifted Workers North', *Toronto Star*, 13 Apr. At: www.thestar.com/article/617487. (28 June 2010)

30. Monsebraaten, L. 2008. 'Toronto at a Crossroads: Why the Time for Change Is Right Now', *Toronto Star*, 7 Oct., R1, R7.

31. Newbold, K.B., and M. Foulkes. 2004. 'Geography and Segmented Assimilation: Examples from the New York Chinese', *Population, Space and Place* 10, 1: 3–18.

32. Paperny, A.M. 2010. 'Failure to Tap into Immigrants' Skills Costs Billions', *Globe and Mail*, 10 June, A1, A17.

33. Pendakur, K., and R. Pendakur. 2004. *Colour My World: Has the Majority–Minority Earnings Gap Changed over Time?* Working Paper 04–11. Vancouver: RIIM–BC Metropolis Centre.

34. Picot, G., and A. Sweetman. 2005. *The Deteriorating Economic Welfare of Immigrants and Possible Causes.* Analytical Studies Research Paper 262. Ottawa: Statistics Canada.

35. Portes, A., and R.G. Rumbaut. 1990. *Immigrant America: A Portrait.* Berkeley: University California Press.

36. Raijman, R., and M. Semyonov. 1995. 'Modes of Labour Market Incorporation and Occupational Cost among New Immigrants to Israel', *International Migration Review* 29, 2: 375–93.

37. ——— and M. Tienda. 1999. 'Immigrants' Socioeconomic Progress post-1965: Forging

Mobility or Survival?', in J. Dewind, C. Hirschman, and S. Castles, eds, *The Handbook of International Migration: The American Experience*. New York: Russell Sage Foundation, 239–56.

38. Reitz, J.G. 1998. *Warmth of the Welcome: The Social Causes of Economic Success for Immigrants in Different Nations and Cities*. Boulder, Colo.: Westview Press.

39. ———. 2002. 'Immigrant Success in the Knowledge Economy: Institutional Change and the Immigrant Experience in Canada, 1970–1995', *Journal of Social Issues* 57, 3: 579–613.

40. ———, R. Breton, K.K. Dion, and K.L. Dion. 2009. *Multiculturalism and Social Cohesion: Potentials and Challenges of Diversity*. New York: Springer.

41. Singer, A. 2008. 'Twenty-First-Century Gateways: An Introduction', in A. Singer, C. Bretell, and S. Hardwick, eds, *Twenty-First Century Immigrant Gateways: Immigrant Incorporation in Suburban America*. Washington: Brookings Institution, 3–30.

42. Statistics Canada. 2006. *Low Income Cut-offs for 2005 and Low Income Measures for 2004*.

Income Research Paper Series. Ottawa: Statistics Canada.

43. US Census Bureau. 2009. 'American Community Survey: 2009 Survey Definitions'. At: www.census.gov/acs/www/data_documentation/documentation_main.

44. Waldinger, R. 1994. 'The Making of an Immigrant Niche', *International Migration Review* 28: 3–30.

45. Wilson, F.D. 2003. 'Ethnic Niching and Metropolitan Labour Markets', *Social Science Research* 32, 3: 429–66.

46. Wright, R., and M. Ellis. 2002. 'The Ethnic and Gender Division of Labour Compared among Immigrants to Los Angeles', *International Journal of Urban and Regional Research* 24, 3: 567–92.

47. Yoo, C. 2007. 'Why Microsoft Loves Richmond, B.C.', UBC Journalism News Service. At: thethunderbird.ca/2007/11/07/why-microsoft-loves-richmond-bc/.(28 June 2010)

48. Zietsma, D. 2010. 'Immigrants Working in Regulated Occupations', Statistics Canada. At: www.statcan.gc.ca/pub/75-001-x/2010102/pdf/11121-eng.pdf. (8 June 2010)

CHAPTER 7

HOW GENDER MATTERS TO IMMIGRATION AND SETTLEMENT IN CANADIAN AND US CITIES

Brian Ray and Damaris Rose

INTRODUCTION

In 2010, the gendered qualities of immigration discourse seem inescapable. In Canada, immigrant women's access to public services has garnered substantial media attention, particularly after the Quebec government, ostensibly for reasons of **gender** equality, proposed legislation in the spring of 2010—in a milder version of policies recently implemented in certain European nations—that would restrict women's access to public services if they wear a full face veil. Likewise, in the United States, Arizona's 2010 immigration bill, which gives police broad powers to question and detain anyone suspected of being an undocumented migrant, also has had widespread media coverage. Advocates for migrant women's rights fear that this legislation will discourage women who experience spousal abuse or other violence from approaching public institutions for protection for fear of losing their immigration status or ability to care for their US-born children. These two issues have encouraged debate about the limits of multiculturalism in Canada and civil rights and identity in the US. They also illustrate a complicated reality when it comes to immigration. Most debates about immigration ignore the fact that immigrants are women and men whose roles and responsibilities are at least partially defined by gendered societal expectations and norms; at best, gender is used as a descriptive category without questioning *how* migration and settlement experiences are gendered (Donato et al., 2006). This erasure renders masculine and feminine roles, responsibilities, and identities invisible, even though gender so fundamentally shapes the human experience. At some moments, however, policy, media, and academic debates cast a spotlight on the intersections between immigration and gender in ways that leave immigrants, especially immigrant women, 'overexposed' and pathologized.

In this chapter, we examine the relationship between immigration and gender, with an emphasis on comparing and contrasting the experiences of women and men in the US and Canada. How does gender status affect the ease of settlement in a new society? How do institutions, organizations, and individuals in the US and Canada put in place structures that influence the migration and post-arrival experiences of women and men?

Our intent is to show how a gendered analysis can reveal nuances embedded in the immigration systems, institutional structures, and built environments of each country that contribute to differences in women and men's settlement experiences. At the same time, we do not overplay these differences, as the two countries in fact have some inescapable similarities with respect to immigration history, the management of migrants, and the social, economic, and political status of women and men. They share a legacy of weaving migration stories into their national ideologies, each projecting a seemingly unassailable image of itself as a 'nation of immigrants' (Li, 2003). Improving the circumstances of immigrants has never been the first priority of New World immigration and settlement policy (Troper, 2003, discusses the case of Canada). Instead, immigration discourse and its politics are shaped largely by pragmatic interests in furnishing labour ('human capital') for economic growth, as well as ensuring population growth in 'frontier' or declining regions (Zolberg, 2000). In this respect, the labour of immigrant men in 'strategic' industrial sectors often takes centre stage, while immigrant women's contributions raising families and working in equally essential but low-status employment remain largely invisible (Iacovetta, 1992; Gabaccia, 1994).

Key historical and contemporary realities also link the experiences of women in both countries irrespective of immigration status. Both struggled with political disenfranchisement until well into the twentieth century and have been equally vigorous in pressing for full political, social, and economic rights over and above achieving the vote. Notwithstanding the many achievements of those movements, the gendered qualities of employment and economic power still leave women and men in very different social positions. In both countries, women who work on a full-time basis experience an average income gap relative to men of approximately 20 per cent.[1] The gap narrows, but never disappears, when age and education are considered—the difference in earnings between highly educated female and male new entrants to the labour force is quite small but persists (Statistics Canada, 2008: 19–20). This fundamental income reality, coupled with higher rates of part-time employment and a greater likelihood of leading lone-parent families, fundamentally structures the experiences of all women—immigrants and non-immigrants—in both Canada and the US.

In examining migration and settlement experiences it is important to recognize that the immigration systems often accord a different status to women and men. The immigration systems in both countries are organized around three pillars: family reunification; the entry of highly educated and skilled 'economic class' migrants; and honouring humanitarian commitments by settling refugees. Women and men, however, are far from equally represented in each category. In Canada, women dominate the family reunification category (59.3 per cent), while the largest proportion of principal applicants in the economic class are men (61.9 per cent) (Figure 7.1). For those individuals who enter under the economic class but are either spouses or dependants, the majority are women (56.9 per cent) and are largely invisible in most research about social and economic integration (Lan-Hung, 2008). Although the composition of the categories is slightly different in the US, similar gender inequalities exist (Figure 7.1). The significance of these classes and their gender composition lies well beyond simply organizing migration flows to each country—policy and media discourses tend to represent economic immigrants as self-sufficient, skilled individuals who contribute

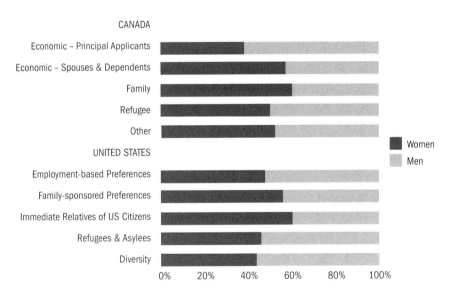

Figure 7.1 Gender Composition of Permanent Residents by Admission Class, Canada and the US, 2008

Sources: Citizenship and Immigration Canada, *Facts and Figures 2008: Immigration Overview— Permanent and Temporary Residents* (Ottawa, 2009), at: www.cic.gc.ca/english/pdf/research-stats/ facts2008.pdf (16 Feb. 2010); US Department of Homeland Security, *Yearbook of Immigration Statistics, 2008* (Washington, 2008), Table 9: 'Persons Obtaining Legal Permanent Resident Status by Broad Class of Admission and Selected Demographic Characteristics: Fiscal Year 2008', at: www.dhs. gov/files/statistics/publications/LPR08.shtm (24 June 2010).

to economic growth but construct those who enter via family reunification as dependent and as more likely to use social and immigrant settlement programs (Li, 2003). As Abu-Laban (1998: 200) argues, 'the effect . . . is to re-assert gender, class and ethnic biases and hierarchies through a renewed emphasis on finding "skilled" independent immigrants.' This kind of discourse reinforces the historical invisibility of immigrant women by discounting the economic contributions of family-class immigrants as paid and unpaid workers and as taxpayers.

Critical study of newcomers' migration and settlement experiences also demands going beyond simply celebrating ethnic, racial, linguistic, and cultural diversity. Social relations between newcomers and long-established groups are grounded in unequal power relations, and it is crucial not to lose sight of the ways in which gender intersects with other aspects of identity as well as with racialization and discrimination. Racism involves processes that are highly context- and place-specific (Kobayashi and Peake, 2000). This means that immigrants of colour have multiple experiences of racialization constructed in part by place-based social relations and conflicts. Existing research too often either ignores intersections between gender and ethnicity or religion, or exploits them in such a way as to render immigrants' femininity or masculinity exotic and threatening. For example, recent discourse about Muslim women who wear a niqab or burka or even a hijab (headscarf) in public space has constructed the practice as a highly contested signifier around which issues of belonging, integration, citizenship, nationalism, and gender are woven (e.g., Shirazi and Mishra, 2010).[2]

Women's and men's lives are shaped by structural inequalities and histories of discrimination whose effects reverberate to the everyday lives of immigrants and non-immigrants both materially and symbolically (Ray and Rose, 2000). In the following sections, we highlight some of the ways that a gendered perspective is valuable for understanding and analyzing migration and urban settlement processes. We begin by examining why gender matters with respect to crossing international borders, and then consider gender with respect to social welfare reform policies and the ways in which immigrants construct new lives in North American cities.

THE GENDERED CHARACTER OF MIGRATION DECISIONS AND ACCESS TO BORDERS

In the broadest of terms, the prospect of betterment for oneself and/or for one's family motivates all who seek to immigrate to Canada or the US, whether under conditions of choice, extreme duress, or somewhere in-between. Gender relations and identities exert powerful influences on why people leave their countries of origin, what draws them to North America, and under what circumstances they are allowed to cross those borders.

The US and Canada share broadly similar approaches to the management and regulation of immigration, although with some important differences shaped by domestic economic and political priorities (see also Chapter 1). Both systems also have similar gendered qualities that differentiate the outcomes of migration for women and men (Boyd and Pikkov, 2005). There are exceptions, however. Notably, same-sex couple relationships are treated very differently by the US and Canadian immigration systems. In both countries, individuals can make an asylum claim based on persecution for sexual orientation in their country of origin, although some important differences exist between Canada and the US as to when such claims can be filed (Ray and Kobayashi, 2005). Only in Canada, however, can gay and lesbian individuals formally sponsor their spouse or partner under family-reunification provisions. This recent and significant change to a decades-old set of sponsorship rules based on heterosexual norma-tivity came about due to LGBT (people who identify as lesbian, gay, bisexual, or transgendered) and feminist organizing and court challenges based on human rights principles. The change has encouraged some American gay men and lesbians to migrate to Canada, particularly if their partner is non-American and living in the US on a temporary basis or illegally.

The reunification of close family members is the conceptual point of depar-ture for both Canadian and US immigration policy. Both countries have a long tradition of women immigrating to join husbands already established in their destination country. The massive nineteenth-century programs to recruit whole families (from the dispossessed to the entrepreneurial) to build up agricultural enterprises gave way in the twentieth century to targeted programs to recruit male migrants to fill labour force gaps in key industries (Kelley and Trebilcock, 1998). But governments, religious leaders, and social policy advocates viewed the persistence of settlements where 'unattached' men predominated as undesir-able and encouraged the immigration of co-ethnic women (Ramirez, 1981). Today, in contrast, bureaucratic backlogs and heightened security checks mean the process of family reunification can entail years of forced separation.

For most women this has always meant that life chances and choices hinge on the status of someone else's application and decision to migrate, and ultimately involves dependence on a spouse or male family member who is the sponsor for between three and 10 years. Today, this supposedly 'easier' route to an immigrant visa is granted on condition that individuals will not claim social welfare benefits and that the family in essence will provide the safety net. The assumption here is that those not selected as immigrants on the basis of their labour market characteristics are a potential burden to the state. Such an assumption ignores the many assets that individuals, most especially women, bring with them in terms of the unpaid work of **social reproduction**—from child care to volunteer work in community organizations (Echaveste, 2009). Moreover, sponsorship can be risky for both parties. Since many, if not most, sponsored immigrant women do not manage to obtain jobs that make them economically self-sufficient, it becomes very difficult for them to leave an abusive spouse or escape dependency on an overly controlling extended family (Côté et al., 2001). Sponsorship also increases the stress for the male sponsor if he falls into precarious employment and becomes unable to fulfill his breadwinner role.

Increasingly, though, urban, educated women are migrating alone to North America from countries of the global South, in spite of the material and ideological barriers in their way. Some do succeed in establishing their credentials as skilled, even professional workers but, as in the past, this mainly happens in the context of targeted recruitment programs that reproduce and reinforce existing patterns of labour market segmentation by gender. A particularly strong illustration is the nursing profession. Although labour shortages in many countries have created an internationally competitive market opening up many employment opportunities for nurses willing to migrate, these jobs rarely lead to career advancement into positions with more recognition and better pay (Collins, 2004; Lowell and Gerova, 2004).

Perhaps the most important examples of targeted migrant recruitment in order to fill labour market shortages are programs recruiting live-in paid domestic workers from overseas, which have been a characteristic of Canadian immigrant and labour market policy since the 1920s (Harzig, 2003; Pratt and Philippine Women Centre, 2005). Earlier programs creating a special immigration category for domestic workers and caregivers were replaced by programs that made admission as a permanent resident contingent on fulfilling a very restrictive labour contract for two years while living in an employer's home. The current Canadian program, the Live-In Caregiver Program (LICP), creates major dilemmas for the often overqualified women who fill its ranks. Not only are their skills grossly undervalued; they frequently experience exploitation and in many instances are squeezed between low pay and obligations to support nuclear and extended family members through remittances.

The transnational extension of immigrant women's responsibilities for social reproduction is not a new phenomenon (Gabaccia and Iacovetta, 2002), but the increasing dependency of extended families, villages, and indeed whole countries on women's earnings is referred to as 'the gendering of survival'. In both Canada and the US, male foreign-born workers are also deeply enmeshed in webs of *transnational* economic support, based in part on gendered expectations to fulfil a breadwinner role for the nuclear and extended family, as well as the 'community' (Johnson and Stoll, 2008). Women, however, because of traditional gender roles and ideologies, often have the added responsibility and stress of transnational caregiving for sick or elderly kin 'back home' (Bernhard et al., 2009).

For many of these reasons, in any given year thousands of individuals cross international borders without valid immigration documents or lapse into an 'undocumented' status. While such migrants undoubtedly constitute an important share of migration flows in both Canada (Magalhaes et al., 2010) and the US, it is only in the US that significant research attention has been given to clandestine border crossing and its gendered aspects. It has been estimated that men comprised three-fifths of the 10.4 million adult undocumented immigrants living in the US in 2008 (Passel and Cohn, 2009), which strongly suggests that gendered roles and/or responsibilities both enable and encourage a disproportionate number of men to enter the US without legal authorization. Research also reveals, however, that with increasing levels of enforcement at land borders, women who undertake clandestine crossings have become more vulnerable relative to men. Women from Mexico are more likely than their male counter-parts to rely on the costly assistance of paid smugglers—exposing them to the risks of exploitation and abandonment—whereas men are much more likely to attempt an illegal border crossing independently (Donato et al., 2008). Women are also more likely to be apprehended on their first trip to the US (Cornelius, 2001).

Since the early 1990s, undocumented workers have increasingly filled the demand for temporary low-skilled migrant labour in the US economy (from gardening for the wealthy to working in hazardous industries), and the majority of these individuals are men. The undocumented lead both US-born males and those with legal immigrant status in labour force participation (94 per cent), while the corresponding rate for female undocumented migrants is relatively low (58 per cent) (Passel and Cohn, 2009: 12). Many employers may be unwilling to hire women for the kinds of physically demanding jobs in which undocumented immigrants are typically employed. Finally, the gendered qualities of undocu-mented migration have repercussions for living arrangements. Among men, a large proportion are separated from their nuclear families—only 53 per cent of men live with a spouse or partner compared to 83 per cent of undocumented women; moreover, most undocumented women have children living with them (ibid., 6). This is important because men who live alone with little familial support report elevated levels of depression and loneliness (Lackey, 2008), and some men form US-based 'second families' during the years when they live away from their original wives and children (Berger, 2004).

Refugees are a small but important component of the flow of migrants to both countries. Women on their own or supporting families without a spouse have tended to be under-represented both among Geneva Convention refugees and those who seek asylum at a border. Women's claims, however, have a higher rate of acceptance than those of men. The gendered nature of refugee selection comes into stark relief when one considers that the UN estimates women are over-represented worldwide among displaced refugee populations (UNHCR, 2001). This may stem in part from the fact that even among refugees, the receiving country determines who will be accepted on a permanent basis depending to a large extent on their 'human capital'. Perhaps more importantly, politically motivated persecution may be conjugated with gendered repression and violence, making exit for women all the more difficult. The motives for seeking asylum, of course, also are gendered: in addition to escaping gendered political violence, women seek refuge from the human trafficking industry (which deals mainly, although not only, in marginalized women) and also from domestic violence. The treatment of refugee and humanitarian claims based on gender-specific persecution is uneven around the world (Freedman, 2009), including between the US and Canada. After a long struggle waged by women's rights activists within the refugee fields (Walton-Roberts, 2004), Canada became one of only a few countries in the world to recognize domestic violence and, more recently, state repression, as legitimate motives for seeking asylum.[3]

In sum, both the Canadian and US immigration systems are implicitly gendered, which in turn influences who attains legal entry relatively easily, their social status upon arrival, and, in certain cases, their access to services. Once in the US or Canada, immigrant women and men begin to learn how their gender intersects with other institutions, public policies, and political priorities to affect further the social and spatial qualities of settlement and integration. This is especially true in terms of the state's involvement in settlement assistance services and its selective treatment of different categories of immigrants and immigrant families.

THE CHANGING BALANCE OF STATE AND FAMILY IN SETTLEMENT AND INTEGRATION

The US and Canada share a substantial history of immigrants gaining access to mainstream social services. In addition, depending on the time frame examined, state agencies or private philanthropic organizations have often sought to offer specialized settlement assistance to immigrant newcomers. The underlying reason is tied to the historical reliance of New World societies on immigration as a vehicle for demographic and economic growth, which generates a strong vested interest in fostering immigrants' allegiance to their adopted country by treating them as a matter of principle as 'future citizens'.

The social welfare programs that immigrants encountered were often paternalistic. It was not unheard of for corporations to intervene directly in immigrant family life by providing instruction to mothers about appropriate child-rearing and social activities, as well as to men concerning the crucial 'breadwinner' role, frugality, and sobriety (Zunz, 1982). Outreach programs for newcomers have often been designed not only to give newcomers access to basic services but also to socialize and 'modernize' the outlook of mothers (Iacovetta, 2006). In the US prior to the mid-1990s, it was illegal for state governments to discriminate between citizens and legal resident aliens (the official term for immigrants who have not obtained US citizenship) in access to education, health, and welfare benefits. Beginning in the 1990s, however, major changes in the philosophy and delivery of social welfare have occurred in both countries, and these changes have had particular impacts on immigrant families and women.

In order to grasp the significance of these changes, we need to recall the key premises of the welfare states that were built up in all Western market-based societies by the mid-twentieth century. First, it was accepted that the dependence of individuals and families on waged labour exposed them to certain 'social risks' that were not their fault (such as unemployment), and that the cost of social protection mechanisms (e.g., unemployment insurance) should be shared via taxation and payroll deductions. Second, there was broad consensus around a shared societal responsibility for individuals' basic well-being (e.g., education, health care, shelter) regardless of ability to pay. Within these broad parameters, welfare states vary in terms of the types of protection offered and in their underlying philosophy (Esping-Andersen, 1990). Importantly, welfare state models also diverge in their normative orientations about family and gender roles, especially in terms of the relative importance given to women as either breadwinners or caregivers or both. An important distinction, for instance, is the degree to which caregiving responsibilities for elders and children remain the unpaid responsibility of women (O'Connor et al., 1999).

There are some notable differences between the social welfare policies of the US and Canada. Social democratic principles have inflected some aspects of Canada's welfare system (most importantly, universal health care), while in the US greater expectations have historically been placed on private philanthropy and charity to fund and deliver social programs. The US federal government has never subsidized settlement services for immigrants, except those for Convention refugees.[4] In contrast, the Canadian federal government has a long tradition of financially supporting various programs that initiate newcomers into Canada's economy and culture, and some provinces also run such programs. While all categories of immigrants are eligible for these programs, it used to be the case that family-class immigrants were excluded from training allowances to attend official-language classes on the (often false) assumption that they were not destined for the labour market. This implicit gender bias was doubly problematic in that women entering Canada are less likely than men to possess official-language skills upon arrival (Boyd and Pikkov, 2005), and yet they need these skills to deal with schools, health-care providers, and so on even if they are not seeking employment. Thanks to campaigns by women's groups and advocates for social justice for immigrants, this distinction was scrapped in 1992.[5] The current language-training programs are only offered to immigrants with less than five years of Canadian residency; however, many longer-established immigrant women have never had the opportunity to learn English or French because of their caring responsibilities, cultural norms, or their type of employment. Research has shown that gender equity in this linchpin of successful integration would be enhanced by more flexible language programs sensitive to the varying circumstances of immigrant women depending on age, origins, family circumstances, and employment situation (Kilbride et al., 2008).[6]

Since the 1990s, American and Canadian social welfare systems have been subject to pressures rooted in a broad neo-liberal turn in political philosophy and social policy (Brenner and Theodore, 2002; Harvey, 2005). Neo-liberalism entails significant privatization of social services and major cuts to 'unproductive' state expenditures for the sake of competitiveness in a globalized economy. Many services that cannot be provided commercially are pushed back onto the family or are devolved to the 'local community' (i.e., private agencies, municipalities, and/or the non-profit sector). Neo-liberalism also entails a retreat from the concept that society should bear a responsibility for individuals' well-being, especially for the most vulnerable, in favour of an emphasis on individual 'self-reliance' (Brodie, 2007). The impacts of neo-liberalism are not gender-neutral: notably, on the one hand, mothers of young children are enjoined—and coerced through welfare reforms—to be 'breadwinners', while, on the other, state supports to child care are eroded (Bakker, 2003).

Welfare reform implemented in the US in 1996 (the Personal Responsibility, Opportunity and Work Reconciliation Act [PRWORA]) illustrates these trends dramatically. Individuals' access to federal benefits was restricted to no more than five years over a lifetime. Income-support programs targeted to needy families (of which racialized minority female-headed families constituted major recipients) were devolved to state governments and were usually made contingent on joining the workforce or job-training programs. New distinctions were introduced between citizens and non-citizens, with non-citizens being unable to access even basic social welfare services such as food stamps and Supplementary Security Income (SSI), although states could provide Temporary Assistance for Needy Families (TANF) and Medicaid benefits after five years of residence in the country.[7] Legal immigrants were caught in the citizenship distinction, with those who arrived after 1996 (plus many of those who had been in the US less than five years when the law was adopted) being disallowed from federal means-tested benefits for at least the first five years after arrival.[8] Undocumented immigrants, but not their US-born citizen children, were also barred from receiving benefits.[9] An intended outcome of this policy change was a decrease in federal government costs associated with a growing number of working immigrants living in very precarious circumstances due to low wages and insecure employment. As a consequence of PRWORA, approximately 935,000 non-citizens lost benefits and their level of poverty increased significantly (Fix and Passel, 2002). The measures most certainly affected both women and men, but women have borne a disproportionate degree of the pain. Many women turned to, and become trapped in, low-wage employment in order to feed and clothe their children. They have also been forced into difficult trade-offs between where good jobs can be found versus the time demands associated with the care of dependants—a situation made worse by longer journeys to work, limited investments in suburban public transit, and shortages of affordable child care.

In Canada and its provinces, reforms to welfare and related social programs have been somewhat less draconian than in the US and have not discriminated on the basis of citizenship. The neo-liberal approach to funding non-profit settlement assistance organizations, however, has reinforced their status as essentially subcontractors to the state (Richmond and Shields, 2005), with funding often being short-term and project-based. Among the implications of this approach is an increase in the administrative workload and precariousness of the agencies. Welfare reform generally has been structured in such a way that welfare recipients become staff for non-profit organizations through 'workfare' programs. These women and men do gain valuable Canadian work experience (Rose et al., 1998), but the low wages associated with the placements lead to high turnover, creating further difficulties for the agencies.

Cutbacks to 'mainstream' social services (i.e., not targeted to immigrants) can have particularly serious repercussions on immigrant women. It has been repeatedly shown that living among co-ethnic family and friends in an institutionally complete minority community brings many advantages to newcomers in terms of practical assistance and cultural support (Iacovetta, 1992); however, the much smaller body of research viewing settlement experiences through a gendered lens has shown that immigrant women may benefit from ethnoculturally diversified social networks. As well as offering what the literature calls 'bridging ties' to 'mainstream' resources (Rose et al., 1998), access to such networks can enable immigrant women to transgress gendered cultural norms in their own communities should they choose or need to do so. In the extreme but not uncommon event of domestic violence (Menjivar and Salcido, 2002), immigrant women may need to turn to organizations or institutions outside their co-ethnic network because their own community's leaders may be in denial about this problem. Consequently, as Walton-Roberts (2008) has pointed out, when the British Columbia government implemented massive cuts to publicly funded women's centres, a valuable source of outreach to minority women facing abuse was lost.

Finally, in both countries, responsibilities for funding key aspects of social welfare and social infrastructure, as well as programs supporting refugee and immigrant settlement and integration, have increasingly been devolved to state/provincial and local levels of government. One consequence is a growing geographical unevenness in service provision and access, especially between the traditional large gateway cities of immigrant reception and the mid-sized and smaller communities where growing numbers of refugees and immigrants are now settling (Singer, 2004). In the next section of this paper, we touch on how this new pattern is affecting immigrant women and men as they go about settling and integrating into different types of urban places.

EVERYDAY LIFE FOR IMMIGRANTS IN TRADITIONAL AND NEW GATEWAY CITIES

To a very significant degree, the settlement and integration experiences that immigrant women and men have in the US and Canada depend on where they live. Appreciation of the complexities involved in building a new life means critically examining cities—large and small—with long-established or more recent histories of immigrant settlement. It also means that attention must be paid to local social and economic circumstances that either encourage or impede employment, access to housing, and education opportunities. The social, economic, and political environments of cities where immigrants settle differ enormously—a variation we acknowledge here simply by drawing a crude distinction between 'traditional' and 'new' **gateway cities**. Large cities with long histories of immigrant settlement are very different urban environments, with distinct opportunity structures and constraints, compared to smaller cities where the built environment is often less complex (e.g., spatial extent, housing options, and transportation systems) and the history of settling newcomers more recent. It is also in the urban geographies of these different types of environments that gender relations play out—where women's and men's roles, responsibilities, and hopes are spatialized.

Many traditional gateway cities, such as New York and Montréal, rank among the largest urban areas in their respective countries and, in addition to decades of history in settling newcomers, they are distinguished by a legacy of strong investment in public infrastructure and services such as transportation, social services, health clinics, and education (from primary to post-secondary). While none of these services was created specifically for immigrants, cities rich in such resources doubtless have a positive influence on the ease with which immigrants construct new lives. These are also cities where generations of immigrants have settled in neighbourhoods proximate to the downtown core and where neighbourhood-based services and institutions have helped newcomers settle and build a new life. But these cities and their older neighbourhoods are changing, with consequences for where and how today's immigrant newcomers live, work, and socialize. In cities such as Toronto, Vancouver, New York, and Chicago widespread gentrification of many inner-city neighbourhoods has meant the disappearance of immigrant enclaves that in the past proved highly supportive for newcomer families. Growing inner-city affluence and the accompanying shrinkage of the affordable housing sector have contributed to relatively poor immigrant households increasingly concentrating in older (and often declining) inner suburbs (Li, 2009; see also Chapter 3). Although housing may be more affordable, the costs for immigrant women can be significant if public transit, in particular a subway line, is less accessible and when the relatives they rely on for everyday support are no longer close at hand (Hirsch, 1998). The suburbanization of employment also has created new geographies of immigrant settlement in traditional gateway cities, but in the context of labour markets that remain highly segmented by gender the different ways that immigrant women and men grapple with these new geographies are still imperfectly understood.

The ability to move easily across urban space, especially within low-density suburban environments, is a major challenge. In the older gateway cities with extensive and dense public transit networks, immigrants may well have considerable choice in terms of employment, housing, and potential education opportunities for children (Truelove, 2000). The simple availability of good public transportation is especially important for women as they are less likely than men to have consistent access to private transportation for employment-based commuting, and they are more likely to juggle employment and the unpaid responsibilities of caring for children and elders, in addition to taking on responsibility for settling their families (dealing with state institutions, finding health care, etc.). Research on social network composition among immigrant women from four different ethnocultural communities living in Montréal found that the ability to move easily through urban space and use public transit enabled many women to feel that they knew the city and 'belonged' (Ray and Rose, 2000). In contrast, settling in the new outer suburbs of the older gateway cities can pose major challenges for newcomer women, even those in middle-class families. A study of Hong Kong immigrants living in exurban Toronto found that the obligations associated with maintaining a large home and the amount of time needed to travel across the city, even if private transportation was available, left many women feeling somewhat isolated or at least unable to pursue the kinds of employment and socializing that had characterized their lives before migration (Preston and Man, 1999). Their access to official-language training was also impeded by travel constraints. Moreover, much as women draw social and material support from the strong ties of kin-based networks, gaining control over their own daily spatial mobility is key to escaping the confines of cultural codes of the ethnocultural community of origin in regard to gender and sexuality, and to facilitating access to the city's spaces of 'unregulated and unwatched' behaviours (Avenarius, 2009: 34).

Since the 1980s ever more immigrants have been living in 'non-traditional' or 'new' gateway cities in the US (Singer, 2004), and to a more limited extent the same phenomenon is occurring in Canada (Leloup and Radice, 2008). In part this 'regionalization' has been built on the practices of state agencies that arrange for government-sponsored Convention refugees to be settled in mid-sized or even small cities. In addition, in cities such as Nashville, Greensboro, Las Vegas, and Salt Lake City, strong labour demand, especially until 2006, encouraged thousands of new legal and undocumented immigrants to settle outside of the traditional gateways, while in Canada labour demand and government policy encouraged more immigrants to settle outside of the three largest cities. While all of these new gateways have or had relatively strong labour demand, they also share limited experience in providing the kinds of resources that are influential in helping immigrant women and their families become established (Ray, 2004).

In many of these new gateway cities, newcomers face a number of challenges. For instance, although many Convention refugees have positive resettlement experiences in smaller cities and towns, others suffer from the absence of a critical mass of immigrants with similar cultural backgrounds and from limited outreach and miscomprehension by the local receiving communities. These newcomers may resort to frequent and costly trips to larger urban centres in order to access appropriate services and maintain their networks, and often end up moving to a larger city (Vatz Laaroussi, 2001). For all immigrants living in smaller cities, public transportation can be much more limited, both in terms of availability and timing/frequency. These smaller, less congested cities may be extraordinarily functional for people with access to private transportation, but they are challenging places to live and work for people, especially women, who depend on buses and/or sharing rides. The new gateway cities also are distinguished by having relatively little affordable rental housing, particularly in suburban neighbourhoods. As a consequence, many women face the challenge of raising children in single-detached suburban houses that are shared with one or more other families (Pruegger and Tanasescu, 2007). Extreme crowding is a problem, with consequences for children's health and education.

While a more even distribution of newcomers between cities may yield many benefits for immigrants and the receiving society, insufficient attention has been given to the effects of regionalization on immigrant women's and men's sense of belonging. Living in cities and suburbs without easy transportation access can enhance women's marginalization and isolation. Live-in domestic caregivers and other service workers (most of them female) who must spend most of their time working in affluent suburbs can easily find themselves deprived of social support and on the margins of the information networks of their own communities (Hagan, 1998; Pratt and Philippine Women Centre, 2005). Yet the social integration experiences of women and men who live in new gateway suburbs are significantly influenced by social class, gender identity, and normative expectations around co-ethnic socializing. Avenarius (2009) found that in Orange County, California, affluent Taiwanese immigrant women with good access to private transportation feel a strong degree of personal contentment in low-density, ethnically diverse suburbs primarily because they experience fewer obligations to socialize with co-ethnic friends and relatives than in more traditional immigrant enclaves. On the other hand, many Taiwanese immigrant men living in low-density suburbs regret the diminished possibilities for interactions with other Taiwanese and in their eyes they have lost opportunities to build social recognition and status. The transnational practices to which we referred earlier also have particular implications for the lives of middle-class women in immigrant suburbs. In **astronaut families** of Pacific Rim immigrants, the male breadwinner maintains both his livelihood and his social status by frequent and prolonged business trips to Asia. In contrast, the suburban lives of wives and mothers may entail a significant degree of isolation. There is considerable variation in experiences; the especially great responsibility placed on these women for settling their families leads some to develop new social networks and greater personal autonomy (Waters, 2002; Zentgraf, 2002).

More research is needed to examine critically the settlement experiences of immigrant women and men in new gateway cities, but it is already clear that these gateways are distinctive urban environments where the experiences that newcomers historically have had in the older big-city gateways are not simply being reproduced on a smaller scale. The gendered roles and responsibilities of immigrant newcomers are spatialized and given meaning in such new destinations as well as in the traditional gateway cities, making it important to examine carefully how class, race, and ethnicity intersect with gender to shape settlement experiences in different kinds of destination cities.

CONCLUSION

This chapter has emphasized that understanding of migration processes and the ways in which immigrants build new lives in cities demands serious consideration of gender. Put simply, the reasons why women and men migrate are frequently different. For instance, women are much more likely to migrate to escape gendered violence, while men are more likely to move—whether legally or illegally—in order to meet family 'breadwinner' obligations. Moreover, women and men are not treated in the same way by systems that regulate who gets into countries of settlement such as Canada and the US, and they do not necessarily experience life in a new city of settlement in the same way. Fundamentally, the roles, responsibilities, and obligations that are the essence of what it is to be a woman or a man, and that are reinforced, practised, and sometimes contested in families, schools, workplaces, neighbourhoods, and recreational spaces, are integral to everyday life and its spatiality. In the US and Canada, the gendered qualities of migration and settlement too often are treated as an invisible part of the immigrant experience. Even less consideration is given to how other aspects of identity, such as ethnicity, race, language, or sexuality, intersect with gender to affect the settlement experiences of women and men. Analyzing the gendered qualities of migration and settlement in a way that incorporates the specificities of cities, neighbourhoods, and workplaces is fundamental if we want to understand why people migrate and what happens to them when they arrive in immigrant-receiving societies.

There are many strong similarities between Canada and the US in regard to the regulation of immigration and the kinds of constraints and opportunities that women and men face. As we have pointed out, women in both countries experience virtually the same degree of wage penalty in the labour market relative to men. There are, however, important differences that need to be incorporated into research and policy. First, racism is a part of Canada's past and present, but the legacy and degree of systemic social discrimination and spatial segregation of African Americans shape integration experiences in US cities for immigrant women and men of colour in particular ways. Second, while both Canada and the US are liberal welfare states, recent neo-liberal changes in welfare policy in the US have targeted immigrants and made the lives of women considerably more difficult. Finally, settlement experiences for women and men are strongly affected by urban form and access to public resources. New landscapes of settlement are being created in both countries by suburbanization and the regionalization of newcomers, and this presents new integration challenges for immigrant women and men. Broad similarities between immigration systems and societal structures in both countries are undeniable, but differences in policy and city form do influence immigrants' integration experiences. To understand the social, economic, and political effects of immigration on cities and their people requires that we consider the complex roles, responsibilities, and identities of women and men, as well as how these are negotiated and produced in the spaces of everyday life.

QUESTIONS FOR CRITICAL THOUGHT

1. US and Canadian governments have made significant changes to social welfare programs since the 1990s. Identify one such policy change in each country and trace its effects—intended or not—on immigrant women and their children. What do such welfare reforms tell us about differences in immigrant integration strategies in the two countries?

2. The need to support children and other dependants often encourages women and men to migrate to Canada or the US while leaving their families behind in their home countries. What kinds of strategies do these migrants adopt to

make raising and supporting a family across borders possible, and what difference does gender make to these strategies?

3. Bearing in mind the gendered dimensions of unpaid and paid work, what would an 'ideal' neighbourhood that promoted integration and a sense of belonging among immigrant women look like in terms of its built environment and social qualities? What kinds of housing, services, and amenities should be available?

SUGGESTED READINGS

1. Benhabib, Seyla, and Judith Resnik, eds. 2009. *Migrations and Mobilities: Citizenship, Borders, and Gender.* New York: New York University Press. This interdisciplinary collection examines the gendered qualities of migration, with an emphasis on leading theories about citizenship and borders in an age of transnationalism.

2. Dobrowolsky, Alexandra. 2008. 'Interrogating "Invisibilization" and "Instrumentalization": Women and Current Citizenship Trends in Canada', *Citizenship Studies* 12, 5: 465–79. This paper examines the ways in which immigrant women's lives have been influenced by a greater emphasis on security and human capital characteristics of migrants in Canadian immigration policy. Dobrowolsky argues that the Canadian state renders women invisible while using them strategically, especially to address labour market shortages.

3. Silvey, Rachel. 2006. 'Geographies of Gender and Migration: Spatializing Social Difference', *International Migration Review* 40, 1: 64–81. This article reviews the contribution made by feminist geographers to the study of gender and migration, with an emphasis on the gendered qualities of place and the socio-political constructions of borders.

NOTES

1. In Canada, the overall gender income gap is 21 per cent compared to 20 per cent in the US (Conference Board of Canada, 2010).

2. Canadian daily newspaper coverage spiked significantly after Bill 94 was introduced in Quebec, especially in Quebec's francophone press (where a media scan turned up over 300 articles in a few months, compared to over 60 in the English-Canadian press). The reasons for such intense interest in this debate are complicated and in part related to language, religion, and fear in a post-9/11 world (Hébert, 2010). See McAndrew (2010: 48) for a discussion of a related earlier (mid-1990s) controversy.

3. The US asylum policy only began to move formally towards in this direction after the Obama administration came to power.

4. This said, some US state and city governments historically have funded immigrant settlement services. In the present context of state constitutional amendments that prohibit borrowing to cover deficits and debt, however, funding for such programs has become more limited and precarious.

5. The reform was a Pyrrhic victory, however, in that it abolished the training allowances altogether.

6. Interestingly, in the US the proportion of legal recent immigrants reporting no knowledge of the official language is higher than in Canada but there is no difference by gender (Fry, 2006:

appendix tables).
7. The PRWORA and its subsequent amendments have established an extraordinarily complicated patchwork of eligibility rules for legal permanent residents across the US. For instance, in 2002 the Farm Security and Rural Investment Act restored immigrants' access to food stamp benefits. This complexity, combined with equally complicated provisions around citizenship, has discouraged many immigrants from accessing the social welfare services for which they are eligible (Kretsedemas, 2003).
8. Convention refugees and recognized asylum claimants were the only exceptions to these rules in that they could receive benefits for seven years after their date of entry.
9. Non-immigrants and undocumented immigrants were only eligible for public health and emergency services, as well as programs deemed by the US Attorney General as being necessary for the protection of life and safety.

REFERENCES

1. Abu-Laban, Yasmeen. 1998. 'Welcome/Stay Out: The Contradiction of Canadian Integration and Immigration Policies at the Millennium', *Canadian Ethnic Studies* 30, 3: 190–211.

2. Avenarius, Christine B. 2009. 'Immigrant Networks in New Urban Spaces: Gender and Social Integration', *International Migration*. At: dx.doi.org/10.1111/j.1468-2435.2009.00511.x.

3. Bakker, Isabella. 2003. 'Neo-liberal Governance and the Reprivatization of Social Reproduction: Social Provisioning and Shifting Gender Orders', in Isabella Bakker and Stephen Gill, eds, *Power, Production, and Social Reproduction: Human In/security in the Global Political Economy*. New York: Palgrave Macmillan, 66–82.

4. Berger, R. 2004. *Immigrant Women Tell Their Stories*. Binghamton, NY: Haworth Press.

5. Bernhard, Judith K., Patricia Landolt, and Luin Goldring. 2009. 'Transnationalizing Families: Canadian Immigration Policy and the Spatial Fragmentation of Care-giving among Latin American Newcomers', *International Migration* 47, 2: 3–31.

6. Boyd, Monica, and Deanna Pikkov. 2005. *Gendering Migration, Livelihood and Entitlements: Migrant Women in Canada and the United States*. Geneva: United Nations Research Institute for Social Development, Occasional Paper 6. At: www.unrisd.org/unrisd/website/document.nsf/8b18431d75 6b708580256b6400399775/9a49929849ceb 521c125708a004c328b/$FILE/OP6pdf.pdf. (11 Nov. 2009)

7. Brenner, Neil, and Nik Theodore. 2002. 'Cities and the Geographies of "Actually Existing Neoliberalism"', *Antipode* 34, 2: 349–79.

8. Brodie, Janine. 2007. 'Reforming Social Justice in Neoliberal Times', *Studies in Social Justice* 1, 2: 93–107.

9. Collins, E. 2004. 'Career Mobility among Immigrant Registered Nurses in Canada: Experiences of Caribbean Women', Ph.D. thesis, University of Toronto.

10. Conference Board of Canada. 2010. *Society: Gender Income Gap*. At: www.conferenceboard. ca/hcp/details/society/gender-income-gap. aspx#countries. (4 June 2010)

11. Cornelius, W.A. 2001. 'Death at the Border: Efficacy and Unintended Consequences of US Immigration Control Policy', *Population and Development Review* 27, 4: 661–85.

12. Côté, Andrée, Michèle Kérisit, Marie-Louise Côté, and Table féministe francophone de concertation provinciale de l'Ontario. 2001. *Sponsorship . . . For Better or For Worse: The Impact of Sponsorship on the Equality Rights of Immigrant Women*. Ottawa: Status of Women Canada, 246. At: epe.lac-bac. gc.ca/100/200/301/swc-cfc/sponsorship-e/010504-0662296427-e.pdf. (11 Feb. 2010)

13. Donato, Katharine M., Donna Gabaccia, Jennifer Holdaway, Martin Manalansan IV, and Patricia R. Pessar. 2006. 'A Glass Half Full? Gender in Migration Studies', *International Migration Review* 40, 1: 3–26.

14. ———, B. Wagner, and E. Patterson. 2008. 'The Cat and Mouse Game at the Mexico–US Border: Gendered Patterns and Recent Shifts', *International Migration Review* 42, 2: 330–59.

15. Echaveste, Maria. 2009. 'Invisible yet Essential: Immigrant Women in America', in Heather Boushey and Ann O'Leary,

eds, *The Shriver Report: A Woman's Nation Changes Everything*. Washington: Center for American Progress, 114–19.

16. Esping-Andersen, Gøsta. 1990. *The Three Worlds of Welfare Capitalism*. Princeton, NJ: Princeton University Press.

17. Fix, M., and Jeffrey S. Passel. 2002. *The Scope and Impact of Welfare Reform's Immigrant Provisions*. Washington: Urban Institute.

18. Freedman, Jane. 2009. 'Protecting Women Asylum Seekers and Refugees: From International Norms to National Protection?', *International Migration*. At: dx.doi.org/ 10.1111/j.1468-2435.2009.00549.x.

19. Fry, Richard. 2006. *Gender and Migration*. Pew Hispanic Center Reports, 64. Washington: Pew Hispanic Center. At: pewhispanic.org/ reports/report.php?ReportID=64. (13 Nov. 2009)

20. Gabaccia, Donna R. 1994. *From the Other Side: Women, Gender and Immigrant Life in the US, 1820–1990*. Bloomington: Indiana University Press.

21. ——— and Franca Iacovetta, eds. 2002. *Women, Gender and Transnational Lives: Italian Workers of the World* Toronto: University of Toronto Press.

22. Hagan, Jacqueline Maria. 1998. 'Social Networks, Gender, and Immigrant Incorporation: Resources and Constraints', *American Sociological Review* 63, 1: 55–67.

23. Harvey, David. 2005. *A Brief History of Neoliberalism*. Oxford and New York: Blackwell.

24. Harzig, Christiane. 2003. 'MacNamara's DP Domestics: Immigration Policy Makers Negotiate Class, Race, and Gender in the Aftermath of World War II', *Social Politics* 10, 1: 23–48.

25. Hébert, Chantal. 2010. 'The Quebec Media Have Distinct Power to Stir Niqab Debate', *Toronto Star*, 5 Apr., 7.

26. Hirsch, Jennifer. 1998. 'En el Norte la Mujer Manda: Gender, Generation, and Geography in a Mexican Transnational Community', *American Behavioral Scientist* 42, 9: 1332–49.

27. Iacovetta, Franca. 1992. *Such Hardworking People: Italian Immigrants in Postwar Toronto*. Montreal and Kingston: McGill-Queen's University Press.

28. ———. 2006. *Gatekeepers: Reshaping Immigrant Lives in Cold War Canada*.

Toronto: Between the Lines.

29. Johnson, Phyllis J., and Kathrin Stoll. 2008. 'Remittance Patterns of Southern Sudanese Refugee Men: Enacting the Global Breadwinner Role', *Family Relations* 57: 431–43.

30. Kelley, Ninette, and Michael Trebilcock. 1998. *The Making of the Mosaic: A History of Canadian Immigration Policy*. Toronto: University of Toronto Press.

31. Kilbride, Kenise Murphy, Vappu Tyyskä, Mehrunnisa Ali, and Rachel Berman. 2008. *Reclaiming Voice: Challenges and Opportunities for Immigrant Women Learning English*. CERIS Working Paper Series, 72. Toronto: CERIS–Ontario Metropolis Centre. At: ceris.metropolis.net/Virtual%20Library/ WKPP%20List/WKPP2008/CWP72.pdf. (16 Feb. 2010)

32. Kobayashi, Audrey, and Linda Peake. 2000. 'Racism out of Place: Thoughts on Whiteness and an Anti-Racist Geography in the New Millennium', *Annals, Association of American Geographers* 90, 2: 392–403.

33. Kretsedemas, Philip. 2003. 'Immigrant Households and Hardships after Welfare Reform: A Case Study of the Miami-Dade Haitian Community', *International Journal of Social Welfare* 12, 4: 314–25.

34. Lackey, G.F. 2008. '"Feeling Blue" in Spanish: A Qualitative Inquiry of Depression among Mexican Immigrants', *Social Science and Medicine* 67, 2: 228–37.

35. Lan-Hung, Nora Chiang. 2008. '"Astronaut Families": Transnational Lives of Middle-Class Taiwanese Married Women in Canada', *Social & Cultural Geography* 9, 5: 505–18.

36. Leloup, Xavier, and Martha Radice, eds. 2008. *Les nouveaux territoires de l'ethnicité*. Québec: Presses de l'Université Laval.

37. Li, Peter S. 2003. *Destination Canada: Immigration Debates and Issues*. Toronto: Oxford University Press.

38. Li, Wei. 2009. *Ethnoburb: The New Ethnic Community in Urban America*. Honolulu: University of Hawai'i Press.

39. Lowell, B.L., and S.G. Gerova. 2004. 'Immigrants and the Healthcare Workforce: Profiles and Shortages', *Work and Occupations* 31, 4: 474–98.

40. McAndrew, Marie. 2010. 'The Muslim Community and Education in Quebec: Controversies and Mutual Adaptation',

Journal of International Migration and Integration 11, 1: 41–58.

41. Magalhaes, Lilian, Christine Carrasco, and Denise Gastaldo. 2010. 'Undocumented Migrants in Canada: A Scope Literature Review on Health, Access to Services, and Working Conditions', *Journal of Immigrant and Minority Health* 12, 1: 132–51.

42. Menjivar, Cecilia, and Olivia Salcido. 2002. 'Immigrant Women and Domestic Violence: Common Experiences in Different Countries', *Gender and Society* 16, 6: 898–920.

43. O'Connor, Julia S., Ann Shola Orloff, and Sheila Shaver. 1999. *States, Markets, Families: Gender, Liberalism and Social Policy in Australia, Canada, Great Britain and the United States*. New York: Cambridge University Press.

44. Passel, Jeffrey S., and D'Vera Cohn. 2009. *A Portrait of Unauthorized Immigrants in the United States*. Washington: Pew Hispanic Center. At: pewhispanic.org/reports/report.php?ReportID=107. (11 Feb. 2010)

45. Pratt, Geraldine, and Philippine Women Centre. 2005. 'From Migrant to Immigrant: Domestic Workers Settle in Vancouver, Canada', in Lise Nelson and Joni Seager, eds, *A Companion to Feminist Geography*. Malden, Mass.: Blackwell, 123–37.

46. Preston, Valerie, and Guida Man. 1999. 'Employment Experiences of Chinese Immigrant Women: An Exploration of Diversity', *Canadian Woman Studies* 19, 3: 115–22.

47. Pruegger, Valerie, and Alina Tanasescu. 2007. *Housing Issues of Immigrants and Refugees in Calgary*. Calgary: City of Calgary, United Way of Calgary and Area, and Poverty Reduction Coalition. At: www.calgary.ca/docgallery/bu/cns/homelessness/housing_issues_immigrants_refugees.pdf.

48. Ramirez, Bruno. 1981. 'Montreal's Italians and the Socio-economy of Settlement, 1900–1930: Some Historical Hypotheses', *Urban History Review* 10, 1: 38–48.

49. Ray, Brian. 2004. *Building the New American Community Project: Encouraging Immigrant Integration and Social Inclusion in Lowell, Nashville and Portland*. Washington: Migration Policy Institute. At: www.migrationpolicy.org/news/BNAC_REPT_SUM.pdf.

50. ——— and Audrey Kobayashi. 2005. 'Negotiating the Nexus of Immigration, Sexuality and Citizenship: Canadian and American Comparisons', *Canadian Issues/Thèmes canadiens* (spring/printemps): 13–16.

51. ——— and Damaris Rose. 2000. 'Cities of the Everyday: Socio-Spatial Perspectives on Gender, Difference, and Diversity', in Trudi Bunting and Pierre Filion, eds, *Canadian Cities in Transition: The Twenty-First Century*. Toronto: Oxford University Press, 502-24.

52. Richmond, Ted, and John Shields. 2005. 'NGO–Government Relations and Immigrant Services: Contradictions and Challenges', *Journal of International Migration and Integration* 6, 3: 513–26.

53. Rose, Damaris, Pia Carrasco, and Johanne Charbonneau. 1998. *The Role of "Weak Ties" in the Settlement Experiences of Immigrant Women with Young Children: The Case of Central Americans in Montréal*. Toronto: CERIS–Toronto Centre of Excellence for Research on Immigration and Settlement, Working Papers. At: ceris.metropolis.net/Virtual%20Library/community/Rose1/rose1.html. (11 Feb. 2010)

54. Shirazi, F., and S. Mishra. 2010. 'Young Muslim Women on the Face Veil (Niqab): A Tool of Resistance in Europe but Rejected in the United States', *International Journal of Cultural Studies* 13, 1: 43–62.

55. Singer, Audrey. 2004. *The Rise of New Immigrant Gateways*. Washington: Brookings Institution, Center on Urban and Metropolitan Policy.

56. Statistics Canada. 2008. *Earnings and Incomes of Canadians Over the Past Quarter Century, 2006 Census*. Catalogue no. 97-563-X. Ottawa: Statistics Canada.

57. Troper, Henry. 2003. 'To Farms or Cities: A Historical Tension between Canada and Its Immigrants', in Jeffrey G. Reitz, ed., *Host Societies and the Reception of Immigrants*. La Jolla: University of California San Diego Press, 509–31.

58. Truelove, Marie. 2000. 'Services for Immigrant Women: An Evaluation of Locations', *Canadian Geographer* 44, 2: 135–51.

59. United Nations High Commissioner for Refugees (UNHCR). 2001. *Women, Children and Older Refugees: The Sex and Age Distribution of Refugee Populations, with a Special Emphasis on UNHCR Policy Priorities*.

Geneva: UNHCR, Population and Geographic Data Section. At: www.womenscommission. org/pdf/unhcr.pdf. (24 Feb. 2010)

60. Vatz Laaroussi, Michèle. 2001. *Le familial au coeur de l'immigration: Stratégies de citoyenneté des familles immigrantes au Québec et en France.* Collection Espaces Interculturels. Paris: L'Harmattan.

61. Walton-Roberts, Margaret. 2004. 'Rescaling Citizenship: Gendering Canadian Immigration Policy', *Political Geography* 23: 265–81.

62. ———. 2008. 'Weak Ties, Immigrant Women and Neoliberal States: Moving beyond the Public/Private Binary', *Geoforum* 39: 499–510.

63. Waters, Johanna L. 2002. 'Flexible Families? "Astronaut" Households and the Experiences of Lone Mothers in Vancouver, British Columbia', *Social and Cultural Geography* 3, 2: 117–34.

64. Zentgraf, Kristine M. 2002. 'Immigration and Women's Empowerment: Salvadorans in Los Angeles', *Gender and Society* 16, 5: 625–46.

65. Zolberg, A.R. 2000. 'The Politics of Immigration Policy: An Externalist Perspective', in Nancy Foner, R.G. Rumbaut, and S.J. Gold, eds, *Immigration Research for a New Century.* New York: Russell Sage Foundation, 60–8.

66. Zunz, Oliver. 1982. *The Changing Face of Inequality: Urbanization, Industrial Development, and Immigrants in Detroit, 1880–1920.* Chicago: University of Chicago Press.

CHAPTER 8
IMMIGRATION, HEALTH, AND HEALTH CARE

Lu Wang, Elizabeth Chacko, and Lindsay Withers

INTRODUCTION

The multicultural urban settings and the contrasting health-care systems in Canada and the US provide the backdrop for this chapter, which addresses the important issue of health and health-care access among immigrant populations who have come to these countries from diverse countries and regions in the last several decades. Canada has a predominantly publicly financed health-care system that aims to provide reasonable access to medically necessary health services for all residents, whereas the health-care system in the US is one of the most privatized in the developed world. The passage of the $940 billion overhaul of the US health-care system in 2010 is supposed to gradually change the character of the US health-care system by expanding coverage to 32 million uninsured and the most economically disadvantaged Americans. But in both countries, underutilization of health-care systems and deteriorating health status among immigrants are frequently reported in the literature.

In this chapter, we draw insights from the existing literature and datasets on health, ethnicity, and immigration to highlight the general trends in health and health care among immigrant populations. Attention is focused on the patterns of access to and utilization of health-care services and the health status of immigrants as examined in various related theories, such as the healthy immigrant theory. Comparisons are made among the general population, immigrants, non-immigrants, and major racial and ethnic groups with respect to health outcomes, incidence of selected chronic diseases, and health-care utilization indicators in Canada and the US. The roles of the health-care system and immigration policy are examined.

The chapter first provides an overview of the literature on immigrant health and the health-care systems in Canada and the US. It then focuses on the differences and commonalities in health outcomes and health-care utilization among immigrants and foreign-born populations in the two countries. The chapter also offers two in-depth case studies of the health experiences, health outcomes, and health-care use among Chinese immigrants in Canada and Latinos in the US at the national level and also in Toronto, 'the most ethnically diverse city in the world' (Hoernig and Walton-Roberts, 2006), and Washington, DC, an increasingly important gateway city for incoming US immigrants. In the case studies we discuss geographic, linguistic, cultural, and economic barriers to formal and alternative health services experienced by the two groups. The chapter yields important policy implications for developing health programs, addressing issues related to foreign-trained physicians, and enhancing the delivery of primary care relevant for immigrants.

BACKGROUND AND CONTEXT

Needs related to health care for new immigrants are critical as they directly affect well-being. Despite the availability of a publicly funded, universal system in Canada, immigrants underutilize health services (Newbold, 2005a) and face strong barriers to health-care access (Leduc and Proulx, 2004). In the US lack of universal health care and welfare reforms have reduced the ability of immigrants to access and use government-sponsored medical care while facing cultural, financial, and other barriers (Akresh, 2009).

In the body of literature that examines the complex relationship among immigration, ethnicity, health, and health-care use (Meadows et al., 2001; Dyck, 2004; Cunningham et al., 2008), one common focus is ethnic inequality and ethnic variations in health status (Nazroo, 1998; Hunt et al., 2004). Differences between immigrants and the dominant population in terms of mortality and life expectancy rates have been noted (Chen et al., 1996; Singh and Hiatt, 2006; Eschbach et al., 2007). According to the population health perspective, socio-economic factors are the basic determinants of health and illness since they guide and rationalize people's health-seeking behaviours (Dunn and Dyck, 2000). According to Brown (1995), however, ethnic residential segregation can lead to poor health and in some cases to **healthy immigrant effects** (Noh and Kaspar, 2003) and **ethnic density effects** (Karlsen et al., 2002). The healthy immigrant effect refers to the health advantage of the foreign-born upon arrival over the US- or Canadian-born due in part to strict immigration selection criteria; however, deterioration of immigrant health and a convergence of immigrant and the US- or Canadian-born in health status have also been observed, likely due to a combination of factors including settlement stress and challenges, adoption of the poor aspects of a North American lifestyle, and aging (Noh and Kaspar, 2003; McDonald and Kennedy, 2004; Abraído-Lanza et al., 2005; Dey and Lucas, 2006). Ethnic residential segregation can be positively associated with minority health as it promotes a sense of community and is protective of health (i.e., the ethnic density effect) (Smaje, 1995), can be negatively related to health because of environmental hazards present in some areas of ethnic concentration (Lanphear et al., 1996), or can have no effect on self-assessed health (Karlsen et al., 2002).

The importance of ethnicity in health-care utilization is further stressed by Prehn et al. (2002) and Gulliford (2003). Cultural and linguistic differences between providers and users of care may create barriers to effective communication and treatment through misinterpretation of symptoms and the difficulty of explaining Western medical concepts and approaches to members of ethnic groups (Zhang and Verhoef, 2002). Immigrant populations have a strong preference for same-language family physicians and patient–physician language discordance in immigrant-receiving cities is a problem (Wang, 2007; Wang et al., 2008). The notion of **cultural competency** is discussed as a strategy to improve quality of care and eliminate ethnic disparities in health and health care (Betancourt et al., 2003). Organizational barriers (e.g., under-representation of minorities among medical school faculties and physicians), structural barriers (e.g., lack of interpreter services), and clinical barriers (e.g., different socio-cultural health beliefs between patient and physician) may lead to a lack of cultural competency in health-care delivery.

The spatial dispersion of recent immigrants in many metropolitan areas in Canada and the US adds to the barriers immigrants face in accessing health care. The points of entry for immigrants were traditionally located in the city core (e.g., Chinatown), where hospitals and medical services are concentrated, although many recent immigrants settle in the suburbs, for example, in 'ethno-burbs' (Li, 1998), without ever experiencing life in a traditional immigrant reception area. The uneven distribution of health-care services and immigrants produces a spatial mismatch between the provision of and demand for culturally sensitive health care. Such is the social and spatial context of the proposed study.

HEALTH-CARE POLICIES AND CHARACTERISTICS IN THE CONTEXT OF IMMIGRATION

Canada has a predominantly publicly financed and administered health insurance system that is essentially an interlocking set of ten provincial and three territorial health insurance plans. The overall goal of the system is to provide reasonable access for all Canadian residents and citizens to medically necessary hospital and physician services on a prepaid basis, without direct charges at the point of service. The five criteria set by the Canada Health Act—public administration, comprehensiveness, universality, portability, and accessibility—ensure that the health-care system will provide universal coverage for medically necessary services on the basis of need rather than the ability to pay. However, there remain a number of unaddressed equity issues on health-care provision, particularly concerning immigrant populations. First, although 'medically necessary' services are largely covered, a range of uninsured hospital, physician, and surgical-dental services, including ambulance services and drug expenses that are usually provided by employers' extended health plans, are deemed 'medically unnecessary'. Unemployed and self-employed immigrants would not have this employment-related health coverage. Immigrants also may find it difficult to get jobs suitable to their skills due to the lack of mechanisms and regulations to recognize foreign credentials (Preston et al., 2003; Lo and Wang, 2005). Second, there is in general a national shortage of physicians in Canada, partly due to the migration of Canadian physicians to the US. In Ontario, approximately 6 per cent of the population does not have a regular family physician (OMA, 2005). This physician shortage has a potential negative impact on recent arrivals who need to secure a family physician in their settlement process. Third, new immigrants may be subject to a waiting period by a province or territory before they are entitled to receive insured health services—the current waiting time in Ontario is three months (Canadian Health Act). The general physician shortage, compounded by settlement stress, labour market integration difficulties, and previously discussed linguistic and cultural barriers, produces a greater challenge for immigrants to access primary care services and to maintain health than is faced by their Canadian-born counterparts.

Unlike the Canadian government's universal health care, the US health-care system is one of the most privatized health-care systems in the developed world. Health care is expensive in the country and having adequate health insurance is critical to accessing good health care. Health insurance in the US can be broadly divided into private and government-sponsored. Private health insurance may be provided by an employer or a union or directly purchased by an individual as a plan from a private company. Government health care includes social insurance programs such as Medicare, Medicaid, and the State Children's Health Insurance Program (SCHIP) (Kovner and Knickman, 2008).

Medicare and Medicaid are particularly important as they are designed to facilitate health-care access and use for more vulnerable sections of the US population. Medicare is a federally funded program that focuses on assisting persons aged 65 and older to pay for health care. Medicaid, on the other hand, is a state-administered program designed to help persons in financial need (the poor and near-poor), particularly the aged, the disabled, and families with dependent children; however, individual states can have different criteria regarding who is eligible for Medicaid and for how long (ibid.).

Immigrants in the US face a number of economic and institutional barriers to accessing health-care services. The economic downturn that began in 2008 has resulted in some counties closing clinics and cutting services for undocumented immigrants (Gorman, 2009). Several legislative acts, such as the 1996 Illegal Immigration and Immigrant Responsibility Act (IIRIRA) and the 1996 Personal Responsibility and Work Opportunity Act (PRWORA), have been passed to curb public benefits to undocumented immigrants. According to original PRWORA provisions, except in the case of emergencies, immigrants who arrived in the United States after 22 August 1996 were not eligible for Medicaid until they had been residents in the country for at least five years. However, *legal* immigrant children and pregnant women may now receive assistance through Medicaid and Children's Health Insurance Program (CHIP) funds (US Agency for Healthcare Research and Quality, 2010).

Community health centres act as a safety net providing affordable health care for those who lack private health insurance and do not qualify for government-sponsored care. They form part of a national matrix of more than 8,000 federally funded health centre delivery sites, serving over 18 million patients, about 40 per cent of whom do not have health insurance and one-third of whom are children (HRSA, 2010a, 2010b). The uninsured users of community health centres in 2006 had increased by more than 50 per cent since 2001 (Wolf, 2008).

In March 2010, US President Barack Obama signed into law a comprehensive health reform law that effectively made targeted changes to a Senate health reform bill passed in December 2009. The new law strives to ensure access to quality, affordable health insurance for all Americans. Included in the Senate bill are provisions to increase access for legal immigrants, but immigrant advocacy groups fear that individual costs will remain high and that undocumented immigrants will experience continued low access to health-care services. The bill, however, contains provisions to significantly increase community health centre funding, expand the collection of data (including race, ethnicity, and language) among patients of federally funded health services, and diversify the health-care system workforce (Sanchez, 2009; Democratic Policy Committee, 2010a, 2010b).

ETHNIC VARIATIONS IN CANADIAN AND US HEALTH OUTCOMES

Both Canada and the US have their own national population health surveys: in Canada, the Canadian Community Health Survey (CCHS) and Canada's National Population Health Survey (NPHS); in the US, the National Health Care Surveys and National Health Care Interviews administered by the US National Center for Health Statistics (NCHS). These surveys typically report aggregated data for racial and ethnic groups such as 'white', 'black', 'Hispanic', and 'Asian', but they largely lack information about specific groups based on countries of origin. Table 8.1 provides some comparable statistics of selected key health indicators among the general population, immigrant population, non-immigrant population, and two major racial groups (native- and foreign-born white and black) in the two countries.

As revealed in Table 8.1, when simply comparing the health outcomes among the general, non-immigrant, and foreign-born populations, contrasting patterns with regard to the 'healthy immigrant effect' emerge. In the US, the foreign-born overall have a superior status on many key health indicators such as high

Table 8.1 Ethnic Variation in Health Status, Canada and the US

	General Population (%)		Non-Immigrants (%)		Immigrants (%)		Non-Immigrant White (%)		Immigrant White (%)		Non-Immigrant Black (%)		Immigrant Black (%)	
	Canada	US	Canada	US	Canada	US	Canada	US	Canada	US	Canada	US	Canada	US
High blood pressure	14.4	22.1	13.5	24.3	17.7	19.8	13.9	22.9	22.5	20.1	5.3	34.7	15.7	26.7
Diabetes	4.6	6.1	4.3	6.1	5.8	6.0	4.3	5.3	6.1	4.4	0.7	10.1	*	8.2
Heart disease	5.0	6.7	4.9	7.6	5.1	5.7	5.0	7.4	7.6	6.9	0.5	8.3	1.8	3.9
Cancer	1.7	7.9	1.7	n.a.	1.6	n.a.	1.7	n.a.	2.4	n.a.	*	n.a.	*	n.a.
Self-perceived health status as fair or poor	11.3	12.1	10.6	11.8	13.7	12.4	10.6	10.3	15.6	9.9	11.7	20.0	9.3	12.0
Self-perceived mental health status as fair or poor**	4.6	2.9	4.4	2.9	5.0	2.9	4.3	2.7	4.1	2.9	6.1	3.3	4.6	1.9

*Cell size less than 5 not reported.

**US data are about serious psychological stress.

Sources: CCHS Cycle 2.1 (Canadian Community Health Survey, 2003) for Canadian data (CCHS data have been weighted); Dey and Lucas (2006) and CDC (2009) for US data.

blood pressure, diabetes, and heart disease and fewer of them self-report fair and poor physical and mental health status. Canada's general population also fares better than that of the US on all indicators in Table 8.1 except for mental health status. Different survey methodologies exist, however, between the cancer rates (current prevalence versus historic, hence the much larger rate in the US), as well as for the perceived mental health status rates, where the US rate reflects those with 'serious psychological stress'.

In Canada, the 'healthy immigrant effect' is less evident. Immigrants fare worse in several categories, including high blood pressure, diabetes, and heart disease, and more of them self-report fair and poor health status as compared to their Canadian-born counterparts (Newbold and Danforth, 2003). After controlling for factors such as age and period of arrival, however, the most recent arrivals were found to report better health (Gee et al., 2004). Lam (1994), Chappell and Lai (1998), and Lai (2004) studied the self-reported health status of Chinese immigrant elders. This group may describe their health as good or excellent because of culture-specific health beliefs that various ailments are only natural as one reaches a certain age (Lam, 1994).

Using the determinants of health framework, several Canadian studies attempt to link socio-economic, demographic, and lifestyle factors with health outcomes (Dunn and Dyck, 2000; Newbold and Danforth, 2003; Newbold, 2005a, 2005b; McDonald and Kennedy, 2004). National studies conducted in the US that compared the mortality, morbidity, and health behaviours of immigrant and US-born groups found that immigrants had lower mortality risks with considerable differentials in mortality for accidents, cancer, and cardiovascular, respiratory, and infectious disease, and also lower morbidity and better health behaviours. In both countries, however, immigrant health deteriorates over time, possibly due to settlement stress, employment challenges, and aging. Other reasons are related to the **acculturation hypothesis**, which states that as immigrants become more 'American' or 'Canadian' they adopt less salubrious aspects of Western lifestyle such as unhealthy diets, greater inactivity, drinking, and smoking, leading to greater health risks such as obesity and hypertension (Noh and Kaspar, 2003). The occurrence of better health indicators for Hispanics than non-Hispanic whites (see Tables 8.1 and 8.3), despite the lower socio-economic status of the former group and lesser access to health care, is sometimes referred to as the **Hispanic or Latino paradox**, and is variously attributed to cultural habits, strong social networks, and the selection processes that bring healthier immigrants to the US; however, prolonged residence in the US and acculturation lead to a negative change in diet, and possibly the loosening of social networks (Abraído-Lanza et al., 2005; Acevedo-Garcia et al., 2005). In terms of mental health, in Canada the 'healthy immigrant effect' was strongest among recent immigrants who are primarily from Asia and Africa. Immigrants in the country for a longer period, who are mostly from Europe, have similar rates of depression as the Canadian-born (Ali, 2002).

The enormous heterogeneity among immigrant populations in Canada and the US points to the need to focus on specific ethnocultural and social groups in specific geographic regions. In several Canadian cities, including Toronto, people of South Asian origin were found to have a higher prevalence of heart disease (11 per cent) than those of European (5 per cent) and Chinese origin (2 per cent) (Anand et al., 2000). Attention also has focused on immigrant women. Ross et al. (2007) found that in Canadian metropolitan areas immigrant men who lived in a neighbourhood with a high proportion of immigrants had lower rates of obesity, but that the same was not true for immigrant women. When compared to Canadian-born women, immigrant women lost health advantage over time and those who had been in Canada more than 10 years were more likely to self-report poor health than were their Canadian-born counterparts (Vissandjee et al., 2004). Qualitative evidence points to the same pattern of deterioration in health status among immigrant women overall (Meadows et al., 2001).

In the US the health status and outcomes for foreign-born women and their infants (particularly Asians/Pacific Islanders and Latinos) are usually more favourable than those of the US-born (Johnson and Marchi, 2009), but variations occur among subgroups. Foreign-born from Puerto Rico have the worst outcomes and those from Cuba and Central and South America the best outcomes in infant mortality (US National Center for Health Statistics, 2009), while one study showed that foreign-born whites and foreign-born Asians had higher risk of low birth weight (<2,500 grams) babies than their US-born counterparts (Acevedo-Garcia et al., 2005). Hispanics also have lower age-adjusted mortality rates than non-Hispanic black and white adults, an advantage that is even greater for foreign-born Hispanics, particularly those from Mexico (Franzini et al., 2001).

Findings at the national level are borne out by those at the level of the city. Foreign-born women in New York City (NYC, 2006) were less likely to give birth to low-birth-weight infants than US-born residents (Kelaher and Jessop, 2002), while foreign-born women in Atlanta experienced fewer pre-term deliveries, less prenatal mortality, and better Apgar scores than US-born women (Forna et al., 2003). In New York City, for all the major causes of death (heart disease, cancer, influenza/pneumonia, stroke, diabetes, chronic lower respiratory disease, AIDS), mortality rates of the foreign-born were lower than or similar to those of the US-born; however, there were marked differences in death rates between foreign-born groups. For example, the mortality rate for heart disease of Jamaican immigrants in New York City, at 304 deaths per 100,000, was more than twice the rate for Mexicans (145 deaths per 100,000 people), although both were below the average US-born rate of 438 deaths per 100,000.

VARIATIONS IN HEALTH-CARE USE

Socio-economic status, health outcomes, and use of the health system are closely related, with lower socio-economic status being associated with poor health and underutilization of health services (Willson, 2009). Place (such as residential location) affects access to healthy and unhealthy environments, health-care resources, social services, and amenities (McLafferty and Chakrabarti, 2009). At a national scale, Table 8.2 reveals that fewer Canadian immigrants used hospital services compared to non-immigrants, more had a regular doctor, and more had visited a doctor in the past 12 months. In the US, however, a smaller percentage of immigrants had regular medical doctors compared to non-immigrant counterparts, possibly due to financial barriers to accessing a doctor under a private health-care system and lack of medical insurance. According to the American Community Survey (US Bureau of the Census, 2008), 45 million people in the United States are uninsured, more than the whole of the Canadian population, comprising 15 per cent of the total US population. The proportion of foreign-born in the United States in 2007 without health insurance was 2.5 times the uninsured US-born population (DeNavas-Walt et al., 2008), and due in part to higher poverty levels, insecure employment, and immigration status, non-citizens are at greater risk of being uninsured (ibid.).

In a public system such as Canada's, income barriers are less significant to accessing health-care services because the costs of using medically necessary services, including visits to primary care physicians and specialists, are covered by government insurance plans. More significant barriers are those related to language and location. Table 8.2 points to the importance of language in physician use in Canada. A smaller proportion of immigrants communicated with physicians in official languages and a larger proportion of immigrants indicated that they could not receive care because of language problems, as compared to non-immigrants. Cultural appropriateness of health services and the intersections of geographical and cultural barriers for specific immigrant population are identified as important (Wang, 2007; McLafferty and Chakrabarti, 2009).

Patterns of health-care utilization also differ among ethnic groups. US Latinos tend to experience the lowest rates of health-care utilization. In 2006 approximately 6.6 per cent of Latinos in the United States did not receive medical care due to the costs involved, compared to 5.6 per cent of non-Latinos (CDC, 2008a). The US National Center for Health Statistics (2007) found that Latino adults diagnosed with diabetes, serious heart conditions, and hypertension were less likely to have a stable source of health care than were other racial/ethnic groups. In Canada, immigrants had a slightly higher overnight hospitalization rate (11.4 per cent) than Canadian-born counterparts (11.0 per cent) (Dunn and Dyck, 2000); however, this could be because immigrants were overall younger than the general population and because younger Canadians were more likely to report an overnight hospitalization, possibly due to accidents and injuries.

Although immigrants had a poorer mental health status compared to the non-immigrant population, as revealed in Table 8.1, they tend to underutilize mainstream mental health services due to unfamiliarity with the services available in the receiving society and the failure of practitioners to understand the mental health-related culture and health beliefs specific to an ethnocultural group (Whitley et al., 2006). Some community-based organizations have begun to work closely with the formal health-care system to offer culturally appropriate mental health services to immigrants, for example, Across Boundaries Ethnoracial Mental Health Centre in Toronto.

Table 8.2   Ethnic Variation in Health-Care Use, Canada and the US

	General Population (%)		Non-Immigrants (%)		Immigrants (%)		Non-Immigrant White (%)		Immigrant White (%)		Non-Immigrant Black (%)		Immigrant Black (%)	
	Canada	US	Canada	US	Canada	US	Canada	US	Canada	US	Canada	US	Canada	US
Use of official languages with doctors	81.1		85.1		67.2		85.5		78.4		84.2		83.9	
Cannot receive care: language problems	0.1		0.02		0.2		0.02		0.11		*		*	
Has regular medical doctor	85.8	82.1	85.4	87.7	87.9	76.4	85.7	88.1	89.6	84.9	85.5	87.1	85.3	82.1
In the past year stayed overnight in a hospital	8.1		8.3	n.a.	7.37		8.3		8.6		8.2		8.4	
No doctor visit in past year	22.64	23.3	22.9	17.6	20.9	29.0	22.7	16.9	18.2	21.6	19.8	19.0	23.4	21.4

*Cell size less than 5 not reported.

Sources: CCHS Cycle 2.1 (Canadian Community Health Survey, 2003) for Canadian data (CCHS data have been weighted); Dey and Lucas (2006) for US data.

CASE STUDIES: HEALTH EXPERIENCE OF CANADIAN CHINESE AND AMERICAN LATINOS

This section explores the patterns in health status and use of health services among Chinese in Canada and Latinos in the United States, the largest and second largest ethnic minority groups in these countries, respectively. Both Toronto and Washington, DC, are important immigrant-receiving cities. The Chinese, including those born in Canada, account for 10.5 per cent of the population in the Toronto metropolitan area, while the Latino population in Washington, including US-born Latinos, comprises 10.6 per cent of the total population. Chinese in Canada came mostly from mainland China and Hong Kong. China has been Canada's top immigrant-sending country since 1997 when Hong Kong, the previous top source country and a former British colony, was returned to China. With a population size of 102,675 (2.0 per cent of the Toronto CMA's population), the Cantonese-speaking Hong Kong-born Chinese represent an established immigrant community, while the Mandarin-speaking Chinese born in China (190,515; 3.8 per cent) are emergent and the fastest-growing immigrant population in the city.

In the Washington metropolitan area, the foreign-born account for over 20 per cent of the total population. Over 60 per cent of Latinos in DC, numbering 48,817 and making up 8.3 per cent of the city's total population, are foreign-born. About 45 per cent of Latinos in the central core came from Central America (primarily from El Salvador), followed by Mexico and other South American countries (US Bureau of the Census, 2005–7). The city has a number of community health-care service providers, including Medicaid and programs for low-income and uninsured populations, aided by several NGOs (District of Columbia Department of Health Finance, 2009).

Table 8.3 offers some key health statistics of these two groups at the national level. In general, the health status of Canadian-born Chinese is superior to that of foreign-born Chinese and the general population. Chinese immigrants have lower self-perceived health and mental health status and have only slightly lower rates of high blood pressure and diabetes than the general population. In the US, US-born Latinos seem to be less healthy than the general population except for self-assessed health status. Foreign-born Latinos have better health status in regard to high blood pressure, heart disease, and self-perceived health and lower status in diabetes and self-perceived mental health. Therefore, there is mixed evidence for a healthy immigrant effect for foreign-born Chinese and Latinos depending on specific chronic conditions examined.

Scholarship on health and immigration, including health outcomes for Chinese Canadians, is limited despite the Chinese being the largest ethnic minority and fastest-growing immigrant group in Canada. Studies by Wang (2007), Wang et al. (2008), and Wang and Roisman (2011) examine the experience of Chinese immigrants in Toronto in using same-language family physicians and alternative health-care providers such as traditional Chinese healers, as well as their transnational health experience (see Box 8.1). Family physicians are 'gatekeepers' to

Table 8.3   Health Status and Health-Care Use of Chinese in Canada and Hispanics in the US

	General Population (%)		Canadian-Born Chinese (%)	Foreign-Born Chinese in in Canada (%)	US-Born Hispanics (%)	Foreign-Born Hispanics in US (%)
	Canada	US				
Health-Care Use Patterns						
Use of official languages with doctors	81.13	n.a.	62.99	26.35	n.a.	n.a.
Cannot receive care: language problems	0.06	n.a.	*	0.93	n.a.	n.a.
Has regular medical doctor	85.82	82.10	84.23	89.76	83.60	69.40
In the past year stayed overnight in a hospital	8.08	n.a	2.51	3.46	n.a.	n.a.
No doctor visit in the past 12 months	22.64	23.30	33.08	21.65	23.50	34.90
Health Status						
High blood pressure	14.38	22.10	3.43	13.58	24.50	19.30
Diabetes	4.60	6.10	0.78	4.59	10.80	7.40
Heart disease	4.96	6.70	0.91	2.52	7.60	5.20
Cancer	1.67	7.9	0.50	1.03	n.a.	n.a.
Self-perceived health status as fair or poor	11.33	24.2	5.94	14.95	18.8	15.8
Self perceived mental health status as fair or poor**	4.55	2.9	3.16	8.97	4.4	3.6

*Cell size less than 5 not reported.

**US data are about serious psychological stress.

Sources: CCHS Cycle 2.1 (Canadian Community Health Survey, 2003) for Canadian data (CCHS data have been weighted); Dey and Lucas (2006); CDC (2009) for US data.

specialists in the Canadian health-care system because, in general, an individual cannot access specialist services without an initial referral by a family practitioner. Although Khan and Bhardwaj (1994) suggest that economic costs are the chief barrier to the consumption of health services, in Canada the less important role of such a market mechanism in a universal health-care system points to the important roles of geography, ethnicity, and language in immigrants' use of health services (see Box 8.1).

DC Latinos, like Latinos nationally, compare favourably on indicators such as life expectancy, birth outcomes, and incidence of heart disease and cancer (excluding breast) diagnoses; however, they fared poorly on other measures such as the incidence of tuberculosis, diabetes, and HIV/AIDS. Moreover, DC

BOX 8.1	Accessing Physicians and Managing Health among Mainland Chinese Immigrants in Toronto

According to the 2006 census, residents of Chinese origin account for 10.5 per cent of the Toronto CMA population. Mainland-born Chinese immigrants, the largest subgroup within the Chinese community, make up 3.8 per cent of the CMA population (2.0 per cent are Hong Kong-born Chinese and 4.7 per cent are Chinese born in Canada and other countries and regions together). Many mainland Chinese immigrants (32 per cent) have no knowledge of Canadian official languages upon arrival and experience linguistic and cultural barriers in utilizing health services in Canada. While the traditional points of entry for immigrants are in the city core (e.g., Chinatown), where hospitals and physicians are clustered, many recent immigrants, despite their socio-economic status, are settling in the suburbs. Physicians in Toronto speak about 100 different non-official languages—and nearly 30 per cent self-reported the use of at least one non-official language. About 6.5 per cent of all family physicians self-reported speaking Chinese. Over half of family physicians are clustered in Toronto's city centre, which has the greatest concentration of hospitals and hospital beds due to the historical development of hospitals in Toronto; only a quarter are located in the suburbs, where immigrants are increasingly concentrated.

Random questionnaires reveal a strong preference among mainland Chinese for same-language family physicians. About 60 per cent expressed a preference for communicating with their family physician in Mandarin—the primary Chinese language spoken by mainland Chinese—but only 40 per cent had been able to secure a Mandarin-speaking physician. Using modified gravity-type and two-step floating catchment area accessibility models in a Geographical Information System (GIS), geographical accessibility scores to family physicians are calculated. The general accessibility, regardless of language and ethnicity, to family physicians decreases from the city core to the suburbs. The accessibility to same-language physicians is low in areas (Chinatown, Scarborough, and Markham) that are heavily populated with mainland Chinese immigrants due to intense competition for same-language physicians among co-ethnics. This produces a spatial mismatch between the residential pattern of mainland Chinese and accessibility to same-language physicians. Using focus groups, Wang et al. (2008) further explore the health management strategies among mainland Chinese in Toronto.

Language and culturally specific health beliefs. All the discussants preferred to see family physicians who have some Chinese proficiency because English 'medical terminologies' were deemed 'difficult' to comprehend and they could communicate more 'effectively' with a Chinese-speaking physician and better understand his/her diagnosis and instructions. Differences between Chinese culture and Western culture in describing and understanding medical symptoms also matter in physician choice. The *yin–yang* (or *cold* and *hot*) balance principle that originates in Taoism is the most fundamental aspect of Chinese culture and traditional Chinese medicine, which is fully accredited in China and coexists with Western medicine. In a similar way, *Chi* (or *Qi*), the vital energy that flows through the human body to maintain good health, is also believed to relate to many clinical symptoms. Some Chinese physicians, although trained in Western medicine, were able to understand these cultural concepts, which was appreciated by some participants.

Health management strategy. Transnational use of health resources was observed. Self-diagnosis and self-treatment using medicines brought from China and hot–cold balanced homemade meals were especially common in managing health for those who immigrated to Canada more recently. The medications brought from overseas included a large proportion of herb-based Chinese medicines. Some participants also self-medicated themselves with antibiotics brought from overseas because it is 'too much trouble to go to a doctor in Canada'.

Underutilization of Canadian health-care system. Some subjects tended to underutilize the Canadian health-care system due to perceived difficulty of getting timely treatment and linguistic barriers. Some others chose to use alternative and complementary medicines such as herbal medicines. The alternative medicine use, on the one hand, seemed to be voluntary as Chinese traditional medicines are thought to be 'safer' than Western medicines and cure the 'cause' of a disease. On the other hand, the use of alternative therapies also related to barriers to accessing Western-style care; often in the focus groups, it was deemed 'too much trouble' to use family physician care (for further details and illustrations, see Wang, 2007; Wang et al., 2008; Wang and Roisman, 2011).

Latinos reported worse overall health than other DC residents (McClure and Jerger, 2005). Other health issues that immigrants in Washington face include cancer and teenage pregnancy. The District of Columbia, which has one of the highest rates of HIV/AIDS in the country, also has the highest prevalence of HIV among Latinos (2,178 per 100,000) of any city in the US: the average rate is 290 per 100,000 for Latinos living in 40 large US cities (NACCHO, 2007). Tuberculosis (TB) is another infectious disease often associated with new immigrants in the US. In 2007, the rate of TB in the District of Columbia was more than double the national rate and Latinos made up a quarter of all cases in DC (CDC, 2008b). High rates of breast and prostate cancer also persist among the DC Latino population. The rate of breast cancer diagnoses in 2004 among DC Latinas, at 3 per cent, was high compared to US-born Latinas (0.6 per cent) and US whites (1.3 per cent) (McClure and Jerger, 2005). Higher rates of diagnoses could be a result of a larger percentage of DC Latinas over age 40 getting mammograms (American Cancer Society, 2005; McClure and Jerger, 2005). In the District of Columbia in 2002, the prevalence rate for diabetes mellitus, compared to an average of 6.7 per cent nationally, was 11.4 per cent for Latinos, 10.6 per cent for blacks, and 2.2 per cent for whites (District of Columbia Department of Health, 2002). One study of DC Latinos also found that 61 per cent of the respondents were overweight or obese and that Latina women in the District had a high incidence (1 in 5 pregnant Latinas) of gestational diabetes (CDC, 2001; McClure and Jerger, 2005).

In the District of Columbia, prenatal care use among Latinas, at only 67 per cent, is a full 10 percentage points below the national average for the ethnic group and 25 per cent below the rate of prenatal care use among non-Hispanic whites in DC (US National Center for Health Statistics, 2009). Withers (2009) explores some geographic elements of community health-care use among Latinos in the Washington metropolitan area (see Box 8.2).

BOX 8.2   Geographic Perspectives on Community Health-Care Use among Latinos in the Washington Metropolitan Area

Through both quantitative and qualitative methods, including the compilation of data on local community health-care centres and local policies, as well as face-to-face, street-level, semi-structured interviews of self-identified Latino residents in Latino neighbourhoods, this study highlights trends in community health-care utilization.

Community health-care distribution and utilization. Although the number of Latino patients using community health-care services in the Washington metropolitan area is more evenly distributed across the three areas within the study region (the District of Columbia and surrounding counties in Maryland and Virginia) than the general patient population, the number using services within the District of Columbia itself remains higher. This is particularly interesting given that the number of Latino residents (including the number living below the poverty level) is higher and most concentrated outside of DC proper. This mismatch first might suggest that Latino populations outside Washington, DC, use community health services less than DC Latino residents, but in fact, particularly after speaking with health centre staff and examining the histories of the health centres used most by the Latino population, the mismatch suggests that community health-care patients tend to go to well-established centres that are known among such communities. This illustrates the importance that social relationships and networks play in these Latino communities. Further, the issue of community health-care policy and funding, particularly when paired with the Latino utilization mismatch, plays an additional role in explaining access. As is seen in the case of health centres located in neighbouring counties or states, the existence of political boundaries and the funding resources or channels available within those boundaries potentially limit the amount of funding provided when patients use sites outside of their own county or state of residence.

Latino perspectives on utilizing DC area community health care. Two Latina immigrants (country of origin unknown) living in Alexandria, Virginia, cited lack of money as having deterred them in the past from using health services and both said they had not used any health services for themselves in the previous two years. One of the women, however, was aware of the nearest hospital and the nearest community clinic, although she only knew the name of the hospital. The woman who was aware of the nearest hospital and clinic had a young child and had been in the Washington area for three years (two years longer than the other female). A male immigrant from Guatemala living in the Langley Park area of Maryland verified the utility of health centre outreach, explaining that he became aware of the health services available at a local health centre after its staff came to visit his local church. On the other hand, three other Guatemalan male immigrants living in the Langley Park area reported no knowledge of any health services (including hospital or community services) and cited their lack of job security, and therefore no health insurance, as being a deterrent to seeking medical care when it was needed. These anecdotal findings suggest that even if Latino residents are aware of a nearby health centre, they may not be aware of the relatively inexpensive services they offer, or of health-care programs available to low-income and uninsured residents (see Withers, 2009, for further discussion and details).

Although transnationalism and its effects on health care have received little attention in the literature, it is evident that DC Latino immigrants and Toronto Chinese immigrants have enduring social ties to their home countries. Chinese immigrants in Canada reported using medicines brought from China and consulting people in China for medical advice. Similarly, 5 per cent of surveyed DC Latina mothers reported having returned to their sending country for prenatal care despite the high cost of international travel (McClure and Jerger, 2005). Transnational activities of Korean immigrants in Toronto included using medical services in their homeland (Suh, 2009).

CONCLUSION

The 'healthy immigrant effect', which posits that the health status of foreign-born immigrants is superior to that of the US- or Canadian-born, is more applicable to US immigrants, particularly those of Latino origin. In Canada, immigrants report poorer health status than that of the Canadian-born, but the reverse is true for US immigrants. Although the data are from national surveys, the sample sizes of some subgroups, such as the black population in Canada, are relatively small and therefore the data should be interpreted with caution. In both countries, the newest arrivals often have the best health status for key indicators of morbidity related to acute and chronic conditions; however, the health of immigrants deteriorates over time and with each succeeding generation. This decline in health status is partly attributed to the 'acculturation hypothesis', whereby immigrants in time adopt certain unhealthy lifestyle attributes of the receiving society such as eating fast foods and becoming less physically active, as well as to stresses related to aging, employment, and ongoing adjustments to an alien cultural milieu.

We also have found that substantial overlap exists in race/ethnicity, immigration/citizenship status, and language skills when addressing the issue of health-care access in both the US and Canada. Health-care coverage and access to care among non-white immigrant populations in both countries, particularly for new arrivals and for the uninsured (in the US), are much more limited than for the general population. In the US Latinos are least likely to have health insurance and hence are most challenged in accessing health care, a problem compounded by their immigration status and limited English proficiency. Even within Canada's system of universal care, lack of language skills can impact access and use by negatively affecting immigrants' ability to gain accurate information on the availability of health care and to communicate with their health-care providers. Furthermore, as was shown in both countries, there is often a spatial mismatch in the largely suburban location of the immigrants and the predominantly central-city location of health-care providers who cater to immigrant groups, making culturally appropriate health care difficult to access. In addition, language discordance between provider and patient makes it difficult for some foreign-born to access and use preventive care, leading to a worsening of existing health problems.

A multi-pronged approach is necessary to overcome the barriers that prevent immigrants from maintaining good health. The ethnic/racial variations in health status described in this chapter are greatly exacerbated by the immigrants' socio-economic conditions and cultural isolation. The restoration of legal immigrants' access to quality health care in the US and special efforts to reach new immigrants who are unaware of their rights and the workings of the health system of the receiving country could go a long way towards achieving and maintaining good health. Greater language assistance in health-care settings in both countries could help bring new immigrants closer to parity with those who are fluent in the language(s) of the receiving country. The employment of physicians who are fluent in the languages spoken by immigrant communities, particularly in areas with large immigrant populations, is one strategy that would increase culturally appropriate and competent care. Lastly, as immigrant populations increasingly tend to be scattered rather than concentrated, greater collaboration between government health-care units and community organizations and institutions can be effective in reaching diverse immigrant groups with culturally appropriate health messages and information and advice on accessing preventive and curative health care.

QUESTIONS FOR CRITICAL THOUGHT

1. What are the relative advantages and disadvantages of urban and suburban living for the health of immigrants?

2. How and why does immigrants' access to health care differ in the US and Canada?

3. What can city/municipal governments do to assist recent immigrants and try to ensure that they retain their health advantages?

SUGGESTED READINGS

1. Dey, A.N., and J.W. Lucas. 2006. 'Physical and Mental Health Characteristics of US- and Foreign-Born Adults: United States, 1998–2003', *Advance Data from Vital and Health Statistics* No. 369 (1 Mar.): 1–19. Hyattsville, Md: US National Center for Health Statistics. This report provides estimates of health-related data for the US foreign-born (data which are less available and/or more difficult to access than health-related data by race) from the 1998–2003 National Health Interview Surveys.
2. Gatrell, A.C. 2009. *Geographies of Health: An Introduction*. Malden, Mass.: Blackwell. This book provides an excellent introduction to the geography of health (e.g., differences in health outcomes and disease diffusion among populations) and the geography of health care (e.g., provision of health care and access to health services) with numerous recent examples from the literature.
3. Noh, S., and V. Kaspar. 2003. 'Diversity and Immigrant Health', in P. Anisef and M. Lanphier, eds, *The World in a City*. Toronto: University of Toronto Press, 316–51. The authors address the immigrant health in Canada, comparing the health status of various immigrant groups with that of the non-immigrant population. The healthy immigrant effect is discussed using Canadian data and examples.
4. Offri, D. 2010. *Medicine in Translation: Journeys with My Patients*. Boston: Beacon Press. Dr Danielle Offri offers readers a first-hand perspective on the health-care experiences of US immigrants and the important role that cultural barriers and competency can play in those experiences.

REFERENCES

1. Abraído-Lanza, A.F., M.T. Chao, and K.R. Flórez. 2005. 'Do Healthy Behaviors Decline with Greater Acculturation? Implications for the Latino Mortality Paradox', *Social Science and Medicine* 61: 1243–55.
2. Acevedo-Garcia, D., M.-J. Soobader, and L.F. Berkman. 2005. 'The Differential Effect of Foreign-born Status on Low Birth Weight by Race/Ethnicity and Education', *Pediatrics* 115: e20–30.
3. Akresh, I.R. 2009. 'Health Service Utilization among Immigrants to the United States', *Population Research and Policy Review* (Mar.) (online).
4. Ali, J. 2002. 'Mental Health of Canada's Immigrants', *Health Report* 13 (supplement). Statistics Canada Catalogue no. 82–003.
5. American Cancer Society. 2005. *Cancer Prevention & Early Detection: Facts & Figures Series*.
6. Anand, S.S., S. Yusuf, V. Vuksan, S. Devanessen, K.K. Teo, et al. 2000. 'Differences in Risk Factors, Atherosclerosis, and Cardiovascular Disease between Ethnic Groups in Canada: The Study of Health Assessment and Risk in Ethnic Groups (SHARE)', *The Lancet* 356: 279–84.
7. Betancourt, J.R., A.R. Green, J.E. Carrillo, and O. Ananeh-Firempong. 2003. 'Defining Cultural Competence: A Practical Framework for Addressing Racial/Ethnic Disparities in Health and Health Care', *Public Health Reports* 118, 4: 293–302.
8. Brown, P. 1995. 'Race, Class and

Environmental Health: A Review and Systematization of the Literature', *Environmental Research* 69: 15–30.

9. CCHS. 2003. Canadian Community Health Survey Cycle 2.1. Ottawa: Statistics Canada.

10. Centers for Disease Control and Prevention (CDC). 2000, 2001, 2002, 2003, 2006. *Behavioral Risk Factor Surveillance System Survey Data*. Atlanta: US Department of Health and Human Services.

11. ———. 2008a. *Summary Health Statistics for the US Population: National Health Interview Survey, 2007*. At: www.cdc.gov/ nchs/products/pubs/pubd/hus/hispanic. htm#population. (10 Aug. 2009)

12. ———. 2008b. *Reported Tuberculosis in the United States, 2007*. Atlanta: US Department of Health and Human Services.

13. ———. 2009. *Summary Health Statistics for US Adults: National Health Interview Survey, 2008*. Vital and Health Statistics, Series 10, Number 242. Hyattsville, Md: US Department of Health and Human Services.

14. Chappell, N., and D. Lai. 1998. 'Health Care Service Use by Chinese Seniors in British Columbia, Canada', *Journal of Cross-Cultural Gerontology* 13: 21–37.

15. Chen, J., E. Ng, and R. Wilkins. 1996. 'The Health of Canada's Immigrants in 1994-1995', *Health Reports* 7, 4: 33–45.

16. Cunningham, S.A., J. Ruben, and K.M. Venkat Narayan. 2008. 'Health of Foreign-born People in the United States: A Review', *Health & Place* 14: 623–35.

17. Democratic Policy Committee. 2010a. The Patient Protection and Affordable Care Act. At: dpc.senate.gov/dpcdoc-sen_health_care _bill_archive_as_passed.cfm. (15 Jan. 2010)

18. ———. 2010b. H.R. 4872, The Healthcare and Education Reconciliation Act of 2010. At: dpc.senate.gov/dpcdoc.cfm?doc_name= lb-111-2-42. (7 Apr. 2010)

19. DeNavas-Walt, C., B.D. Proctor, and J.C. Smith. 2008. *Income, Poverty, and Health Insurance Coverage in the United States: 2007*. US Bureau of the Census, Current Population Reports, P60–235. Washington: US Government Printing Office.

20. Dey, A.N., and J.W. Lucas. 2006. 'Physical and Mental Health Characteristics of US- and Foreign-Born Adults: United States, 1998–2003', *Advance Data from Vital and Health Statistics* No. 369 (1 Mar.): 1–19. Hyattsville, Md: US National Center for Health Statistics.

21. District of Columbia Department of Health. 2002. Diabetes Data and Statistics. At: app. doh.dc.gov/services/special_programs/ diabetes/pdf/final_data_and_stat_diabetes. shtm. (30 Aug. 2009)

22. District of Columbia Department of Health Finance. 2009. DC Department of Health Finance: Health Care Services for the Uninsured. At: dhcf.dc.gov/dhcf/site/ default.asp. (13 Oct. 2009)

23. Dunn, J.R., and I. Dyck. 2000. 'Social Determinants of Health in Canada's Immigrant Population: Results from the National Population Health Survey', *Social Science and Medicine* 51: 1573–93.

24. Dyck, I. 2004. *Immigration, Place and Health: South Asian Women's Accounts of Health, Illness and Everyday Life*. Research on Immigration and Integration in the Metropolis RIIM Working Paper Series. Vancouver: Vancouver Centre of Excellence.

25. Eschbach, K., J.P. Stimpson, Y.F. Kuo, and J.S. Goodwin. 2007. 'Mortality of Foreign-Born and US-Born Hispanic Adults at Younger Ages: A Re-examination of Recent Patterns', *American Journal of Public Health* 97: 1297–1304.

26. Forna, F., D.J. Jamieson, D. Sanders, and M.K. Lindsay. 2003. 'Pregnancy Outcomes in Foreign-Born and US-Born Women', *International Journal of Gynecology and Obstetrics* 83: 257–65.

27. Franzini, L., J.C. Ribble, and A.M. Keddie. 2001. 'Understanding the Hispanic Paradox', *Ethnicity & Disease* 11: 496–518.

28. Gee, E.M., K.M. Kobayashi, and S.G.. Prus. 2004. 'Examining the Healthy Immigrant Effect in Mid- to Later Life: Findings from the Canadian Community Health Survey', *Canadian Journal on Aging* (supplement): S55–63.

29. Gorman, A. 2009. 'California Counties Cut Healthcare to Illegal Immigrants', *Los Angeles Times*, 27 Apr. At: articles. latimes.com/2009/apr/27/local/me-immighealth27. (20 Nov. 2009)

30. Gulliford, M. 2003. 'Equity and Access to Health Care', in M. Gulliford and M. Morgan, eds, *Access to Health Care*. London: Routledge, 36–60.

31. Health Resources and Services Administration (HRSA). 2010a. Health Care Service Delivery Sites Detail. At: datawarehouse. hrsa.gov/sitesdetail.aspx. (04 Jan. 2011)

32. ———. 2010b. Health Center Data. 2010b.

At: www.hrsa.gov/data-statistics/health-center-data/index.html. (04 Jan. 2011)
33. Hoernig, H., and M. Walton-Roberts. 2006. 'Immigration and Urban Change: National, Regional, and Local Perspectives', in T. Bunting and P. Filion, eds, *Canadian Cities in Transition: Local Through Global Perspectives*, 3rd edn. Toronto: Oxford University Press, 408–18.
34. Hunt, L.M., S. Schneider, and B. Comer. 2004. 'Should "Acculturation" Be a Variable in Health Research? A Critical Review of Research on US Hispanics', *Social Science and Medicine* 59: 973–86.
35. Johnson, M., and K.S. Marchi. 2009. 'Segmented Assimilation Theory and Perinatal Health Disparities among Women of Mexican Descent', *Social Science and Medicine* 69: 101–9.
36. Karlsen, S., J. Nazroo, and P. Stephenson. 2002. 'Ethnicity, Environment and Health: Putting Ethnic Inequalities in Health in Their Place', *Social Science and Medicine* 55: 1647–61.
37. Kelaher, M., and D.J. Jessop. 2002. 'Differences in Low Birthweight among Documented and Undocumented Foreign-Born and US-Born Latinas', *Social Science and Medicine* 55: 2171–5.
38. Khan, A.A., and S. Bhardwaj. 1994. 'Access to Health Care: A Conceptual Framework and Its Relevance to Health Care Planning', *Evaluation & the Health Professions* 17: 60–76.
39. Kovner, A.R., and J.R. Knickman, eds. 2008. *Jonas & Kovner's Health Care Delivery in the United States*. New York: Springer.
40. Lai, D.W. 2004. 'Health Status of Older Chinese in Canada: Findings from the SF-36 Health Survey', *Canadian Journal of Public Health* 95, 3: 193–7.
41. Lam, L. 1994. 'Self-assessment of Health Status of Aged Chinese-Canadians', *Journal of Asian and African Studies* 1, 2: 77–90.
42. Lanphear, B.P., M. Weitzman, and S. Eberly. 1996. 'Racial Differences in Urban Children's Environmental Exposures to Lead', *American Journal of Public Health* 86, 10: 1460–3.
43. Leduc, N., and M. Proulx. 2004. 'Patterns of Health Services Utilization by Recent Immigrants', *Journal of Immigrant Health* 6, 1: 15–27.
44. Li, W. 1998. 'Anatomy of a New Ethnic Settlement: The Chinese Ethnoburb in Los Angeles', *Urban Studies* 35, 3: 479–501.
45. Lo, L., and L. Wang. 2005. 'Economic Status of the Chinese Population in the Toronto CMA: Immigrant Origin and Landing Period', *Journal of International Migration and Integration* 5, 1: 107–40.
46. McClure, H., and K. Jerger. 2005. *The State of Latino Health in the District of Columbia*. Washington: Council of Latino Agencies.
47. McDonald, J.T., and S. Kennedy. 2004. 'Insights into the "Healthy Immigrant Effect": Health Status and Health Service Use of Immigrants to Canada', *Social Science and Medicine* 59: 1613–27.
48. McLafferty, S., and R. Chakrabarti. 2009. 'Locating Diversity: Race, Nativity and Place in Health Disparities Research', *GeoJournal* 74: 107–13.
49. Meadows, L.M., W.E. Thurston and C. Melton. 2001. 'Immigrant Women's Health', *Social Science and Medicine* 52: 1451–8.
50. National Association of County and City Health Officials (NACCHO). 2007. *Big Cities Health Inventory*. Washington: NACCHO.
51. Nazroo, J.Y. 1998. 'Genetic, Cultural or Socio-economic Vulnerability? Explaining Ethnic Inequalities in Health', *Sociology of Health & Illness* 20, 5: 710–30.
52. Newbold, K.B. 2005a. 'Health Status and Health Care of Immigrants in Canada: A Longitudinal Analysis', *Journal of Health Services Research Policy* 10, 2: 77–83a.
53. ———. 2005b. 'Self-rated Health within the Canadian Immigrant Population: Risk and the Healthy Immigrant Effect', *Social Science and Medicine* 60: 1359–70.
54. ——— and J. Danforth. 2003. 'Health Status and Canada's Immigrant Population', *Social Science and Medicine* 57: 1981–95.
55. New York City Department of Health and Mental Hygiene (NYC). 2006. *The Health of Immigrants in New York City*. At: www.nyc.gov/html/doh/downloads/pdf/episrv/episrv-immigrant-report.pdf. (15 Aug. 2009)
56. Noh, S., and V. Kaspar. 2003. 'Diversity and Immigrant Health', in P. Anisef and M. Lanphier, eds, *The World in a City*. Toronto: University of Toronto Press, 316–51.
57. Ontario Medical Association (OMA). 2005. 'The Ontario Physician Shortage 2005: Seeds of Progress, but Resources Crisis Deepening', position paper. Toronto.
58. Prehn, A.W., B. Topol, S. Stewart, et al. 2002. 'Differences in Treatment Patterns for Localized Breast Carcinoma among Asian/

Pacific Islander Women', *Cancer* 95: 2268–75.

59. Preston, V., L. Lo, and S. Wang. 2003. 'Immigrants' Economic Status in Toronto: Stories of Triumph and Disappointment', in P. Anisef and M. Lanphier, eds, *The World in a City*. Toronto: University of Toronto Press, 192–262.

60. Ross, N.A., S. Tremblay, S. Khan, D. Crouse, et al. 2007. 'Body Mass Index in Urban Canada: Neighborhood and Metropolitan Area Effects', *American Journal of Public Health* 97, 3: 500–9.

61. Sanchez, R. 2009. 'Evaluation of the "Patient Protection Affordable Care Act"', National Council of La Raza. At: www.nclr.org/content/publications/detail/61094/. (15 Jan. 2010)

62. Singh, G..K., and R.A. Hiatt. 2006. 'Trends and Disparities in Socioeconomic and Behavioural Characteristics, Life Expectancy, and Cause-Specific Mortality of Native-Born and Foreign-Born Populations in the United States 1979–2003', *International Journal of Epidemiology* 35, 4: 903–19.

63. Smaje, C. 1995. 'Ethnic Residential Concentration and Health: Evidence for a Positive Effect?', *Policy and Politics* 13, 3: 251–69.

64. Suh, S.M. 2009. 'The Self-Perceived Health Status and Access to Health Care Services among Korean Immigrants in Ottawa', Master's research paper, Ryerson University.

65. US Agency for Healthcare Research and Quality. 2010. 'CHIPRA Children's Health Care Quality Measurement and Improvement Activities', Nov. At: www.ahrq.gov/chipra/. (16 Nov. 2010)

66. US Bureau of the Census. 2005–7. American Community Survey, Three-Year Estimates. B05006, 'Place of Birth for the Foreign-Born Population', and S0201, 'Selected Population Profile in the United States'.

67. ———. 2008. American Community Survey, One-Year Estimates. B27001, 'Health Insurance Coverage Status by Age for the Civilian Non-institutionalized Population'.

68. US National Center for Health Statistics (NCHS). 2007. *Health, United States, 2007, with Chartbook*. Hyattsville, Md.

69. ———. 2009. *Health, United States, 2008, with Chartbook*. Hyattsville, Md.

70. Vissandjee, B., M. Desmeules, Z. Cao, S. Abdool, and A. Kazanjian. 2004. 'Integrating Ethnicity and Migration as Determinants of Canadian Women's Health', *BMC Women's Health* 4: S1–32.

71. Wang, L. 2007. 'Immigration, Ethnicity, and Accessibility to Culturally Diverse Family Physicians', *Health and Place* 13: 656–71.

72. ——— and D. Roisman. 2011 (forthcoming). 'Modelling Spatial Accessibility of Immigrants to Culturally Diverse Family Physicians', *Professional Geographer*.

73. ———, M. Rosenberg, and L. Lo. 2008. 'Ethnicity, Accessibility, and Utilization of Family Physicians—A Case Study of Mainland Chinese Immigrants in Toronto, Canada', *Social Science and Medicine* 67, 9: 1410–22.

74. Whitley, R., L.J. Kirmayer, and D. Groleau. 2006. 'Understanding Immigrants' Reluctance to Use Mental Health Services: A Qualitative Study from Montreal', *Canadian Journal of Psychiatry* 51: 205–9.

75. Willson, A.E. 2009. 'Fundamental Causes of Health Disparities: A Comparative Analysis of Canada and the United States', *International Sociology* 14, 1: 93–113.

76. Withers, L. 2009. 'Community Health Care Service Awareness and Use among the Washington Metropolitan Area Latino Population', Master's thesis, George Washington University.

77. Wolf, R. 2008. 'Rising Health Care Costs Put Focus on Illegal Immigrants', *USA Today*, 21 Jan. At: www.usatoday.com/news/washington/2008-01-21-immigrant-healthcare_N.htm. (20 Dec. 2009)

78. Zhang, J., and M.J. Verhoef. 2002. 'Illness Management Strategies among Chinese Immigrants Living with Arthritis', *Social Science and Medicine* 55, 10: 1795–1802.

CHAPTER 9

IMMIGRANT POLITICAL INCORPORATION IN AMERICAN AND CANADIAN CITIES

Els de Graauw and Caroline Andrew

INTRODUCTION

In North America, immigrants' incorporation into the political life of their adoptive country has long been and continues to be an urban phenomenon.[1] Cities are the places where immigration and integration policies meet, and few institutions have a more direct impact on immigrant well-being than municipal governments. Cities also offer immigrants valuable opportunities to participate actively in the political life of the receiving country. Naturalized citizens, and sometimes non-citizens, can participate in local elections. Also, all immigrants, regardless of documentation and citizenship status, can participate in informal politics, including protests. Cities, therefore, are key loci for immigrants' political integration into the receiving country.

In this chapter, we compare the political incorporation of immigrants in American and Canadian cities. Immigrant political incorporation refers to the process through which immigrants, over time, become part of arenas of democratic voice and choice. We focus on five key indicators of immigrant political incorporation: (1) *naturalization*, or the percentage of immigrants in American and Canadian cities who have acquired national citizenship in the receiving country; (2) *voting*, or the percentage of immigrants who vote in national and local elections; (3) *demographic representation*, or the percentage of foreign-born individuals and their descendants who are elected to city council; (4) *substantive representation*, or the extent to which cities have created institutional spaces specifically aimed at facilitating immigrants' interaction with local government officials; and (5) *civic engagement*, or the extent to which immigrants participate in civic organizations such as community-based non-profit organizations, religious institutions, and labour unions.

We compare five American cities with five Canadian cities. For the United States, we discuss New York City, Chicago, and San Francisco, three established gateway cities that have experienced large-scale immigration since the nineteenth century, and Los Angeles and Houston, two cities that have attracted large numbers of immigrants mainly since World War II. For Canada, we consider the established gateways of Toronto, Vancouver, and Montreal and the newer immigrant destinations of Ottawa and Calgary. By comparing the American and Canadian cities on the five incorporation indicators listed above, we can determine where immigrants are more politically incorporated: in US or Canadian cities, and in established or newer gateway cities.

We then discuss three sets of factors that explain immigrant political incorporation in the United States and Canada. First, we consider socio-economic or demographic characteristics as individual-level determinants of political incorporation. Second, we consider institutional factors and government policies that shape immigrants' political incorporation. Finally, we consider the role of civic organizations, including local political parties and other community-based organizations, which can mediate between the micro-level of individuals and the macro-level of larger political communities. Determining the fit of these explanatory factors with our urban-level incorporation indicators enriches our understanding of the ways in which immigrants have, or have not, been incorporated into the political systems of the two countries.

THE CHALLENGES OF A US–CANADA COMPARISON

Five important differences between the United States and Canada make it challenging to compare immigrant political incorporation in American and Canadian cities: (1) municipal boundaries; (2) the extent to which political incorporation serves as an indicator of immigrants' overall integration into the receiving society; (3) the roles of political parties at the local level; (4) the presence (or absence) of a substantial population of undocumented immigrants; and (5) the roles and funding sources of community-based immigrant organizations.

Differences in Municipal Boundaries

Municipal boundaries are different in the two countries. City government jurisdictions in the United States tend to be bounded by the limits of central-city areas. City government in Canada, in contrast, covers not only the city centre, but also large parts of outlying suburban areas. In addition, much of the Canadian population data are collected not for the unit of the city as it is understood in the United States, but rather by **census metropolitan area** (CMA), a socio-economic entity including an urban core population of at least 10,000 and an overall population of at least 100,000. To complicate matters further, Canadian cities vary considerably in the population size of the central city relative to the CMA of which that central city is part. Consequently, a comparison of urban cores of various Canadian cities can entail measuring different geographic areas or different populations.

These differences in municipal boundaries are of methodological consequence because they make it difficult to compare similar units of municipal government in the two countries. In comparing American with Canadian central cities, we have to contend with substantial variation among the Canadian central cities. In comparing American central cities with Canadian CMAs, we can even out differences among Canadian cities, yet we cannot compare American and Canadian central cities. A key objective for us is to compare American and Canadian central cities and discuss the importance of the urban city environment in the process of immigrant political incorporation. For that reason, we opted to use central-city data for both countries. A drawback of this focus is that in some cases we have only limited data for the Canadian cities.

The differences in municipal boundaries also matter empirically because urban and suburban areas likely provide different contexts for immigrant political incorporation. Central cities often are more diverse and more politically progressive; therefore, we can expect central cities to be more inclusive towards newcomers and facilitative of their incorporation. Suburban areas, on the other hand, have traditionally been more homogeneous and politically conservative (Oliver, 2001). These differences between central cities and suburbs as loci of immigrant political incorporation are another reason why we chose to compare US and Canadian central cities, rather than US central cities with Canadian CMAs.

Different Conceptions of Immigrant Integration

The United States is a republican-inspired[2] country and understandings of what it means to be an American draw heavily on notions of patriotism and civic participation. Civic and political incorporation consequently serve as important indicators of immigrant integration. Canada's social democratic tradition, on the other hand, places more emphasis on the socio-economic dimension of integration, and tends to focus on employment and economic outcomes to assess immigrants' integration. The Canadian statist tradition, furthermore, means that the administrative or bureaucratic dimension of the state tends to overshadow the electoral one.

The different conceptions of immigrant integration matter for two reasons. First, Canada's lesser emphasis on political incorporation goes hand in hand with a dearth of available data on immigrants' political incorporation in Canadian cities. Despite a growing interest in political incorporation dynamics in Canada today, there are more data on immigrants' political activities in the United States than in Canada. Differences in political culture and understandings of immigrant integration likely have something to do with this difference in research interests and data availability. Second, the different conceptions of immigrant integration likely also affect the incorporation outcomes that American and Canadian cities will pursue. Given Canada's more bureaucratic tradition, likely more pressure is on municipal governments to expand and improve services to immigrant communities rather than to mobilize immigrants' electoral participation.

Different Roles of Local Political Parties

While the Democratic and Republican parties have a notable presence in the politics of many American cities, national and provincial political parties have no recognized role at the municipal level in Canada, and only in British Columbia and Quebec do a shifting array of purely local parties contest local elections, as is the case in Vancouver and Montreal. This difference makes it difficult to compare and contrast the role of political parties as agents of immigrant political incorporation in American and Canadian cities. As we discuss below, however, local political parties in the United States have lost ground in recent decades. The two countries, therefore, may be more similar on this point today than they were in the past when local parties were key engines of immigrant incorporation in various American gateway cities.

Differences in the Size of the Undocumented Population

A fourth difference between the two countries is the absence of a substantial population of undocumented immigrants in Canada. There is no credible estimate of the size of the undocumented population in Canada today, but the latest calculations put the number in the United States at 11.9 million (Passel and Cohn, 2009). Undocumented immigrants are heavily concentrated in the central cities of large US metropolitan areas and they tend to be poor, uneducated, and not proficient in English (ibid.). With minimal resources, undocumented immigrants have correspondingly few opportunities to engage in the political process. We may, therefore, observe overall lower levels of immigrant political incorporation in American than in Canadian cities.

Differences in Community-Based Immigrant Organizations

Finally, the role and funding structure of community-based immigrant organizations in the two countries differ. Canadian cities count many organizations that receive funding from the federal and provincial governments to deliver settlement services to immigrants and refugees. These organizations are significant players at the local level, yet the close ties and accountability arrangements with their government funders have produced speculation that the focus on service provision has driven out advocacy within the settlement sector. American cities likewise include many immigrant organizations that provide essential social services to newcomer populations. The American organizations, however, draw financial support from both public and private sources and many are dual-goal organizations that combine service provision with advocacy around immigrant issues (Hung, 2007; de Graauw, 2008). Immigrant organizations in the American context, therefore, likely enjoy more independence from government and likely engage in more advocacy and immigrant mobilization than their Canadian counterparts.

INCORPORATION INDICATORS[3]

Naturalization

For classic countries of immigration like the United States and Canada, **naturalization**, the legal process enabling people to acquire citizenship after birth, has played an important role in building their populations and economies (Bloemraad, 2006). Naturalization also is an important indicator of immigrant political incorporation. It signals immigrants' identification with the receiving society, confers the right to vote and hold any public office in the country (except that of US President), and communicates immigrants' readiness to participate actively in the political affairs of the country (Johnson et al., 1999; Jones-Correa, 2001). Naturalization also is important, especially in the United States today, because it gives immigrants access to government-funded social services on par with the population as well as immigration benefits for family members.

In the past, the United States and Canada had racial and gender restrictions on citizenship. Today, however, their naturalization policies are among the most open and inclusive in the world (Brubaker, 1989). Currently, adult immigrants in the United States are eligible for citizenship after a minimum of five years of legal permanent residence in the country (or three years if married to a US citizen), and if they can demonstrate knowledge of English as well as of American history and government. Applicants for naturalization also need to pay an application fee of US$675, take an oath of allegiance, and renounce their previous citizenship. Legal permanent residents in Canada need to wait only three years before applying for citizenship and pay a lower application fee of Cdn$200; however, they also need to prove a basic knowledge of Canada and of either English or French and take an oath of allegiance. Canada, furthermore, has legally recognized dual citizenship since 1977; immigrants do not have to renounce their previous citizenship to become Canadians. In sum, the two countries have similar naturalization requirements, but immigrants in Canada can become naturalized citizens more quickly, naturalization is more affordable, and they are legally allowed to hold dual citizenship upon naturalization. The benefits to naturalization are similar, and naturalized citizens enjoy legal and political equality with the non-immigrant population in both countries.

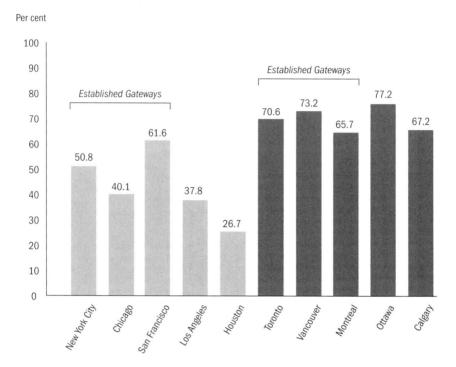

Figure 9.1 Naturalization Rates, Selected US and Canadian Cities

Note: All data are for central cities.

Sources: American Community Survey, three-year estimates, 2005–7; Statistics Canada, *Census of Canada, 2006.*

Only a minority of contemporary immigrants in the United States are naturalized citizens. Data from the 2005–7 American Community Survey show that 42.5 per cent of the nation's foreign-born were naturalized citizens. As Figure 9.1 shows, the naturalization rates were higher than the US national average in the established gateway cities of New York City (50.8 per cent) and San Francisco (61.6 per cent); however, they were lower in Chicago (40.1 per cent) and in the newer gateways of Los Angeles (37.8 per cent) and Houston (26.7 per cent). Naturalization rates are notably higher in Canada. In 2006, 72.7 per cent of the foreign-born had acquired Canadian citizenship (Statistics Canada, 2006). Slightly different from the pattern in the United States, only the established gateway city of Vancouver (73.2 per cent) and the newer gateway city of Ottawa (77.2 per cent) had naturalization rates exceeding the national average. The naturalization rates were slightly lower than the national average in Toronto (70.6 per cent) and even lower in Calgary (67.2 per cent) and Montreal (65.7 per cent). As Figure 9.1 indicates, differences in naturalization rates among Canadian cities are smaller than among US cities: an 11.5 per cent difference between the cities with the highest and lowest naturalization rates in Canada, compared to a gap of almost 35 per cent in the United States.

Voting

Immigrants' participation in electoral politics, as measured by voter registration and the number of immigrants turning out to vote, is a second common indicator of political incorporation (Cain et al., 1991; DeSipio, 1996; Ramakrishnan and Espenshade, 2001; Tam Cho, 1999). In this section, we discuss immigrant voting in both national and local elections in the United States and Canada.

Voting in National Elections

Current Population Survey data show that naturalized citizens in the United States, with the exception of those who emigrated from Canada, are less likely to vote than US-born citizens (see Table 9.1). The effect of nativity is most pronounced during mid-term, or congressional, elections, when the gap in voter registration and voting rates between naturalized and US-born citizens often is larger than it is during presidential election years. Overall, naturalized citizens from Asia have had among the lowest registration and voting rates, followed by naturalized citizens from Latin America, Africa, Europe, and then North America. Table 9.1 also shows that the gap in voter registration and voting rates between US-born citizens and naturalized citizens from Europe has widened in the two most recent elections. Finally, a record high percentage of naturalized citizens from Africa voted during the presidential election of 2008, which likely was because Barack Obama (a mixed-race candidate whose father was from Kenya) ran for President that year.

Data on immigrant voting are relatively limited in Canada, and the two kinds of data available through Elections Canada do not suggest a clear pattern about immigrants' propensity to vote—a degree of ambiguity confirmed by emerging scholarship on the topic (e.g., White et al., 2006). However, data for federal elections since 2004 show that turnout in the 16 electoral districts with the highest percentages of foreign-born residents in 2001 had turnout rates consistently lower than the Canadian average turnout rate for the three most recent federal elections. Also, in its analysis of Statistics Canada's 2002 Ethnic Diversity Survey, Elections Canada provides self-reported rates of voting in federal elections. Table 9.2 shows that immigrants born in Europe report voting at rates *higher* than Canadian-born citizens, yet immigrants from Asia (with the exception of Indian immigrants) reported voting at rates *lower* than the Canadian-born population. Data from this study also show that second and later generations of immigrants (with the exception of those of Japanese descent) tend to report voting less than first-generation immigrants (Table 9.3).

The numbers in Tables 9.1–9.3 suggest that immigrants in Canada are more likely to participate in national elections than are immigrants in the United

States. We need to remember, though, that the two countries use different formulas to calculate turnout. In the United States, turnout is expressed as a percentage of the voting age population (i.e., people 18 years and older, regardless of whether they are eligible to vote). In Canada, turnout is expressed as a percentage of eligible voters (i.e., those 18 years and older whose names appear on the voting roll). With a much larger denominator, it is not surprising that immigrant turnout rates in the United States are consistently lower than in

Table 9.1   Reported Voter Registration and Voting among US-Born and Naturalized Citizens in the US, by Region of Origin, 1996–2008

Year (Election)	US-Born Citizens	Naturalized Citizens by Region of Origin				
		Europe	Asia	North America	Latin America	Africa
1996 (Presidential Election)						
Total population						
18 yrs and over	171,713,000	2,247,000	2,560,000	274,000	2,824,000	146,000
% Registered	71.3	68.4	59.2	77.7	61.6	57.8
% Voted	58.6	57.6	46.9	67.3	53.8	52.1
1998 (Congressional Election)						
Total population						
18 yrs and over	173,862,000	2,198,000	2,998,000	316,000	3,691,000	195,000
% Registered	67.8	65.7	46.1	71.8	54.8	40.1
% Voted	45.7	49.5	28.6	55.8	38.1	24.4
2000 (Presidential Election)						
Total population						
18 yrs and over	175,679,000	2,384,000	3,329,000	339,000	4,000,000	185,000
% Registered	70.2	65.3	52.5	74.3	59.3	44.0
% Voted	60.0	57.3	45.0	63.9	52.0	38.0
2002 (Congressional Election)						
Total population						
18 yrs and over	180,473,000	2,390,000	4,055,000	302,000	4,740,000	261,000
% Registered	67.3	63.8	49.2	75.0	53.5	52.2
% Voted	46.8	45.5	30.8	57.0	35.2	38.8
2004 (Presidential Election)						
Total population						
18 yrs and over	183,880,000	2,524,000	4,561,000	228,000	5,075,000	348,000
% Registered	72.9	69.7	55.9	81.1	60.7	65.1
% Voted	64.5	63.2	48.3	75.3	52.7	59.8
2006 (Congressional Election)						
Total population						
18 yrs and over	187,132,000	2,552,000	4,837,000	286,000	5,428,000	323,000
% Registered	68.6	62.4	52.0	69.4	51.7	53.7
% Voted	48.6	45.4	33.2	53.6	34.9	36.5
2008 (Presidential Election)						
Total population						
18 yrs and over	190,683,000	2,725,000	5,340,000	255,000	6,304,000	637,000
% Registered	71.8	62.8	56.6	81.2	61.7	67.5
% Voted	64.4	57.3	47.8	76.2	56.4	61.1

Source: Current Population Survey (Nov. supplement), 1996-2008.

Canada. These differences in calculating turnout make it difficult to compare immigrant voting in the two countries.

To learn about immigrants' propensity to vote in national elections, it is more instructive to compare the turnout rates of the foreign-born with those of the non-immigrant population in the same country, which leads us to conclude that fewer immigrants to the United States, with the exception of those originally from Canada, vote than do individuals born in the United States. The Canadian

Table 9.2   Self-Reported Rates of Voter Participation by Selected Place of Birth, 2000 Canadian Federal Election

Place of Birth	Eligible Voters Surveyed	% Reported Voting
Canada	27,028	78.8
Italy	504	94.0
Netherlands	224	93.7
United Kingdom	846	86.4
Germany	370	85.9
Poland	301	81.6
India	360	81.1
Portugal	243	79.8
United States	391	78.0
Philippines	291	75.6
People's Republic of China	373	70.5
Hong Kong	428	64.7

Source: Ethnic Diversity Survey, Statistics Canada and Department of Canadian Heritage, 2002.

Table 9.3   Self-Reported Rates of Voter Participation, Selected Visible Minorities by Place of Birth, 2000 Canadian Federal Election

Group	% Canadian-Born Reported Voting	% Foreign-Born Reported Voting
Not Visible Minority	80.9	83.6
Visible Minority		
South Asian	60.4	78.2
Filipino	59.6	75.6
Arab	55.9	75.1
Black	53.2	74.6
Chinese	62.8	65.8
Japanese	77.1	60.8

Source: Ethnic Diversity Survey, Statistics Canada and Department of Canadian Heritage, 2002.

data show a mixed bag. More immigrants of European descent vote than Canadian-born individuals, while comparatively fewer immigrants from Asia vote. Canadian data also suggest that more first-generation immigrants participate in national elections than do successive generations of immigrants from the same country of origin.

Voting in Local Elections

Few hard data are available on the voting behaviour of naturalized citizens in American cities. Nativity questions are almost never included in surveys conducted by news organizations, including media exit polls. In recent years, however, immigrant advocacy groups have collaborated with academic institutions to conduct multilingual non-partisan exit polls of immigrant voters in a number of American cities, including New York City, Chicago, Los Angeles, and Houston.[4] Exit polls survey only the voters in these cities, making it impossible to compare the registration and turnout rates of naturalized citizens vis-à-vis American-born citizens. With exit poll data now available for several election years and for several cities, however, it is possible to see the trend that immigrants are becoming an increasingly important bloc of voters in large gateway cities. In New York City, over 275,000 immigrant voters have been added to the city's voter rolls since 1998, and today naturalized citizens make up more than one million of the city's 3.8 million registered voters (NYIC, 2009; Santos, 2008). Similarly, immigrants in Los Angeles now constitute one-fifth of the city's electorate and they are also an important driving force behind the expansion of the city's new voters (CARECEN, 2006).

City-level data on immigrant voting are virtually absent in Canada. A study of the 2003 election in Toronto, though, indicated that the city's electoral districts with the highest percentages of immigrant residents had lower levels of turnout compared to the overall turnout rate in the city (Siemiatycki, 2009). Siemiatycki also is completing a study of the 2006 election in Toronto, and results appear to be less clear than they were for the 2003 municipal election (personal communication with author).

Voting in US national elections is a right reserved to citizens and only registered voters can participate, but a handful of American municipalities allow non-citizens to vote in local elections. The best known of these is Takoma Park, Maryland, which introduced the practice in 1992. Among the US gateway cities we focus on in this chapter, Chicago currently allows non-citizens to vote in school board elections, and New York City did so until elected school boards were abolished in 2003. Since the 1990s, Los Angeles and San Francisco have unsuccessfully considered proposals to expand the franchise to non-citizens. In Canada, the issue of non-citizen voting has surfaced in Toronto, where the 'I Vote Toronto' campaign seeks to give non-citizens the right to vote locally (ibid.). While only a few North American cities allow non-citizens to vote in local elections today, the fact that campaigns for non-citizen voting are most pronounced in cities underscores the importance of the urban context in the process of immigrants' incorporation into the political life of the United States and Canada.

Demographic Representation

The extent to which immigrants and their immediate descendants have won elective seats in legislative bodies of government is another indicator of political incorporation. At the national level, immigrants enjoy notably more **demographic representation** in Canada than they do in the United States. In Canada's 40th Parliament (2008–10), 38 of 304 members sitting in the House of Commons are foreign-born, compared to only 8 out of 535 members in the 111th US Congress (2009–11). In both countries, however, the foreign-born today are under-represented in the national legislatures: immigrants make up 19.8 per cent of the population in Canada (2006 census), but only 12.5 per cent of Parliament. Likewise, the foreign-born constitute 12.5 per cent of the US population (ACS three-year estimates, 2005–7), but only 1.5 per cent of the US Congress. One reason why the percentage of immigrant legislators lags considerably behind the immigrant share of the population is that it takes considerable time for first-generation immigrants to become citizens, vote, and run for and win political office. Immigrants' under-representation in the US Congress is further explained by the high cost associated with running for federal elective office in the United States.

As at the national level, immigrants today are under-represented on city councils in both countries, although they have relatively more electoral success in Canadian cities. As Figure 9.2 illustrates, the foreign-born enjoy close to proportional representation in Ottawa, where they constitute 18 per cent of all municipal councillors and 22 per cent of the central-city population. Immigrants also enjoy significant representation on the city councils of Toronto and Vancouver, where approximately a third of all municipal representatives and nearly half of each city's population are foreign-born. Immigrants enjoyed the least representation in Montreal and Calgary, where they made up 7 per cent of municipal councils, but 33 and 21 per cent of the overall populations, respectively.

Figure 9.2 also shows that the foreign-born enjoy some representation on city councils in the American cities we studied, but this representation is well below immigrants' share of the cities' populations. The representation picture improves when we include **ethnic and racial minorities**, or city council members who are black or the direct descendants of immigrants from Asia and Latin America. San Francisco stands out as the only city where ethnic and racial minorities were slightly over-represented among the city's legislators in 2009. Non-Hispanic whites remain over-represented and Asians and Latinos have yet to gain a stronger foothold on the city councils of New York City, Los Angeles, Chicago, and Houston. Yet, these cities have come a long way in recent decades. In New York City, no council members were of immigrant origins at the end of the 1980s, but today legislators with Asian or Hispanic roots make up 24 per cent of the city's 51-member council. Voters in Houston selected the

A. Per cent FB Population and Per cent FB City Councillors, 2009

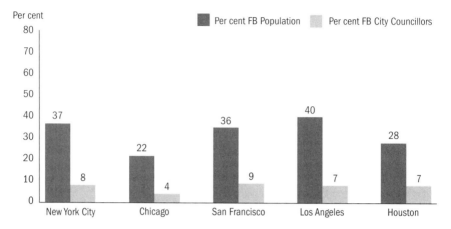

B. Per cent ERM Population and Per cent ERM City Councillors, 2009

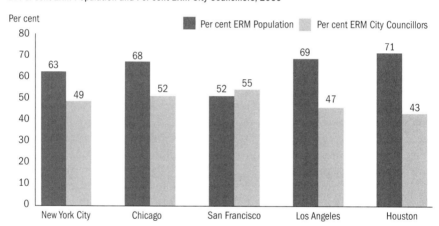

C. Per cent FB Population and Per cent FB City Councillors, 2002

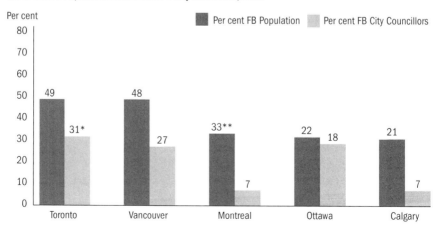

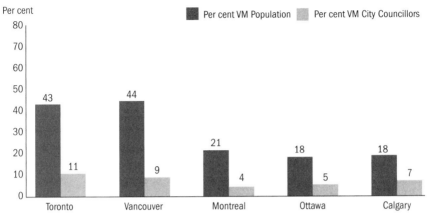

D. Per cent VM Population and Per cent VN City Councillors, 2002

Figure 9.2 Foreign-Born and Ethnic and Racial/Visible Minority Representation on City Councils, Selected US and Canadian Cities

Notes: FB = foreign-born; ERM = ethnic and racial minority; VM = visible minority.

*This is an estimate and should be interpreted with caution. We do not have data on the precise number of foreign-born elected officials on the Toronto city council in 2002. We use an estimate based on the total number of foreign-born elected officials in the Toronto area for all three levels of government (federal, provincial, and municipal) combined. City elected officials normally make up about 50 per cent of all elected officials in Toronto. Therefore, we estimate that about 50 per cent of all foreign-born elected officials at all three levels of government combined were Toronto city officials in 2002.

**Foreign-born population of 15 years and over.

Sources: Population statistics for American cities are from the American Community Survey, three-year estimates, 2005–7; data for American city councils are for 2009 and, with the exception of New York City, they are the authors' compilation of city council records available on the Internet; New York City data are from John Mollenkopf (personal communication), with updates by the authors; population statistics for Canadian cities are from Statistics Canada, *Census of Canada, 2001*; data for Canadian city councils are for 2002 and come from Andrew et al. (2008).

first Mexican American for city council in 1979 and the first Asian American in 1993. In contrast to American cities, visible minorities have notably lower representation on the municipal councils of Canadian cities. The differences in demographic representation between American and Canadian cities raise interesting questions about the significance of race and ethnicity, but also about how accessible local elective bodies are to the foreign-born versus the American-born and Canadian-born descendants of immigrants in the two countries.

Substantive Representation

Another indicator of immigrants' political incorporation is the extent to which their interests are reflected in the adoption and implementation of public policies at the local level. Few studies consider substantive representation as an indicator of immigrant political incorporation, which is understandable given the difficulty in determining what immigrants' collective political interests are.[5] Immigrants are diverse in terms of their origins, immigration status, and linguistic and socio-economic backgrounds. As a result, their political interests are equally diverse.

To get a sense of immigrants' substantive representation, we collected data on institutional spaces that city governments have created to address the issues that are especially salient to immigrant communities. We focused specifically on city agencies that seek to integrate immigrants into various aspects of local civic life. Conceptually, such agencies provide an important indicator that cities are willing to use municipal resources to provide programs of particular value to immigrants. Empirically, they are relatively rare (Ramakrishnan and Lewis, 2005). Thus, if a city has an immigrant-focused agency, this is a good indicator that the city is taking proactive steps to integrate its foreign-born residents.

Four of the five American gateway cities have immigrant affairs offices. They are recent creations and all are executive-level agencies within the mayor's office. Houston created its Mayor's Office of Immigrant and Refugee Affairs in 2001, the same year in which New York City established the Mayor's Office of Immigrant Affairs as a charter agency. Los Angeles created a Mayor's Office of Immigrant Affairs in 2004 and San Francisco established the Office of Civic Engagement and Immigrant Affairs in 2009. These offices engage in similar activities and have as their mission the promotion of the well-being of immigrant communities. Chicago is the only one of the five cities that does not have a specific immigrant affairs office, but its Commission on Human Relations includes an Advisory Council on Immigrant and Refugee Affairs, which dates back to 1990.

BOX 9.1   Immigrant Integration in San Francisco

In recent years, San Francisco has adopted a number of policies aimed at improving immigrant integration. In 2001, the city's legislators adopted the Equal Access to Services Ordinance, which requires city departments to offer information and services to residents in a number of foreign languages, including Spanish and Chinese. In 2003, San Francisco voters approved the Minimum Wage Ordinance, which mandates that all low-wage workers in the city be paid a minimum wage that is higher than the federal and state minimums. Since immigrants are over-represented in the city's low-wage service sector, they in particular have benefited from the higher minimum wage. Finally, in 2009, the city started issuing local ID cards to undocumented immigrants, who can use the card to access city services, identify themselves with local police, and open bank accounts. These three policies are good examples of the substantive representation of immigrants. Because all three policies have been enacted with strong advocacy of immigrant organizations, they also show that immigrant civic engagement can have an impact on policy outcomes (see de Graauw, 2008).

At the local level in Canada, the government response to immigrant integration is less institutionalized and none of the five Canadian cities we studied has an immigrant affairs office. Toronto comes closest with its Diversity Management and Civic Engagement Unit, housed in the City Manager's Office. This unit oversees the immigration and settlement policy framework that the city council adopted in 2001 with the goal of improving immigrant integration in the city (Graham and Phillips, 2007). In 2006, Toronto also signed the Canada–Ontario–Toronto Memorandum of Understanding on Immigration and Settlement, by which the city pledges to collaborate with the federal and provincial governments in meeting the needs of the city's foreign-born residents.

In Canada, the local government response to immigrants also takes a different form in that integration issues tend to be addressed through public–private partnerships that rely on significant input from community-based organizations. Vancouver, for example, has created neighbourhood houses (akin to yesteryear's settlement houses in American gateway cities) in partnership with local community organizations. The Ottawa Police Department, furthermore, collaborates with immigrant and visible minority communities through COMPAC, a community policing initiative established in 1999 (Andrew, 2007). The cities of Calgary and Ottawa have similar public–private initiatives. A weak mayoral system in Canadian cities, where mayors rarely have the political power to initiate policies by themselves, is one reason why these public–private partnerships have emerged. The partnerships, however, also result from the absence of local-level political parties. Montreal and Vancouver have civic parties, but the other three cities do not have any formal role for local-level political parties, let alone a role for parties in immigrant integration.

Civic Engagement

A final indicator of political incorporation we considered is participation in communal activities through community-based organizations such as non-profit organizations, religious institutions, and labour unions. One important reason for considering immigrants' **civic engagement** is that it provides them with relatively low-threshold access to aspects of public life. Immigrants, for example, do not need to be citizens, or even legal residents, to participate in community organizations. At the same time, immigrants might be more inclined to participate in civic affairs because they are afraid to participate in politics or because they perceive traditional politics to be dirty or corrupt. We also want to consider immigrants' civic engagement given the rich literature that shows a clear link between civic engagement, on the one hand, and formal political participation and policy influence at the local level, on the other (e.g., Verba et al., 1995). Civic engagement can foster immigrants' political engagement by making them more aware of current events, increasing their sense of political efficacy, and teaching them skills that carry over into an active political life (Ramakrishnan and Bloemraad, 2008).

It is difficult to quantify immigrants' civic engagement, but recent political events suggest that immigrants in American urban centres are quite active in the civic sphere. The immigration protests in the spring of 2006 serve as illustration. Immigrants and their supporters took to the streets in over 160 American cities to protest the anti-immigrant legislation that members of the US Congress were considering at the time. In Los Angeles, an estimated 500,000 people protested, and about 300,000 did so in Chicago (Bloemraad et al., 2011; Bernstein, 2006). These demonstrations were not spontaneous outbursts of frustration. Rather, they had been staged by immigrant organizations, labour unions, churches, and ethnic media outlets, which drew on the active involvement of immigrants to craft and implement public education campaigns, advocacy initiatives, and community mobilization efforts (Cordero-Guzmán et al., 2008).

Immigrant civic engagement takes slightly different forms in Canadian cities. Because the federal, provincial, and municipal governments all deliver public services through third parties, service organizations that cater to immigrants and visible minorities dominate Canada's civil society sector (Biles, 2008). These organizations' heavy dependence on the public purse has produced debates about the ability of the immigrant settlement sector in Canadian cities to perform advocacy in addition to service provision (ibid.). There are immigrant advocacy organizations in Canadian cities, but they simply are not as numerous and not as well organized as those in the United States. In sum, immigrant organizations in Canadian cities provide newcomers with essential services that improve their socio-economic well-being, but they likely are less significant in providing newcomers with a formative advocacy experience.

Some data, however, show immigrants' involvement in local civic life in Canada. Scott, Selbee, and Reed (2006), for example, document that the

BOX 9.2   Immigrant Integration in Ottawa

The City for All Women Initiative (CAWI) is a public–private partnership between the city of Ottawa and representatives from local women's groups, many representing recent immigrants to Canada. A recent CAWI project, called the Equity and Inclusion Lens, focuses on providing questions and generating data that will help city employees and staff with community-based organizations work more inclusively in all stages of the policy-making process. The project covers the five marginalized groups included in Ottawa's policy on diversity: visible minorities, women, Aboriginal people, persons with disabilities, and lesbian, gay, bisexual, and transgender (LGBT) people. Added to these groups were: immigrants, people living in poverty, rural populations, youth, the elderly, and francophones. This project is noteworthy for its participatory nature. It also is a strong attempt at 'mainstreaming' immigrant concerns into the city's decision-making processes (see: www.cawi-ivtf.org).

foreign-born volunteer in numbers similar to Canadian-born individuals, although they do so more with religious organizations. Furthermore, Tossutti (2003) argues that active membership in voluntary associations, rather than just volunteering, improves immigrants' civic engagement. These findings have sparked a debate in Canada about the connection between volunteering and immigrants' political incorporation. They also have resulted in local initiatives, such as the one undertaken by the Maytree Foundation in Toronto, aimed at increasing immigrant membership on the boards of mainstream organizations (Maytree Foundation, 2007). Such efforts to create more inclusive community structures at the local level suggest that also in Canada civic engagement is seen as an important avenue for immigrants' political incorporation.

EXPLAINING IMMIGRANT POLITICAL INCORPORATION

The Role of Socio-economic Status (SES)

North American scholarship on immigrant political incorporation tends to emphasize that immigrants' individual characteristics are key to understanding their participation in various aspects of political life. This scholarship, following in the footsteps of the well-established literature on political behaviour (e.g., Verba and Nie, 1972; Wolfinger and Rosenstone, 1980), draws prominently on survey data to measure the effect that immigrants' age, gender, race, educational attainment, country of origin, income, language ability, home ownership, marital status, and length of stay in the receiving country have on their likelihood to naturalize, register to vote, develop partisanship, and cast a ballot on election day. Statistical analysis demonstrates that immigrants of higher educational status and those who are more proficient in the receiving country's dominant language, are married, have spent more time in their adoptive country, are homeowners, etc. are more likely to acquire citizenship and participate in formal politics than those who lack such resources and skills (e.g., Bass and Casper, 2001; Cain et al., 1991; Jones-Correa, 1998; Portes and Curtis, 1987; Ramakrishnan and Espenshade, 2001; Yang, 1994). In this context, Canadian immigration laws, which strongly favour immigrants with higher levels of formal education, help explain why naturalization rates are comparatively higher in Canada.

This SES model of immigrant political incorporation also helps explain the differences in naturalization rates in the five American gateway cities we analyzed. In comparison to New York City and San Francisco, Los Angeles, Chicago, and Houston have immigrant populations that are relatively (1) more recent arrivals to the United States, (2) less educated, (3) less proficient in English, and (4) poorer. As Table 9.4 illustrates, these three cities also have notably higher percentages of immigrants who have *not* acquired US citizenship. Among the five American cities, San Francisco has the largest percentages of immigrants who have lived in the country for at least 18 years and have an

Table 9.4   Demographic Characteristics of the Foreign-Born Population, Selected US Cities, 2005–7

	New York City		Los Angeles		Chicago		Houston		San Francisco	
	Estimate	%	Estimate	%	Estimate	%	Estimate	%	Estimate	%
Total Population	8,246,310	100	3,770,590	100	2,740,224	100	2,034,749	100	757,604	100
US-born	5,218,136	63.3	2,260,826	60.0	2,145,383	78.3	1,463,803	79.1	487,123	64.3
Foreign-born	3,028,174	36.7	1,509,764	40.0	594,841	21.7	570,946	28.1	270,481	35.7
Citizenship										
Foreign-born	3,028,174	100	1,509,764	100	594,841	100	570,946	100	270,481	100
Naturalized	1,537,379	50.8	570,657	37.8	238,238	40.1	152,421	26.7	166,504	61.6
Non-citizen	1,490,795	49.2	939,107	62.2	356,603	59.9	418,525	73.3	103,977	38.4
Period of Entry										
Foreign-born	3,028,174	100	1,509,764	100	594,841	100	570,946	100	270,481	100
Entered 2000 or later	628,644	20.8	304,872	20.2	139,921	23.5	171,389	30.0	50,074	18.5
Entered 1990 to 1999	989,611	32.7	439,490	29.1	198,541	33.4	200,013	35.0	75,848	28.0
Entered before 1990	1,409,919	46.6	765,402	50.7	256,379	43.1	199,544	34.9	144,559	53.4
World Region of Birth										
Foreign-born	3,028,174	100	1,509,764	100	594,841	100	570,946	100	270,481	100
Europe	529,930	17.5	96,625	6.4	116,589	19.6	23,980	4.2	36,785	13.6
Asia	775,213	25.6	398,578	26.4	110,046	18.5	90,209	15.8	167,428	61.9
Africa	112,042	3.7	21,137	1.4	20,819	3.5	21,696	3.8	3,516	1.3
Oceania	9,085	0.3	4,529	0.3	1,190	0.2	1,142	0.2	2,434	0.9
Latin America	1,583,735	52.3	975,308	64.6	342,034	57.5	429,922	75.3	56,260	20.8
Northern America	21,197	0.7	13,588	0.9	4,164	0.7	3,997	0.7	4,057	1.5
Educational Attainment										
Foreign-born 25 yrs and over	2,595,009	100	1,273,019	100	495,264	100	454,904	100	245,674	100
No high school degree	713,627	27.5	535,941	42.1	187,210	37.8	222,448	48.9	71,737	29.2
Graduate or professional degree	256,906	9.9	78,927	6.2	42,593	8.6	34,573	7.6	27,024	11.0

	New York City		Los Angeles		Chicago		Houston		San Francisco	
	Estimate	%	Estimate	%	Estimate	%	Estimate	%	Estimate	%
Language Spoken at Home and English Ability										
Foreign-born 5 yrs and over	3,012,244	100	1,501,085	100	590,935	100	566,067	100	268,971	100
Speak English only	753,061	25.0	117,085	7.8	48,457	8.2	40,191	7.1	31,470	11.7
Speak English less than 'very well'	1,472,987	48.9	951,688	63.4	371,698	62.9	382,661	67.6	155,734	57.9
Occupation										
Foreign-born civilian employed pop. 16 yrs and over	1,735,271	100	903,066	100	356,510	100	347,699	100	153,144	100
Management or professional	466,788	26.9	191,450	21.2	68,806	19.3	60,847	17.5	54,060	35.3
Service occupation	489,346	28.2	229,379	25.4	90,910	25.5	89,359	25.7	40,277	26.3
Poverty Status in Past 12 Months										
Foreign-born for whom poverty status is determined	3,003,372	100	1,496,979	100	590,173	100	564,657	100	268,718	100
Below 100% of the federal poverty level	531,597	17.7	306,881	20.5	108,592	18.4	131,565	23.3	34,396	12.8

Source: American Community Survey, three-year estimates, 2005-7.

Table 9.5   Demographic Characteristics of the Foreign-Born Population, Selected Canadian CMAs, 2006

	Toronto (CMA)		Vancouver (CMA)		Montreal (CMA)		Ottawa–Gatineau (CMA)		Calgary (CMA)	
	Estimate	%	Estimate	%	Estimate	%	Estimate	%	Estimate	%
Total Population	5,072,075	100	2,097,965	100	3,588,520	100	1,117,120	100	1,070,295	100
Canadian-born (1)	2,675,590	52.8	1,227,495	58.5	2,806,230	78.2	905,745	81.1	805,645	75.3
Foreign-born (2)	2,396,485	47.2	870,470	41.5	782,285	21.8	211,375	18.9	264,650	24.7
Citizenship										
Foreign-born	2,396,485	100	870,470	100	782,285	100	211,375	100	264,650	100
Naturalized (3)	1,754,355	73.2	632,325	72.6	544,565	69.6	161,720	76.5	178,370	67.4
Non-citizen (4)	642,130	26.8	238,145	27.4	237,720	30.4	49,655	23.5	86,280	32.6
Period of Entry										
Immigrants (5)	2,320,155	100	831,265	100	740,355	100	202,730	100	252,760	100
Entered 2001 or later	447,925	19.3	151,695	18.2	165,345	22.3	35,085	17.3	57,940	22.9
Entered 1991 to 2000	720,185	31.0	290,830	35.0	190,570	25.7	60,000	29.6	67,365	26.7
Entered before 1991	1,152,045	49.7	388,740	46.8	384,440	51.9	107,645	53.1	127,455	50.4
World Region of Birth										
Foreign-born	2,396,485	100	870,470	100	782,285	100	211,375	100	264,645	100
Europe	704,235	29.4	186,525	21.4	265,650	34.0	69,905	33.1	77,545	29.3
Asia and the Middle East	1,172,140	48.9	569,810	65.5	229,270	29.3	82,855	39.2	132,475	50.1
Africa	122,645	5.1	28,020	3.2	115,935	14.8	24,200	11.4	18,785	7.1
Oceania	7,500	0.3	24,810	2.9	1,020	0.1	1,010	0.5	3,580	1.4
Latin America	164,025	6.8	25,930	3.0	70,370	9.0	11,335	5.4	14,230	5.4
Caribbean and Bermuda	177,450	7.4	6,245	0.7	78,690	10.1	12,310	5.8	5,285	2.0
Northern America	46,580	1.9	28,575	3.3	20,340	2.6	9,595	4.5	12,535	4.7
Educational Attainment										
Foreign-born 25 yrs and over	2,012,195	100	723,885	100	655,900	100	174,765	100	221,320	100
No certificate, diploma, or degree	398,610	19.8	123,295	18.8	148,015	20.4	23,845	13.6	38,975	17.6
University degree (6)	583,720	29.0	221,760	30.6	192,905	29.4	71,780	41.1	38,975	32.4

	Toronto (CMA)		Vancouver (CMA)		Montreal (CMA)		Ottawa–Gatineau (CMA)		Calgary (CMA)	
	Estimate	%	Estimate	%	Estimate	%	Estimate	%	Estimate	%
TLanguage Spoken at Home										
Foreign-born 5 yrs and over	2,378,435	100	864,820	100	773,410	100	209,345	100	261,925	100
English only	745,590	31.3	233,985	27.1	81,365	10.5	65,590	31.8	92,575	35.3
French only	4,035	0.2	1,205	0.1	119,000	15.4	8,860	4.2	315	0.1
Occupation										
Foreign-born labour force										
15 yrs and over	1,458,865	100	505,370	100	446,685	100	127,145	100	173,160	100
Management occupations	140,470	9.6	54,530	10.8	42,705	9.6	12,405	9.8	16,675	9.6
Sales & service										
occupations	318,305	21.8	136,755	27.1	107,150	24.0	30,090	23.7	44,445	23.7
Family Income										
Immigrants economic										
families	814,075	100	243,985	100	274,135	100	74,030	100	88,590	100
Below Cnd$25,000 (7)	102,525	12.6	44,635	18.3	50,345	18.4	9,300	12.6	7,655	8.6

Notes: (1) referred to as 'non-immigrants' by Statistics Canada; (2) calculated as the total of the Statistics Canada categories of 'immigrants' and 'non-permanent residents'; (3) calculated as the difference between 'foreign born' (see note #2) and 'non-citizen' (see note #4); (4) referred to as 'not Canadian citizens' by Statistics Canada; (5) does not include 'non-permanent residents'; (6) includes bachelor's, graduate, and professional degrees; (7) $25,000 Canadian is an approximation of the Statistics Canada low-income cut-offs, which are based on family expenditure data.

Source: Statistics Canada, *Census of Canada, 2006*.

advanced educational degree and a professional job, and the fewest immigrants who live in poverty. With an immigrant population that is relatively rich in resources and skills, it is not surprising that San Francisco, at 61.6 per cent, also has the largest percentage of naturalized citizens among the five American cities we studied.

For the Canadian cities, the SES model can account for some, although not all, of the differences in the naturalization rates we reported (also see Table 9.5). In recent years, the province of Alberta has taken an active role in bringing large numbers of low-skilled and temporary workers to Calgary. Low-skilled workers tend to lack the skills and resources that facilitate naturalization, and temporary workers are not eligible for Canadian citizenship. These workers' characteristics help explain the relatively low naturalization rate in Calgary. The SES model, however, does a poor job explaining the naturalization rates in Ottawa and Montreal. The relatively high naturalization rate in Ottawa, the seat of the federal government, likely has to do with the fact that people with Canadian citizenship receive preference for federal government jobs. Changes in municipal boundaries, furthermore, help explain the relatively low naturalization rate in Montreal. In recent years, Montreal municipalities with large English-speaking or allophone populations (i.e., people whose first language is neither English nor French) have supported de-amalgamation from the city of Montreal. As a result, the central city now has a relatively large number of francophone immigrants, including Haitians, who tend to have low naturalization rates also because Haiti does not recognize dual citizenship (Simard, 2008).[6] The de-amalgamated areas, in contrast, have larger numbers of immigrants from South Asia, who tend to have higher naturalization rates. Shifts in municipal boundaries, therefore, help explain why the naturalization rate for the city of Montreal today is almost 4 per cent lower than that for the Montreal CMA.

The Role of Policy and Contextual Factors

Contemporary research also considers policy and contextual factors as explanations for immigrants' political incorporation. The availability of dual citizenship (Jones-Correa, 1998, 2001; Ramakrishnan, 2005), the nature of the naturalization process (North, 1987), characteristics of the home country (Yang, 1994), government support for immigrant communities (Bloemraad, 2006; Reitz, 2003), election rules (Jones-Correa, 2001; Wolfinger and Rosenstone, 1980), home country ties (Rogers, 2006), and immigrants' perceptions of discrimination in the host society (Portes and Curtis, 1987; Ramakrishnan, 2005) are examples of policy and contextual factors that affect immigrants' political incorporation. It is difficult to generalize about their effects, but research reveals that immigrants who make decisions about political participation in contexts amenable to such participation—e.g., easy naturalization requirements, permission of dual nationality, less costly voter registration rules, no purging of non-voters, more government support for immigrant communities—are more likely to acquire citizenship and participate in electoral politics.

Because the federal governments in the United States and Canada determine immigration and citizenship policies, this model for immigrants' political incorporation cannot account for variation in incorporation *within* American or Canadian cities. It does, though, shed light on differences in incorporation *between* American and Canadian cities. In comparison to the United States, Canada legally recognizes dual citizenship, has a naturalization process that is less costly, has an easier voter registration process, and provides more government support to immigrant communities through settlement assistance and an official policy of multiculturalism (Bloemraad, 2006). The overall effect of these differences is that immigrants in Canada have an easier time incorporating into certain aspects of Canadian political life than do comparably situated immigrants in the United States. As our data show, immigrants in Canadian cities are more likely to naturalize than immigrants in US cities. Also, certain immigrant groups in Canadian cities are more likely to vote than Canadian-born citizens, while immigrant turnout in US cities is consistently lower than turnout among US-born citizens.

The Role of Civic Organizations

Scholarship on the role of civic organizations in the incorporation process is making a resurgence, but is still in its infancy. Civic organizations can mediate between the immigrants and the larger political communities they are part of. They are especially important for non-citizen and poor immigrants, who rarely wield power and political influence on their own in the political system. The literature on the early European immigrants in the American context teaches us that civic organizations—including local political parties, churches, immigrant settlement houses, and ethnic mutual aid societies—provided these newcomers with the skills, information, and motivation that facilitated their participation in both formal and informal politics (e.g., Dahl, 1961; Sterne, 2001). Accounts on the role of settlement houses in the incorporation process add that civic organizations served as vehicles that raised the political visibility of the immigrant community and advocated for policies that substantively benefited disadvantaged immigrants in gateway cities such as Chicago and New York City (Davis, 1967; Trolander, 1987).

Local party organizations in the United States today, however, no longer serve the important role they once did. Jones-Correa (1998), Wong (2006), and Rogers (2006) document that local party organizations in New York City and Los Angeles have been passive in helping contemporary immigrants to acquire US citizenship, register to vote, and become members of the party's political clubs. Often, parties' passivity in mobilizing immigrants into the political process is a result of the non-partisan and non-competitive nature of local elections. Local parties do not need the immigrant vote to win an election and, therefore, have little incentive to invest time and resources in reaching out to new voters. In this context, the influence that immigrant-serving non-profit organizations, labour unions, and religious institutions (especially the Catholic Church) have on immigrants' political incorporation has become more notable and more consequential (Andersen and Cohen, 2005; de Graauw, 2008; Wong, 2006).

Although local political parties are virtually absent in Canadian cities, large settlement agencies play an important role in the incorporation process, but their work focuses mostly on service provision and not advocacy. Also, Canadian settlement agencies often are headed by immigrants and have majority immigrant staff, which put them in a position where they can provide an economic ladder and leadership opportunities to many immigrants. Their bureaucratic nature and close ties with government funders, however, make it more difficult for Canadian settlement agencies to serve as advocates for policy change or as agents of immigrant mobilization than is the case for similar non-governmental organizations in US cities. Finally, religious organizations are very active in the settlement process in Canadian cities, but they have received little scholarly attention and their contributions to immigrants' political incorporation are not yet well understood (Bisson, 2009).

CONCLUSION

It is difficult to make generalizations about whether political incorporation has advanced more, or less, in American or Canadian cities because successful political incorporation means slightly different things in the two countries, despite the fact that both are classic countries of immigration. Another reason is that American and Canadian cities vary in how they perform on the five incorporation indicators we documented. Naturalization rates, for example, are higher and the foreign-born enjoy better representation on city councils in Canadian cities. Yet, US cities perhaps provide immigrants with more opportunities for civic engagement and they appear to have taken greater steps to institutionalize their commitment to immigrant integration.

A US–Canada comparison also teaches us that the context of incorporation matters. American and Canadian cities provide different contexts in which immigrants engage in the political life of the receiving country. Differences in municipal boundaries, mayoral power, the historical roles of local political parties, and the sources of funding for immigrant-serving organizations affect immigrants' political incorporation. Yet, it should be noted that different city contexts have not prevented civic organizations from fulfilling important roles in immigrants' incorporation in both countries. We identified some key differences in the role that civic organizations play in American and Canadian cities, but there is an obvious need for additional research that analyzes the precise role that a range of civic organizations have in the incorporation process.

QUESTIONS FOR CRITICAL THOUGHT

1. What can American cities learn from the political incorporation experiences of immigrants in Canadian cities, and vice versa? What would a model of best political incorporation practices look like in American and Canadian cities?

2. Who should have responsibility for improving immigrant political incorporation in the United States and Canada—immigrants themselves, government (local, state/provincial, or federal), or non-governmental organizations?

3. In recent decades, political parties have declined in importance in many American cities, and they have never played a significant role in Canadian cities. What role, if any, should political parties have in incorporating immigrants into aspects of public life?

SUGGESTED READINGS

1. Biles, John, Meyer Burstein, and James Frideres, eds. 2008. *Immigration and Integration in Canada in the Twenty-first Century*. Montreal and Kingston: McGill-Queen's University Press. This edited collection examines various dimensions of immigrant integration and describes the evolution of recent immigration policies across Canada.

2. Bloemraad, Irene. 2006. *Becoming a Citizen: Incorporating Immigrants and Refugees in the United States and Canada*. Berkeley: University of California Press. This book compares the civic and political incorporation in the United States and Canada, with a focus on Portuguese immigrants and Vietnamese refugees in Boston and Toronto.

3. Elections Canada. 2006. 'Electoral Participation of Ethnocultural Communities', *Electoral Insight* 8, 2. At: www.elections.ca/res/eim/pdf/ Insight_2006_12_e.pdf. This edited collection, produced by Elections Canada, analyzes the results of Statistics Canada's Ethnic Diversity Survey and focuses on immigrant voting patterns in Canada.

4. Kasinitz, Philip, John H. Mollenkopf, Mary C. Waters, and Jennifer Holdaway. 2008. *Inheriting the City: The Children of Immigrants Come of Age*. Cambridge, Mass.: Harvard University Press. The authors examine the integration of second-generation immigrants, with a focus on the offspring of Dominican, West Indian, South American, Chinese, and Russian-Jewish immigrants in metropolitan New York.

NOTES

1. We use the term 'political incorporation' (rather than 'political integration') because this is the standard terminology in the North American political science literature to describe the process through which immigrants, over time, become part of the political systems of the United States and Canada. The terms 'political incorporation' and 'political integration', however, are interchangeable and denote the same idea.

2. Republicanism refers to the political value system that has been dominant in the United States since the American Revolution. It emphasizes liberty (or individual freedom) and natural rights and vests political power in the people rather than inherited rulers.

3. For a related discussion on incorporation indicators at the national level in the United States, see de Graauw (2011, forthcoming).

4. Columbia University and the City University of New York (in collaboration with the New York Immigration Coalition; see NYIC, 2008) conduct the New Americans Exit Poll and they have collected exit poll data on immigrant voters in New York City since 2000. The Leavey Center for the Study of Los Angeles at Loyola Marymount University (in collaboration with the Central American Resource Center, the Coalition for Humane Immigrant Rights of Los Angeles, and the National Korean American Service and Education Consortium) has conducted the New Americans Exit Poll of immigrant voters in Los Angeles since 2005. Finally, the Asian American Legal Defense and Education Fund (AALDEF) has been collecting exit poll data on Asian-American voters since 1988. On election day in 2008, AALDEF surveyed 16,665 voters at 113 poll sites in 39 cities across 11 US states.

5. Very few studies consider substantive representation that is clearly *pro*-immigrant. Notable exceptions include Bloemraad (2006), de Graauw (2008), Jones-Correa (2008), and Marrow (2009). Various other studies, however, do consider substantive non-representation, or policies that are decidedly *anti*-immigrant. Included here are studies on immigrants' exclusion from policy benefits, such as California's Proposition 187—a 1994 initiative designed to prohibit undocumented immigrants from using social services, health care, and public education in California—and federal welfare reform in 1996. See, e.g., Kretsedemas and Aparicio (2004).

6. As discussed in the next section, evidence shows that dual citizenship laws encourage immigrants to naturalize. The fact that Haiti does not allow its emigrants to acquire the citizenship of another country while retaining their Haitian citizenship helps explain why naturalization rates are lower for the city of Montreal (where many Haitian immigrants live) than they are for Montreal CMA.

REFERENCES

1. Andersen, Kristi, and Elizabeth F. Cohen. 2005. 'Political Institutions and Incorporation of Immigrants', in Christina Wolbrecht and Rodney E. Hero, eds, *The Politics of Democratic Inclusion*. Philadelphia: Temple University Press, 186–205.

2. Andrew, Caroline. 2007. 'La gestion de la complexité urbaine', *Téléscope* 13, 3: 60–7.

3. ———, John Biles, Myer Siemiatycki, and Erin Tolley. 2008. *Electing a Diverse Canada: The Representation of Immigrants, Minorities, and Women*. Vancouver: University of British Columbia Press.

4. Bass, Loretta E., and Lynne M. Casper. 2001. 'Differences in Registering and Voting between Native-born and Naturalized Americans', *Population Research and Policy Review* 20: 483–511.

5. Bernstein, Nina. 2006. 'In the Streets, Suddenly, an Immigrant Groundswell', *New York Times*, 27 Mar., A-14.

6. Biles, John. 2008. 'Integration Policies in English-Speaking Canada', in John Biles, Meyer Burstein, and James Frideres, eds,

Immigration and Integration in Canada in the Twenty-first Century. Montreal and Kingston: McGill-Queen's University Press, 139–86.

7. Bisson, Ronald. 2009. *État des lieux de l'immigration d'expression française à Ottawa*. Ottawa: Ronald Bisson and Associates.

8. Bloemraad, Irene. 2006. *Becoming a Citizen: Incorporating Immigrants and Refugees in the United States and Canada*. Berkeley: University of California Press.

9. ———, Kim Voss, and Taeku Lee. 2011. 'The Immigration Rallies of 2006: What Were They, How Do We Understand Them, Where Do We Go?', in Kim Voss and Irene Bloemraad, eds, *Rallying for Immigrant Rights*. Berkeley: University of California Press.

10. Brubaker, William R. 1989. 'Citizenship and Naturalization: Policies and Politics', in William R. Brubaker, ed., *Immigration and the Politics of Citizenship in Europe and North America*. Lanham, Md: University Press of America, 99–127.

11. Cain, Bruce E., D. Roderick Kiewiet, and Carole J. Uhlaner. 1991. 'The Acquisition of Partisanship by Latinos and Asian Americans', *American Journal of Political Science* 35: 390–422.

12. Central American Resource Center in Los Angeles (CARECEN). 2006. *CARECEN Notes* 23, 4: 3.

13. City for All Women Initiative. 2010. 'Equity and Inclusion Lens'. At: www.cawi-ivtf.org.

14. Cordero-Guzmán, Héctor R., Nina Martin, Victoria Quiroz-Becerra, and Nik Theodore. 2008. 'Voting with Their Feet: Nonprofit Organizations and Immigrant Mobilization', *American Behavior Scientist* 52, 4: 598–617.

15. Dahl, Robert. 1961. *Who Governs? Democracy and Power in an American City*. New Haven: Yale University Press.

16. Davis, Allen F. 1967. *Spearheads for Reform: The Social Settlements and the Progressive Movement, 1890–1914*. New York: Oxford University Press.

17. de Graauw, Els. 2008. 'Nonprofit Organizations: Agents of Immigrant Political Incorporation in Urban America', in S. Karthick Ramakrishnan and Irene Bloemraad, eds, *Civic Hopes and Political Realities: Immigrants, Community Organizations, and Political Engagement*. New York: Russell Sage Foundation, 323–50.

18. ———. 2011 (forthcoming). 'Thematic Essay: Immigrant Political Incorporation in the United States', in Elliott Barkan, ed., *An Encyclopedia of U.S. Immigration History*. Santa Barbara, Calif.: ABC-CLIO.

19. DeSipio, Louis. 1996. *Counting on the Latino Vote: Latinos as a New Electorate*. Charlottesville: University of Virginia Press.

20. Graham, Katherine A.H., and Susan D. Phillips. 2007. 'Another Fine Balance: Managing Diversity in Canadian Cities', in Keith G. Banting, Thomas J. Courchene, and F. Leslie Seidle, eds, *Belonging? Diversity, Recognition and Shared Citizenship in Canada*. Montreal: Institute for Research on Public Policy, 155–94.

21. Hung, Chi-Kan Richard. 2007. 'Immigrant Nonprofit Organizations in US Metropolitan Areas', *Nonprofit and Voluntary Sector Quarterly* 36, 4: 707–29.

22. Johnson, Hans P., Belinda I. Reyes, Laura Mameesh, and Elisa Barbour. 1999. *Taking the Oath: An Analysis of Naturalization in California and the United States*. San Francisco: Public Policy Institute of California.

23. Jones-Correa, Michael. 1998. *Between Two Nations: The Political Predicament of Latinos in New York City*. Ithaca, NY: Cornell University Press.

24. ———. 2001. 'Institutional and Contextual Factors in Immigrant Naturalization and Voting', *Citizenship Studies* 5, 1: 41–56.

25. ———. 2008. 'Immigrant Incorporation in Suburbia: The Role of Bureaucratic Norms in Education', in Douglas S. Massey, ed., *New Faces in New Places: The Changing Geography of American Immigration*. New York: Russell Sage Foundation, 308–40.

26. Kretsedemas, Philip, and Ana Aparicio. 2004. *Immigrants, Welfare Reform, and the Poverty of Policy*. Westport, Conn.: Praeger.

27. Marrow, Helen B. 2009. 'Immigrant Bureaucratic Incorporation: The Dual Roles of Government Policies and Professional Missions', *American Sociological Review* 74, 5: 756–76.

28. Maytree Foundation. 2007. *Diversity in Governance: A Toolkit for Inclusion on Nonprofit Boards*. Toronto: Maytree Foundation.

29. New York Immigration Coalition (NYIC). 2008. 'New Americans Exit Poll Project: Preliminary Results from the 2008 New York City Voter Exit Poll'. At: www.urbanresearch.org/resources/docs/NYIC_2008_ExitPoll.pdf.

30. ———. 2009. *Annual Report 2008–2009*.

31. North, David S. 1987. 'The Long Grey Welcome: A Study of the American Naturalization Program', *International Migration Review* 21, 2: 311–26.

32. Oliver, J. Eric. 2001. *Democracy in Suburbia*. Princeton, NJ: Princeton University Press.

33. Passel, Jeffrey S., and D'Vera Cohn. 2009. *A Portrait of Unauthorized Immigrants in the United States*. Washington: Pew Hispanic Center.

34. Portes, Alejandro, and John W. Curtis. 1987. 'Changing Flags: Naturalization and Its Determinants among Mexican Immigrants', *International Migration Review* 21, 2: 352–71.

35. Ramakrishnan, S. Karthick. 2005. *Democracy in Immigrant America: Changing Demographics and Political Participation*. Stanford, Calif.: Stanford University Press.

36. ——— and Irene Bloemraad. 2008.

'Introduction: Civic and Political Inequalities', in S. Karthick Ramakrishnan and Irene Bloemraad, eds, *Civic Hopes and Political Realities: Immigrants, Community Organizations, and Political Engagement*. New York: Russell Sage Foundation, 1–42.

37. —— and Thomas J. Espenshade. 2001. 'Immigrant Incorporation and Political Participation in the United States', *International Migration Review* 35, 3: 870–907.

38. ——and Paul Lewis. 2005. *Immigrants and Local Governance: The View from City Hall*. San Francisco: Public Policy Institute of California.

39. Reitz, Jeffrey G. 2003. *Host Societies and the Reception of Immigrants*. La Jolla, Calif.: Center for Comparative Immigration Studies.

40. Rogers, Reuel R. 2006. *Afro-Caribbean Immigrants and the Politics of Incorporation: Ethnicity, Exception, or Exit*. New York: Cambridge University Press.

41. Santos, Fernanda. 2008. 'Upheaval among New York's Voting Blocs', *New York Times*, 13 July, A-25.

42. Scott, Katherine, Kevin Selbee, and Paul Reed. 2006. *Making Connections: Social and Civic Engagement among Canadian Immigrants*. Ottawa: Canadian Council on Social Development.

43. Siemiatycki, Myer. 2009. 'Urban Citizenship for Immigrant Cities: One Resident, One Vote in Municipal Elections', *Plan Canada* 49 (special issue): 75–8.

44. Simard, Carolle. 2008. 'Political Representation of Minorities in the City of Montréal: Dream or Reality?', in Andrew et al. (2008: 70–91).

45. Statistics Canada. 2001, 2006. *Census of Canada*. Ottawa: Statistics Canada.

46. —— and Department of Canadian Heritage. 2003. *Ethnic Diversity Survey: Portrait of a Multicultural Society*. Ottawa: Statistics Canada.

47. Sterne, Evelyn S. 2001. 'Beyond the Boss: Immigration and American Political Culture from 1880 to 1940', in Gary Gerstle and John Mollenkopf, eds, *E Pluribus Unum?*

Contemporary and Historical Perspectives on Immigrant Political Incorporation. New York: Russell Sage Foundation, 33–66.

48. Tam Cho, Wendy. 1999. 'Naturalization, Socialization, Participation: Immigrants and (Non)Voting', *Journal of Politics* 61: 1140–55.

49. Tossutti, Livianna. 2003. 'Does Volunteering Increase the Political Engagement of Young Newcomers? Assessing the Potential of Individual and Group-based Forms of Unpaid Services', *Canadian Ethnic Studies* 35, 3: 70–84.

50. Trolander, Judith Ann. 1987. *Professionalism and Social Change: From the Settlement House Movement to Neighborhood Centers, 1886 to the Present*. New York: Columbia University Press.

51. US Census Bureau. 1996–2008. *Current Population Survey: November Voting and Registration Supplement* Washington: US Census Bureau.

52. ——. 2009. '2005–7 American Community Survey Three-Year Estimates, Tables S0501 and S0502', *American Factfinder*. Washington: US Census Bureau.

53. Verba, Sidney, and Norman H. Nie. 1972. *Participation in America: Political Democracy and Social Equality*. New York: Harper & Row.

54. ——, Kay Lehman Schlozman, and Henry E. Brady. 1995. *Voice and Equality: Civic Voluntarism in American Politics*. Cambridge, Mass.: Harvard University Press.

55. White, Stephen, Neil Nevitte, André Blais, Joanna Everitt, Patrick Fournier, and Elisabeth Gidengil. 2006. 'Making Up for Lost Time: Immigrant Voter Turnout in Canada', *Electoral Insight* 8, 2: 10–16.

56. Wolfinger, Raymond, and Steven J. Rosenstone. 1980. *Who Votes?* New Haven: Yale University Press.

57. Wong, Janelle S. 2006. *Democracy's Promise: Immigrants and American Civic Institutions*. Ann Arbor: University of Michigan Press.

58. Yang, Philip Q. 1994. 'Explaining Immigrant Naturalization', *International Migration Review* 28, 3: 449–77.

PART III

IMMIGRANT GROUPS IN NORTH AMERICAN
CITIES AND SUBURBS

CHAPTER 10

CONTEMPORARY ASIAN IMMIGRANTS IN THE UNITED STATES AND CANADA

Shuguang Wang and Qingfang Wang

Of the 6.9 billion people on earth, 60 per cent live in Asia. In the past century or so, millions of Asians have left their homelands and have migrated to other continents to escape poverty, natural disasters, war, and political turmoil. Today, hundreds of thousands of Asians are still on the move for the same reasons, but also for the economic opportunities afforded to them by the globalization of the world economy. Over the last two decades, the number of Asian immigrants[1] to North America has increased substantially. They have become the largest immigrant group by continent of origin in Canada and the second largest group in the US, next only to those from Latin America. Besides their sheer numbers, the contemporary Asian immigrants possess much greater financial and human capital than their predecessors, and they have made significant imprints on the social, economic, and cultural fabric of North American cities.

Although they come from the same region, Asian immigrants consist of many diverse groups. Furthermore, notable differences exist between the Asian immigrants in the United States and their counterparts in Canada. This chapter is devoted to a comparative analysis of foreign-born Asian groups in the US and Canada in four aspects: the changing patterns of Asian immigration by country of origin; their socio-economic characteristics; their labour market performance; and trends of settlement pattern changes in major US and Canadian cities.

CHANGING PATTERNS BY ORIGIN

Historically, Asian immigration to the US and Canada is closely linked to the economic and geopolitical conditions in the countries of origin, but also is attributed to the changes in the immigration policies of the two countries of destination. For example, Asian immigration to both countries started in the mid-nineteenth century, when large numbers of Chinese labourers were brought in to work in mining and railroad construction. Due to discriminatory immigration laws, however, Asian immigration was very restricted until such laws were repealed in the 1960s, when Asian immigration to the US and Canada began to expand, with significant increases occurring after the early 1990s.

According to the US census, the number of foreign-born Asians has changed from less than half a million in 1960 (5.1 per cent of the total foreign-born) to 7.2 million in 2000 (Gibson and Lennon, 1999; US Bureau of the Census, 2002a). The 2005–7 American Community Survey indicates there are 10 million foreign-born Asians in the US, representing 27 per cent of the country's total immigrants. While they constitute the second largest group of immigrants in the US, they are much smaller in both absolute number and percentage than the immigrants from Latin America (mainly Mexico), who represent 54 per cent of all immigrants in the country.

The total number of Asian immigrants in Canada is much smaller than that in the US; yet they constitute the county's largest immigrant group. The number of Asian immigrants in Canada more than doubled from 1.1 million in 1991 to 2.5 million in 2006. In 1991, Asian immigrants accounted for only 25 per cent of the total immigrants in Canada; by 2006, their representation increased to 41 per cent.

Asian immigrants are far from homogeneous, of course. They consist of many groups that differ by country of origin, language, culture, religion, and type of human capital. Figure 10.1 shows the top 15 groups of Asian immigrants by country of birth in both the US and Canada since the 1990s.

Over the last two decades, the five largest groups in the US consistently have been (mainland) Chinese, Filipinos, Indians, Vietnamese, and Koreans (with Indians having surpassed Vietnamese and Koreans since 1990). Notably, immigrants from the Middle East—particularly Iran, Israel, and Lebanon—have increased since 1990; and those from Pakistan and Bangladesh in South Asia also have been on the rise.

In Canada, the top five groups of Asian immigrants came from mainland China, India, the Philippines, Hong Kong, and Vietnam, with four of these groups being the same as in the US. Immigrants from Hong Kong constitute the fourth largest group in Canada (compared with the ninth rank in the US), but their number has dwindled since 2001 (see Figure 10.1). As in the US, immigrants from the Middle East and such South Asian countries as Pakistan, Bangladesh, and Sri Lanka have also increased since the 1990s.

We explain the huge influx of Asian immigration to North America and the differences between the US and Canada as an outcome of mixed and cumulative

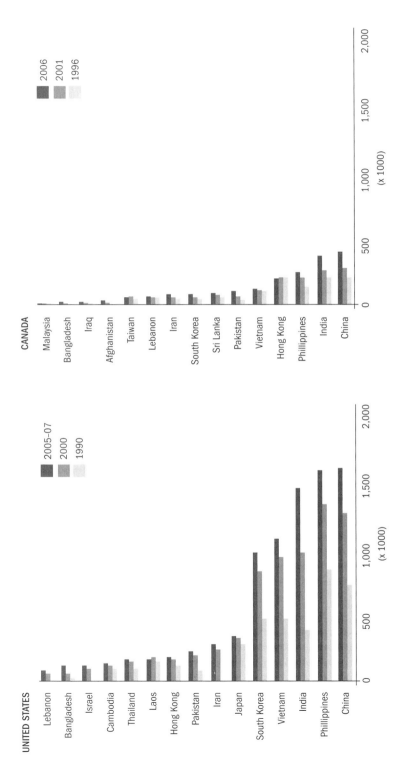

Figure 10.1 Changes in Number of Asian Immigrants in the United States and Canada

Note: Only the top 15 countries/regions of birth are shown.

causes that have played out at the global and national levels, as well as local and personal or household levels. On the one hand, the lack of opportunities and freedom, the frequent occurrences of natural disasters, and prolonged wars in many Asian countries have led to personal dissatisfaction among many Asians, which motivated them to leave their homelands for a better life elsewhere. On the other hand, the much higher standard of living, together with better economic opportunities (or prospects) and the much appreciated political freedom in the US and Canada, has made North America a 'dream land' for most Asian emigrants. In addition, the more open and favourable immigration policies have made the US and Canada much more attractive destinations than Western Europe.

The past political and even military connections between the US and the various Asian countries have shaped the structure of Asian immigration to the US. For example, the US colonization and military presence in the Philippines and the US military presence in South Korea have contributed to the mass migration from these two countries to the US. Likewise, the political connections between the US and South Vietnam and the American military involvement in the Vietnam War were responsible for the high volumes and types of Vietnamese refugees/immigrants to the US over the past three decades. These connections account for the ranking of the Philippines, South Korea, and Vietnam among the top five countries of origin of Asian immigrants to the US (Min, 2006).

Canada has no colonial connections with the Philippines, South Korea, and Vietnam, although it participated in the US-led Korean War in the early 1950s; however, it fulfilled its international obligations in the early 1980s to accept and settle a significant number of refugees from Vietnam (and Cambodia) and, subsequently, their family members and close relatives. As Figure 10.1 shows, immigration from Vietnam has increased very little since the mid-1990s, reflecting the restoration of political stability there and the resultant decrease in outflow of refugees. In contrast, immigrants from the Middle East to Canada have increased due to the destructive wars in the region.

As the landing data reveal, a large number of immigrants from the Philippines were admitted to Canada as live-in caregivers (96 per cent of the total in this category of immigrants to Canada), and most of the Korean immigrants were accepted as independent economic immigrants and entrepreneurs, benefiting from Canada's points system and Business Immigrant Program. Oh (2008) found that, when asked during an interview why they chose Canada as their destination of migration, all participating Korean immigrants answered that it was simply because the process of immigration to Canada was easier compared with other countries. Most of them also mentioned that they first considered the US, but discovered that it was much more difficult due to the longer waiting period and stricter conditions of the immigration process in the US.

The sudden surge of contemporary Chinese immigrants to the US and Canada in the 1990s was first prompted by the suppression of the student protest in Beijing's Tiananmen Square by the mainland Chinese government in 1989; it was later facilitated by the relaxation of the exit policies governing the emigration of Chinese nationals. The uncertainty of Hong Kong's return from a British colony to Chinese rule in 1997 served as a push factor for the large exodus of Hong Kong citizens in the late 1980s and the early 1990s. For the reasons above, the US and, especially, Canada were viewed as preferred destinations. The fact that more Hong Kong immigrants came to Canada than to the US is also attributed to Canada's favourable Business Immigrant Program as well as to the Commonwealth connection (Wang and Lo, 2005). The landing data show that 89 per cent of Hong Kong immigrants who were admitted to Canada in the 26 years from 1980 to 2005 arrived between 1984 (when China and the United Kingdom signed the agreement) and 1997 (when Hong Kong was officially returned to China's sovereignty), and that 26 per cent were admitted as entrepreneurs and investors.

While many Indian immigrants came to the US to pursue higher education or employment opportunities under the H-1B Temporary Visa Program, people from other countries of South Asia came for different reasons. The number of Pakistani and especially Bangladeshi immigrants was relatively low before the 1990s. When employment opportunities in the oil-producing countries of the Middle East declined in the early 1990s, many Pakistanis and Bangladeshis had to look for economic opportunities elsewhere (Kibria, 2006), and subsequently benefited from the Diversity Immigrants Program (commonly known as the **Green Card** Lottery Program) instituted in the American Immigration Act of 1990. The pattern of contemporary Asian immigration to Canada is similar to that of the US. As Singh and Thomas (2004: 2) have noted, 'With limited opportunities for permanent residence coupled with political instability and insecurity in the Middle East, many South Asians are immigrating to Canada. Canada has become the country of choice of South Asians [also] because of a shorter immigration-processing period.' In sum, the above examples illustrate multi-scaled, cumulative, yet differentiated explanations for the large inflows of Asian immigrants to the US and Canada.

CHARACTERISTICS OF CONTEMPORARY ASIAN IMMIGRANTS

The multi-scaled, cumulative process of migration has impacted not only the total volumes but also the composition and socio-economic characteristics of Asian immigrants in both the US and Canada. Among all immigrants (16 years and older) in the US, Asian immigrants seem to possess higher-than-average human capital. Seventy-nine per cent have good spoken English (see Table 10.1), almost 10 per cent higher than the general immigrant population; whereas linguistically isolated households (i.e., non-English-speaking households) of Asian immigrants are 5.6 per cent lower. As well, 44.5 per cent of Asian immigrants (25 years of age and over) possess a bachelor's degree or higher, compared with only 24.4 per cent for the general immigrant population and 25.8 per cent for the total adult population born in the US. Asian immigrants have higher personal incomes than both the general immigrant population and the total US-born population ($35,800 vs $28,400 and $35,100,

Table 10.1 Socio-economic Characteristics of Asian Immigrants in the US, 2005–7, Ages 15+

Socio-economic Characteristics	Non-Immigrants	All Immigrants	Asia	China	Hong Kong	India	Philippines	Korea	Vietnam
Total population 15 years and over (000s)	148,620	36,821	9,677	1,235	210	1,410	1,662	959	1,081
Period of immigration (% of total)									
Before 1996 (>10 years)		65.9	67.2	60.0	79.8	49.0	73.3	69.8	79.9
1996 to 2006 (<10 years)		34.1	32.8	40.0	20.2	51.0	26.7	30.2	20.1
Speak English well or very well (%)	99.4	70.0	79.0	58.4	85.3	90.4	93.1	71.8	61.6
Language-isolated households (%)	0.7	28.8	23.2	41.7	22.3	10.7	10.0	32.5	39.3
With bachelor's or higher degree, 25 years and over (%)	25.8	24.4	44.5	42.1	49.1	70.1	44.9	45.4	22.2
Average personal income ($)	35,100	28,400	35,800	31,200	46,000	49,000	34,200	31,600	28,400
Population at or below poverty line (%)	9.9	14.5	10.4	13.4	8.8	6.4	5.1	11.3	12.7
Population with government welfare income (%)	1.4 / 1.3	1.2	1.1	1.2	0.4	0.5	0.8	0.9	2.0

Source: American Community Survey, 2005–7, three-year sample microdata (US Bureau of the Census, 2010), which is 3-in-100 national random sample of the population.

respectively) and they constitute a lower proportion at or below the poverty line (10 per cent vs 15 per cent). They are also less likely to receive government welfare.

Like their counterparts in the US, Asian immigrants in Canada possess above-average human capital as measured by higher education attainment.[2] In particular, 32 per cent of the Asian immigrants have university educations, compared with 25 per cent for the general immigrant population and only 16 per cent for the Canadian-born population (see Table 10.2), although more of them obtained their higher education outside of Canada prior to their immigration (62 per cent vs 54 per cent for the general immigrant population). A higher percentage of Asian immigrants have higher English proficiency than do members of the general immigrant population (although fewer of them speak French than the immigrants from the other regions). Despite their higher education credentials, however, Asian immigrants have much lower personal income not only than the Canadian-born but also than the general immigrant population ($27,900 vs $36,200 and $33,400, respectively). Accordingly, the prevalence of low-income families among Asian immigrants is much higher: 26 per cent vs 19 per cent for the general immigrant population and 10 per cent for the Canadian-born. Still, the proportion receiving government transfers[3] is slightly lower than for the general immigrant population (see Table 10.2).

Significant subgroup disparities are noted. Of the six major groups of Asian immigrants in the US,[4] those from India, Hong Kong, and the Philippines possess more human capital than the others, and their personal income is higher as well. Specifically, Indian immigrants enjoy the highest personal annual income among all Asian groups, despite their relative newness in the country, 51 per cent having immigrated to the US after 1995. This achievement reflects their higher human capital; for example, 90 per cent speak English well, and 70 per cent have at least a bachelor's degree.

The Philippine immigrants have the highest English proficiency, with 93 per cent speaking English well. Compared with the Indian and Hong Kong immigrants, however, a lower proportion have university degrees (only 45 per cent, compared with 70 per cent and 49 per cent for the Indian and Hong Kong immigrants, respectively); correlatively, their average personal income is much lower ($34,200 for Filipinos vs $49,000 for the Indians and $46,000 for the Hong Kong immigrants). Compared with Hong Kong and Indian immigrants, the proportion of Philippine immigrants who are at or below the poverty line as defined by the US federal government is also lower (at 5.1 per cent); however, the proportion receiving government welfare is higher. The socio-economic characteristics of the Philippine immigrants are closely related to the history of immigration from the Philippines over the past several decades. The Philippines became a US colony in 1898 and did not gain full independence from the US until 1946. During the American colonization period, the US influence permeated not only the Philippines economy, but also its political, social, and cultural systems, including the use of English as an official language. The more recent employment-based immigration policy of the US has further increased the

Table 10.2 Socio-economic Characteristics of Asian Immigrants in Canada, 2006, Ages 15+

Socio-economic Characteristics	Non-Immigrants	All Immigrants	Asia	China	Hong Kong	Other E. Asia (S. Korea, Japan)*	Philippines	India	Other S. Asia (Pakistan, Bangladesh, Sri Lanka)
Total population 15 years and over (000s)	19,592	5,841	2,340	428	209	175	282	419	243
Period of immigration (% of total)									
Before 1996 (>10 years)		72.1	58.0	47.2	82.6	49.0	60.9	53.5	42.4
1996 to 2006 (<10 years)		27.9	42.0	52.8	17.4	51.0	39.1	46.5	57.5
Knowledge of official languages (%)									
English only	65.1	77.3	80.2	67.7	90.0	85.0	96.1	84.6	88.4
French only	15.7	3.9	1.4	1.6	0.1	0.6	0.1	0.1	0.3
English and French	18.8	12.3	7.1	3.6	3.7	4.6	2.9	3.3	5.7
Neither Eng. nor French	0.4	6.4	11.3	27.1	6.3	9.8	0.9	12.1	5.6
Persons 15 years and over with university certificate, diploma, or degree, bachelor's level or above (%)	15.8	25.4	32.2	35.3	32.6	40.6	35.7	33.4	31.0
Persons 15 years and over with post-secondary certificate, diploma, or degree obtained outside Canada (%)	2.1	53.5	62.1	71.3	32.5	66.4	73.2	72.8	68.9
Average income for persons 15 years and over ($)	36,200	33,400	27,900	24,400	34,600	23,300	30,400	30,200	23,900
Prevalence of low income before tax in 2005 for economic family members (%)	9.8	19.3	26.3	30.3	21.9	42.3	12.1	16.2	37.2
Government transfer payments (%)	10.5	13.1	11.1	15.1	4.9	9.9	8.4	11.1	14.1

*According to the landing records (Citizenship and Immigration Canada, 2006), a total of 114,215 immigrants came to Canada from South Korea, North Korea, and Japan (i.e., Other East Asia) between 1980 and 2005. Of this total, 82.6 per cent came from South Korea; 17.2 per cent came from Japan, and only 0.2 per cent came from North Korea.

Source: Statistics Canada (2009).

number of immigrant professionals (especially in the field of health care) from the Philippines (Batalova and Lowell, 2007).

In the US, the immigrants from mainland China, Korea, and Vietnam all possess lower human capital than do Asian immigrants in general. In particular, the proportion of linguistically isolated households ranges from 33 per cent for Korean immigrants to 42 per cent for the mainland Chinese and 39 per cent for the Vietnamese. They also have lower-than-average personal income (see Table 10.1). Compared with immigrants from Hong Kong, the majority of mainland Chinese came in recent years (after the mid-1990s), and they have lower levels of English proficiency, educational attainment, and personal income. In addition to having the lowest personal income among all the Asian groups, Vietnamese immigrants are the most likely to depend on government welfare. The US–Vietnam diplomatic and economic relations were essentially frozen for more than a decade after the 1975 Communist victory in South Vietnam. Re-establishment of relations took major steps forward in the mid-1990s when the two countries reopened embassies in each other's capitals (Manyin, 2005). The level of Vietnamese immigration was very low until the mid-1970s when refugees began to arrive in the US following the 'fall of Saigon', and reached its highest point in the early 1990s. The special status of the Vietnamese refugees/immigrants distinguishes their socio-economic characteristics from the other Asian immigrants.

In Canada, natives of Hong Kong, 82 per cent of whom immigrated to Canada before 1996, comprise the best-established group of contemporary Asian immigrants (Table 10.2). Ninety-four per cent possess knowledge of the Canadian official languages (90 per cent English and 4 per cent both English and French), which is 6 per cent higher than the total among Asian immigrants. While the percentage of Hong Kong immigrants with college or university education is nearly the lowest among the six major Asian groups (33 per cent), most of their degrees (68 per cent) were obtained within Canada—the least likely to be discounted and discriminated against in the Canadian labour market. As a result, the Hong Kong immigrants have the highest personal income ($34,600) and the lowest prevalence of low-income families, and are the least likely to receive government transfers.

The differences in socio-economic status between the mainland Chinese and the Chinese from Hong Kong are more striking in Canada than in the US. Although the immigrants from mainland China have exceeded those from Hong Kong in number, over half of them are recent immigrants, arriving in Canada after 1995 (Table 10.2). Among the six major groups of Asian immigrants, the mainland Chinese have the highest percentage without knowledge of a Canadian official language (27 per cent compared with 6 per cent for Hong Kong immigrants, 10 per cent for other East Asians, 1 per cent for Filipinos, 12 per cent for Indians, and 6 per cent for other South Asians). While 35 per cent of the mainland Chinese immigrants have a college or university education, the majority (71 per cent) of them obtained their higher education from outside of Canada, mainly from their native country of China (59 per cent). Accordingly, their personal income is 29 per cent lower than that of the Hong Kong immigrants, and 20 per cent lower than the average personal income of Filipinos and Indians. Unfortunately, their prevalence of low-income family members (30 per cent) and government assistance recipients are both higher than most other Asian groups (except other Southeast Asians and other South Asians), and the percentage of them receiving government transfers is the highest, at 15 per cent.

Compared with mainland Chinese immigrants, a higher proportion of immigrants from the 'other East Asian' countries (83 per cent of those from South Korea and 17 per cent from Japan) have knowledge of Canadian official languages (90 per cent vs 73 per cent) and college educations (41 per cent vs 35 per cent), and a higher percentage of the Korean and Japanese immigrants have received their post-secondary education within Canada (Table 10.2). Nonetheless, their average personal income is 5 per cent lower than that of those from mainland China, and their prevalence of low-income families is 12 per cent higher, although a lower percentage received government transfers, indicating that there is a wider income gap among the Korean (and the Japanese) immigrants than among the mainland Chinese.

Contemporary Philippine immigrants also have a longer length of residency in Canada, next only to the Hong Kong immigrants (with 60 per cent having lived in Canada for 10 or more years at the time of the 2006 census; see Table 10.2). Almost all Filipinos (99 per cent) possess knowledge of an official Canadian language. Their personal income is the second highest among the major groups of Asian immigrants ($30,400), next only to Hong Kong immigrants. Although 73 per cent of them received their college or university education in their home country, most of their education is likely to have been obtained with English as the language of instruction.

Of all South Asian immigrants, Indians have the longest length of residency in Canada, with 53 per cent arriving in Canada before 1996, compared with only 42 per cent of those from Pakistan, Bangladesh, and Sri Lanka (see Table 10.2). Despite their similarities in English proficiency and source of higher education, Indian immigrants have a 26 per cent higher personal income than the other South Asian immigrants. Accordingly, the prevalence of low-income families is 21 per cent lower, and the percentage receiving government transfers is 3 per cent lower. Immigrants from Pakistan, Bangladesh, and Sri Lanka have the least human capital as measured by college or university education; they also are the most economically disadvantaged, with the lowest personal income and the highest rate of dependency on government transfers.

To summarize, the Asian immigrants in the US and Canada are highly educated and are better educated than the general immigrant population and even than the US- and Canadian-born, although internal variations exist among the subgroups. Overall, Asian immigrants in the US are better educated than their counterparts in Canada, with 12 per cent more having university degrees. This gap may be attributable to the differences in the two countries' immigration policies. For example, the US requires that an applicant already have a job offer before he/she can be admitted; usually the job offer is made on the basis of the applicant's education qualifications, which lead to better job market outcomes. Canada is generous in admitting family members (according to the landing data), and its universal health-care system does not discourage elderly parents from immigrating to Canada.

LABOUR MARKET PERFORMANCE

The socio-economic characteristics of immigrants play a critical role in their adaptation to the receiving society. In this section, we look into the labour market performance of Asian immigrants in terms of their unemployment rates, the distribution of their employment in the various industries, and their work-related earnings.

In the US, immigrants as a whole have a lower unemployment rate than the US-born, 6 per cent vs 7 per cent (see Table 10.3). Asian immigrants have an even lower unemployment rate, at 5 per cent. Compared with the general immigrant population, Asian immigrants are much less likely to work in the primary industries and in utilities/construction/transportation/warehousing: only 0.4 per cent working in the former and 7 per cent in the latter categories, compared with 2 per cent and 16 per cent for the general immigrant population. Instead, they are more likely to work in finance, insurance, and real estate (FIRE), and in professional/technical services, which should have contributed to their higher employment income. Despite their higher levels of education, they still are less likely than their US-born counterparts to work in public administration.

Among the Asian groups in the US, consistent with their higher English proficiency and education attainment, the immigrants from India, Hong Kong, and the Philippines are more likely to work in FIRE and professional/technical services industries; over 50 per cent of the immigrants from these groups work in these sectors. Of the three groups, Hong Kong immigrants are more likely to work in FIRE than are Indian and Philippine immigrants (13 per cent vs 9 per cent and 8 per cent, respectively), whereas the latter two groups are more likely to work in professional/technical services sectors (49 per cent and 45 per cent, respectively, vs 36 per cent). The Indian and Hong Kong immigrants top the list of wage earnings among all Asian groups, although there are also more Hong Kong immigrants working in the low-paid retail/food/accommodation sectors (28 per cent vs 21 per cent of Indians). Filipinos have lower earned income than do Hong Kong immigrants; but both Hong Kong and Philippine immigrants have higher-than-average proportions working in the public administration sector (4 per cent vs the average of 3 per cent for all Asian immigrants).

Lower-than-average proportions of mainland China immigrants work in FIRE (7 per cent) and the professional/technical services industry (35 per cent). Instead, higher-than-average proportions work in retail/food/accommodation services (35 per cent) and manufacturing (15 per cent). As a result, nearly half (49 per cent) earn less than $30,000 from employment. Korean immigrants are similar to mainland Chinese immigrants, with an above-average proportion working in retail/food/accommodation services (44 per cent), and below-average

Table 10.3 Labour Market Performance of Asian Immigrant Labour Force* in the US, 2005–7 (%)

Employment Status and Industry	Non-Immigrants	All Immigrants	Asia	China	Hong Kong	India	Philippines	Korea	Vietnam
Unemployment rate	6.6	5.9	5.1	4.8	5.2	4.6	4.6	4.8	5.5
Self-employment rate	9.5	11.0	12.3	10.7	11.3	10.7	5.7	22.0	12.1
Agriculture, forestry, fishing and hunting, mining and oil and gas extraction	0.4	2.3	0.4	0.3	0.2	0.4	0.5	0.3	0.6
Utilities and construction, transportation and warehousing	4.5	15.6	6.8	6.2	7.7	5.5	7.0	6.7	6.3
Manufacturing	6.1	13.3	14.0	14.9	11.4	12.9	10.5	9.4	25.2
Wholesale and retail trade, accommodation and food services, other services (except public administration)	28.2	30.9	31.8	35.4	27.5	21.4	24.9	43.6	38.6
FIRE	9.5	5.8	7.6	6.5	13.1	9.1	8.0	6.8	5.3
Professional and technical services	46.5	29.9	36.5	34.8	36.1	48.7	44.7	30.8	21.3
Public administration	4.8	2.2	2.9	2.0	4.1	2.1	4.4	2.3	2.8
Average earned income from work ($)	34,500	38,000	49,900	45,200	59,000	65,800	43,500	47,900	38,600
Earned income <$30,000	53.3	56.8	42.3	48.6	32.6	30.7	42.5	43.4	51.5
Earned income $30,000–$49,999	25.8	21.1	22.2	18.7	20.9	16.9	26.3	24.2	24.0
Earned income >$50,000	20.9	22.1	35.5	32.7	46.5	52.3	31.2	32.4	24.6

Note: In the US, 'labour force' refers to persons 16+ years old. Except for unemployment rate, all other measurements are calculated for the civilian employed labour force.

Source: American Community Survey, 2005–7 three-year sample microdata (US Bureau of the Census, 2010), which is 3-in-100 national random sample of the population.

proportions working in FIRE (7 per cent) and in the professional/technical services industry (31 per cent). Nonetheless, their average employment income is higher than that of the mainland Chinese immigrants, even higher than the Philippine immigrants. The Vietnamese immigrants have the highest proportion working in manufacturing (25 per cent) and the second highest proportion (39 per cent) working in retail/food/accommodation services. They also have the lowest proportions working in FIRE and professional services. Accordingly, their employment income is the lowest ($38,600) of the studied groups, with 51 per cent making less than $30,000 a year.

In contrast with the US, both the general immigrant population and the Asian immigrants in Canada experience a higher unemployment than their US counterparts (see Tables 10.3 and 10.4). In Canada, Asian immigrants have a 1.5 per cent higher unemployment rate than non-immigrants. Compared with the general immigrant population, Asian immigrants are under-represented in utilities/construction/transportation/warehousing (by 2 per cent), professional services (by 4 per cent), and public administration (by 1 per cent). On the other hand, they are over-represented in manufacturing (by 2 per cent), and in wholesale/retail/accommodation/food (by 5 per cent). Overall, their employment income is 14 per cent lower than the general immigrant population, and 59 per cent of Asian immigrants made less than $30,000 from work, whereas 53 per cent of the general immigrant population and 52 per cent of non-immigrants earn less than $30,000. The higher percentage of low-income Asian immigrants probably is related to the fact that the largest proportion of them (33 per cent) work in the low-paid wholesale/retail/accommodation/food sector.

As in the US, Hong Kong immigrants in Canada have the highest proportion working in FIRE (15 per cent); they also have one of the highest proportions working in professional/technical services (33 per cent); accordingly, they make the highest employment income ($39,000) and, among all the Asian groups, have the highest proportion (28 per cent) making $50,000 or more annually. The mainland Chinese immigrants in Canada have the second highest unemployment rate (9 per cent) and the second highest percentage (36 per cent) working in wholesale/retail/accommodation/food services (next only to the Koreans). Their employment income is among the lowest, with 65 per cent earning less than $30,000 from work. Koreans are similar to the mainland Chinese in their unemployment rate, average employment income, and the percentage earning less than $30,000. With the highest percentage (45 per cent) working in wholesale/retail/accommodation/food services among all Asians, Koreans have the highest level of self-employment (26 per cent) among all Asian immigrants, an observation that has been confirmed in many other studies. Their average employment income ($28,800) and the percentage making less than $30,000 (66 per cent) are similar to the figures for mainland Chinese immigrants.

Immigrants from the Philippines enjoy the lowest unemployment rate (as low as 5 per cent) among all the Asian immigrants. As well, they have the highest percentage working in the professional services (36 per cent), likely due to their high level of English proficiency (see Table 10.2). Yet, their employment income

Table 10.4 Labour Market Performance of Asian Immigrants in Canada, 2006 (%)

Employment Status and Industry	Non-Immigrants	Immigrants	Asia	China	Hong Kong	Other E. Asia (S. Korea, Japan, N. Korea)	Philippines	India	Other S. Asia (Pakistan, Bangladesh, Sri Lanka)
Unemployment rate	6.4	6.9	7.9	8.5	5.8	8.5	4.9	7.8	10.7
Self-employment rate	11.2	14.3	13.1	14.3	14	26.2	4.5	12.7	10.5
Agriculture, forestry, fishing and hunting, mining, oil and gas extraction	5.1	2.2	1.7	1.3	0.8	1.1	0.8	3.9	0.7
Utilities, construction, transportation and warehousing	12.4	10.7	8.6	5.6	6.4	5.6	5.6	14.8	9
Manufacturing	10.9	15.6	17.7	18.1	10.5	7.3	18.5	20.9	18.7
Wholesale and retail trade, accommodation and food services, other services (except public administration)	27.1	27.9	32.6	35.8	31	44.7	29.6	25.1	33.3
FIRE	5.6	7.1	7.6	6.8	14.5	7.5	7.5	6.6	8
Professional and technical services	32.4	33.2	29.4	30.2	33.4	32	35.5	26.2	28.1
Public administration	6.5	3.3	2.4	2.3	3.3	1.9	2.3	2.4	2.3
Average employment income ($)	36,600	36,600	31,500	29,000	39,000	28,800	31,500	32,200	27,100
Employment income <$30,000	52.2	53.0	58.8	64.6	49.2	66.3	52.6	58.7	65.0
Employment income $30,000–$49,999	22.9	23.5	22.7	18.2	22.7	17.6	31.2	22.5	20.7
Employment income >$50,000	24.9	23.5	18.5	17.3	28.1	16.1	16.2	18.8	14.2

Source: Statistics Canada (2009).

is only about the average for all Asian immigrants (at $31,500), substantially lower than that of the Hong Kong immigrants. More than half (53 per cent) of Filipinos make less than $30,000 from employment.

Indian immigrants have a much higher unemployment rate (8 per cent) than the immigrants from Hong Kong (6 per cent) and the Philippines (5 per cent). This difference may be attributed to their shorter period of settlement in Canada (see Table 10.2). Notably, higher proportions of Indian immigrants work in the primary and secondary sectors than do the other groups of Asian immigrants. They are under-represented in FIRE and professional industries, despite their higher level of English proficiency. Their employment income and the distribution of Indian immigrants in different income brackets are similar to the averages for Asian immigrants. Other South Asian immigrants—those from Pakistan, Bangladesh, and Sri Lanka—as shown in Table 10.4, seem to be the lowest labour market performers. They have the highest unemployment rate (11 per cent), a higher-than-average proportion working in wholesale/retail/accommodation/food services, their employment income is the lowest, and they have the lowest proportion in the high income bracket (only 14 per cent, compared with the average of 19 per cent for all Asian immigrants).

Both in the US and in Canada, Asians are more likely than those in other ethnic minority groups to start and operate their own small businesses. As shown in Table 10.3, the self-employment rate of Asian immigrants in the US is higher than the rates for the general immigrant population and for the US-born labour force (12 per cent vs 11 per cent and 9 per cent, respectively). In Canada, the rate is slightly lower than for the general immigrant population but still higher than that for the Canadian-born. The US Census Bureau's *Survey of Business Owners* (US Bureau of the Census, 2002b) reveals that as of 2002, there were 1.1 million Asian-American-owned businesses, a growth of 24 per cent from 1997. Among all Asian immigrants, Koreans have the highest self-employment rate in both countries, and, in both countries, those from the Philippines have the lowest rate of self-employment. This difference may partially explain why Koreans in the US have higher work-related incomes than those of the Chinese and Philippine immigrants, whose self-employment rates are much lower (at 11 per cent and 6 per cent, respectively).

In sum, Asian immigrants in the US not only have a lower unemployment rate than the general immigrant population and the non-immigrant population, they also have a lower proportion making less than $30,000 in employment earnings and a much higher percentage in the highest income bracket. In contrast, Asian immigrants in Canada experience a higher unemployment rate than that of the general immigrant population and the Canadian-born. A much higher proportion of Asian Canadians make less than $30,000, with a much lower percentage in the highest income bracket. In terms of industrial distribution, the Asian immigrants in Canada are more likely than their counterparts in the US to work in primary and secondary industries, but they are much less likely to work in the professional and technical services sector, possibly because of differences in the two countries' economic structures (see Chapter 6). The US immigration requirement that a work visa applicant must have a job offer before being admitted to the country may also help to explain the higher percentage of Asian immigrants in the professional and technical services sector, who have high education qualifications. Canada no longer has such requirements for independent immigrants. In both countries, Asian immigrants are much less likely than the Canadian- or US-born to work in public administration, indicating clearly the existence of barriers to pursuing a career in the public sector.

SETTLEMENT PATTERNS

Like most other immigrants in the US and Canada, Asians are both urban-oriented and suburbanized. It is noteworthy that the suburbanization of Asian immigrants has led to the formation of distinctive **ethnoburbs** (i.e., ethnic suburbs) (Li, 1998), where ethnic businesses flourish in significantly large clusters (see Box 10.1).

In the US, Asian immigrants concentrate in a small number of metropolitan areas. According to the 2005–7 American Community Survey, 40 per cent of the country's Asian immigrants live in the three metropolitan areas of Los Angeles, New York, and San Francisco. The other principal destination metro areas are Chicago, San Jose, Houston, Washington, DC, Seattle, and San Diego. Within the San Francisco metro area, Asian immigrants make up more than half of the foreign-born population. In the San Jose metro area, the proportion of Asian immigrants is almost 60 per cent. Spatial concentration is much higher in Canada, where 69 per cent of Asian immigrants are concentrated in the three gateway cities of Toronto, Vancouver, and Montreal. In the last decade, Calgary, Edmonton, Ottawa, Hamilton, and Winnipeg also have attracted large numbers of Asian immigrants. Clearly, the trend towards concentration in the above census metropolitan areas (CMAs) has intensified.

Within the metropolitan areas, Asian immigrants definitely are suburbanized, where a number of Asian-dominated ethnoburbs are in the making. A typical example in the US is the San Gabriel Valley in suburban Los Angeles (Li, 1998, 2006b). The first surge of Chinese immigration into Los Angeles occurred after the promulgation of the new immigration law of 1965, followed by new waves in the 1980s and the 1990s, propelled by a series of geopolitical events. Many of the new Chinese immigrants never lived in the inner-city Chinatown; they settled directly into suburbs, especially the San Gabriel Valley in the eastern suburbs of Los Angeles, where Chinese immigrant population growth stimulated expansion of the real estate market. Chinese realtors and investors purchased properties and often converted them into multi-family dwellings for Chinese immigrant occupants. From 1980 to 1990, the San Gabriel Valley gained more than 60,000 Chinese immigrants (mainly from China, Hong Kong, and Taiwan), accounting for 40 per cent of the total increase of Chinese in Los Angeles County (Li, 2006b). By 2000, there were nearly 200,000 Chinese immigrants living in the Valley. In nine municipalities, the Chinese accounted for more than 25 per cent of the total population, with the two highest over 40 per cent.

BOX 10.1   The Concept of Ethnoburb

The term 'ethnoburb' was coined by Wei Li in her influential 1998 paper in *Urban Studies*. The concept was developed in a case study of Los Angeles County to describe a new model of ethnic settlement patterns. By Li's definition, ethnoburb refers to a new type of suburban ethnic concentration: 'suburban ethnic clusters of residential area and business districts in large metropolitan regions' (Li, 1998: 482). They are also multi-ethnic communities, in which one ethnic minority group has a significant concentration, but does not necessarily comprise a majority. This concept provides an alternative framework for the study of the changing social geographies in North American cities.

The definition of 'ethnoburb' suggests two critical defining factors: (1) the size of ethnic populations; and (2) the level of ethnic business concentrations. While the size of an ethnic population can be measured with census data, it is difficult to compile a comprehensive list of ethnic businesses in a metropolitan region, because few governments keep and disseminate business records by ethnicity.

In a Canadian application, Wang and Zhong (2008) used 'visible minority' as the defining indicator and 50 per cent or more as the critical mass of visible ethnic concentration to delineate ethnoburbs in the Toronto census metropolitan area (CMA). From this standpoint, three contiguous ethnoburbs are identifiable in the Toronto CMA: one includes most of Brampton and Mississauga and part of Etobicoke; the second includes most of Markham and parts of Richmond Hill, Scarborough, and North York; a third is located in Pickering and Ajax. Although ethnic businesses are not part of the defining indicators, in fact, many ethnic businesses are located within these ethnoburbs. More importantly, when both 2001 and 2006 censuses are mapped, it is found that the contiguous ethnoburbs have been expanding rapidly over time. If the current trend continues, it is expected that the entire Toronto CMA will become an ethno-metropolitan city in two to three decades. Statistics Canada (2010) has projected visible minorities will comprise the majority of Toronto residents by 2017, and that by 2031 they will account for 63 per cent of the Toronto population. By 2031, also, 50 per cent of Toronto's population is projected to be foreign-born.

In Canada, typical examples are two Asian-dominated ethnoburbs in the Toronto CMA. The ethnoburb that includes most of Brampton, Mississauga, and part of Etobicoke is dominated by South Asians; the one that encompasses most of Markham and parts of Richmond Hill, Scarborough, and North York is dominated by Chinese. According to the 2006 census, 41 per cent of the Toronto CMA's South Asians live in Brampton and Mississauga; and over 50 per cent of the CMA's Chinese live in the Chinese-dominated ethnoburb.

The emergence of ethnoburbs in North American cities not only changed population composition but also altered local economic structures (ibid.). The Asian-dominated ethnoburbs have created an interdependent pool of resources that a prospective ethnic entrepreneur can use. These include a ready source of ethnic labour and a critical mass of ethnic consumers (Kaplan, 1998; Zhou, 1992). In the case of the San Gabriel Valley, ethnic economic activities increased dramatically, especially the roles played by the ethnic Chinese banking sector (Dymski and Li, 2004). In the Toronto CMA, there are now over 60 Chinese shopping centres of various sizes, most of which are located in the Chinese-dominated ethnoburb that includes Markham, Richmond Hill, northwest Scarborough, and northwest North York (Wang, 1999; Lo, 2006; Wang et al., 2011). Such developments have led to conflicts involving several parties—including not only local residents and business owners but also city councils. Similar Asian-dominated ethnoburbs are observed in the metro area of New York (Smith and Logan, 2006) and in the Vancouver CMA (Edgington et al., 2006; Ray et al., 1997).

As Quadeer and Kumar (2003) explain, people of the same ethnic origin concentrate in the same geographic areas as a defence against discrimination, to support each other, to preserve cultural heritage, and even to facilitate political actions/lobbying. Increased political participation and representation also have been observed in ethnoburbs, where the usually silent minorities become more vocal. Ethnic votes are growing in size in ethnoburbs. More minority candidates are running for elections, changing the political landscape. In the 2008 Canadian federal election, three South Asians won seats in the Brampton–Mississauga ethnoburb; two Chinese ran in the Markham–Richmond Hill–Scarborough ethnoburb, although both lost. Similarly, in the San Gabriel Valley and Silicon Valley ethnoburbs, Asian Americans are politically active as candidates, fundraisers, and campaign workers, as well as voters. The first Chinese-American female city mayor in the US mainland, Lily Lee Chen, an immigrant from Taiwan, was elected in Monterey Park in 1984. Since then a number of Asian immigrants have been elected to offices at different levels (Li, 2009; Li and Park, 2006). Politicians of various ethnic backgrounds now see ethnoburbs as important sources of political capital, and party leaders often take time to attend ethnic festivals to solicit votes.

CONCLUSION

Asian immigrants have become the largest sources of legal immigrants for both the US and Canada, and they make important contributions to both countries' socio-economic development. Since the late 1980s, Asians have become a dominant group in the immigration of all professionals in the US and Canada (Kanjanapan, 1995).

Contemporary Asian immigrants possess much improved human capital, including higher levels of educational attainment and official-language ability than their predecessors possessed. They are also better educated than the general immigrant populations and even than the US- and Canadian-born populations, although internal variations exist among the subgroups. They should not, however, be portrayed as 'model minorities' for other racialized groups (see Box 10.2). Most of the Asian immigrants in Canada received their higher education prior to immigration, with little or no North American content. We assume such is also the case for many Asian immigrants in the US. While the majority of the Asian immigrants are reported to have knowledge of the official languages of the receiving countries, their actual language abilities might be lower than indicated in the census because this information is based on self-reporting rather than on any form of testing.

BOX 10.2   The Myth of a 'Model Minority'

According to Lim (2001), Peterson (1966) coined the term 'model minority' in his *New York Times Magazine* article, 'Success Story: Japanese American Style', in which he argues that the Japanese culture with its family values and strong work ethic saved the Japanese from becoming a 'problem minority' in the US. The success of the Japanese Americans is easily generalized across all ethnic Asian groups. Under the 'model minority' framework, Asian Americans viewed as one group are hard-working, technologically competent, and mathematically skilled, and are believed to be capable of achieving a higher degree of success than other ethnic minorities, as measured by income, education, and family stability (Lee and Joo, 2005).

The term 'model minority', however, is often viewed as a myth that amounts to racial stereotyping. One argument against the myth is that the statistics neglect history and the role of selective immigration of Asian Americans. It also ignores the heterogeneity of Asian-American groups and their significantly varied levels of success, hiding the plight of some Asian sub-groups under the high success rate of more established Asian communities. Revisionist critics argue that the model minority myth is not only invalid but also detrimental to the welfare of Asian Americans because it suggests they do not need and could not benefit from affirmative action (Min, 2006).

The positive stereotype of Asian Americans could negatively affect other minority groups as well. Since the efforts of the US civil rights movement to remove institutional, legal, and social disparities between majority and minority groups, political conservatives have pointed to Asian Americans as an exemplar and testimony that the American dream is colour-blind. The message has been loud and clear: 'If Asian Americans can succeed in America, why not blacks, Hispanics, and Native Americans?' This idea implied that the system should not be debilitating for other minorities if Asian Americans succeed, and that perhaps the socio-economic failure of these other groups is due to their inherent laziness and that success could be achieved through hard work and determination. The resentment aroused in other minority groups, as well as mainstream Americans, by the supposed success of Asians has caused tension between Asian Americans and other ethnic groups and has even provoked hate crimes towards to Asian Americans.

Overall, the Asian immigrants are doing better in the US labour market than their counterparts in Canada. This may well be attributed to the variations in their human capital and to the differences in the two countries' immigration policies. In the US, with a much higher proportion in the professional and technical services sector, Asian immigrants have higher personal income than do both the US-born and the general immigrant population; they also have a low prevalence of families at or below poverty lines. In Canada, high proportions of Asian immigrants work in the primary and secondary industries and in retail/accommodation/food services. Asian immigrants have much lower personal incomes than both the Canadian-born and the general immigrant populations, as well as a much higher prevalence of low-income families.

Significant differences exist among the subgroups of Asian immigrants. For example, the Indian and Philippine immigrants in both the US and Canada are the best performers in the labour market in terms of job earnings. In contrast, the Vietnamese in the US are the least successful in the labour market; the immigrants from Pakistan, Sri Lanka, and Bangladesh are at the bottom of the labour market hierarchy in Canada. Even within the Chinese group, those from Hong Kong and those from mainland China have demonstrated significant differences.

To a large extent, the differences in labour market performances and integration experiences among Asian groups are closely related to the level of human capital they possess. In addition, we must realize that the migration patterns and integration experiences of Asian immigration to North America are an outcome of cumulative factors that have played across multiple spatial scales. Historical linkages between the two continents, (im)migration laws, economic and cultural globalization, socio-economic inequalities between origin and destination countries, and differences in the resources at both individual and community levels all have had an impact on who can migrate, when they migrate, and what will be the long-lasting implications for how they adapt in the receiving societies.

In terms of residential settlement patterns, Asian immigrants concentrate in a few metropolitan areas and have 'traditionally been the most urbanized of all the major racial/ethnic groups' (Le, 2007: 141). Asian ethnic enclaves and communities, such as Chinatown, Koreatown, and Little Tokyo, have grown out of the spatial concentration of Asian immigrants in the receiving countries and from persistent institutional discrimination in the late 1800s and early 1900s (e.g., Anderson, 1991). On the one hand, continuing streams of immigration into existing ethnic communities reinforce and perpetuate the urban nature of Asian-American communities; on the other hand, Asian-dominated ethnoburbs have evolved in many metropolitan areas in both the US and Canada, such as the San Gabriel Valley in California and the suburbs in the Toronto CMA. In these instances, ethnic residential concentrations and ethnic businesses support each other to flourish and have become incubators of social, cultural, and political changes in North American cities.

More issues deserve further study. While Asian immigrants in general possess higher levels of educational attainment, most of their higher education was obtained in their home countries and frequently is not recognized in the receiving countries, notably by professional organizations that certify the members of their professions. Related to blocked opportunities due to lower levels of English proficiency among the Asian immigrants and 'glass ceilings' that exist in the high-end labour market, underemployment and underutilization of skills often are disguised by high labour force participation rates. Understanding this labour market process in relation to institutional environments, group characteristics, and individual experiences in both countries and across groups would provide valuable insights. Another trend is that with globalization more and more Asian businesses are leveraging their transnational ties and connections to Asia in expanding their social and economic coverage. As described by Saxenian (2006: 10), the 'new Argonauts', 'the foreign-born engineers, entrepreneurs, managers, lawyers, and bankers who have the linguistic and cultural abilities as well as the institutional knowledge to collaborate with their home-country counterparts', are typical cases. Looking into the transnational activities of Asian immigrants and how different groups gain or lose under both sending and receiving contexts will significantly enrich the existing immigration theories, in particular how the theories apply to Asian immigrants.

QUESTIONS FOR CRITICAL THOUGHT

1. What are the push and pull factors for increased Asian immigration to the US and Canada over the last two decades?

2. While coming from the same continent, the Asian immigrants consist of many diverse groups, with significant internal differences in their composition. Summarize the differential characteristics of the Asian immigrants in the US and Canada.

3. Despite much improved human capital possessed by contemporary Asian immigrants, many still underperform in the US and Canadian labour markets. What are the barriers to their economic success?

SUGGESTED READINGS

1. Anderson, K.J. 1991. *Vancouver's Chinatown: Racial Discourse in Canada, 1875–1980*. Montreal and Kingston: McGill-Queen's University Press. This book contributes to contemporary debates about 'race' and racism by examining the development of the Chinatown district in Vancouver, and highlights the importance of a geographical perspective on the relationship between people and place.

2. Li, W. 2009. *Ethnoburb: The New Ethnic Community in Urban America*. Honolulu: University of Hawai'i Press. This book documents the processes that have evolved with the spatial transformation of the Chinese-American community of Los Angeles and that have converted the San Gabriel Valley into ethnoburbs in the latter half of the twentieth century.

3. Min, P.G. 2006. *Asian Americans: Contemporary Trends and Issues*, 2nd edn. Thousand Oaks, Calif.: Pine Forge Press. While historical information is provided for each group, this book offers a broad overview of issues that impact Asian-American life today, such as economic status, educational achievements, intermarriage, intergroup relations, and settlement patterns.

4. Wang, S., and L. Lo. 2005. 'Chinese Immigrants in Canada: Their Changing Composition and Economic Performance', *International Migration* 43, 3: 35–71. The authors of this article found that despite their increased human capital, Chinese immigrants still experience very different economic outcomes in the Canadian labour market compared to members of the general population.

NOTES

1. 'Asian immigrants' in this chapter denotes all people who self-identify as 'Asian' or as any of several Asian subgroups in either the Canadian or US census.
2. There are inconsistencies in the available census data for the two countries with regard to higher education. In the US data, higher education refers to university degrees (bachelor's or higher); in the Canadian data, higher education includes university certificates and college diplomas for two-year courses of study.
3. The US data single out welfare income as a form of government transfer, whereas the Canadian data include public pensions as well as welfare in government transfers.
4. Hong Kong immigrants are included along with the five largest groups of Asian immigrants in the US because they are the fourth largest group in Canada and will be examined as a separate group in the analysis of Canada's Asian immigrants. Their inclusion in the US analysis allows for a comparison between the two countries.

REFERENCES

1. Anderson, K.J. 1991. *Vancouver's Chinatown: Racial Discourse in Canada, 1875–1980.* Montreal and Kingston: McGill-Queen's University Press.
2. Batalova, J., and B.L. Lowell. 2007. 'Immigrant Professionals in the United States', *Society* 44, 2: 26–31.
3. Citizenship and Immigration Canada. 2006. Landed Immigrant Data System. Ottawa.
4. Dymski, G., and W. Li. 2004. 'Financial Globalization and Cross-Border Co-movements of Money and Population: Foreign Bank Offices in Los Angeles', *Environment & Planning A* 36, 2: 213–40.
5. Edgington, D., M.A. Goldberg, and T.A. Hutton. 2006. 'Hong Kong Business, Money and Migration in Vancouver, Canada', in Li (2006a: 155–83).
6. Gibson, J., and E. Lennon. 1999. *Historical Census Statistics on the Foreign-Born Population of the United States: 1850–1990.* Washington: US Bureau of the Census, Population Division Working Paper No. 29.
7. Kanjanapan, W. 1995. 'The Immigration of Asian Professionals to the United States: 1988–1990', *International Migration Review* 29, 1: 7–32.
8. Kaplan, D. 1998. 'The Spatial Structure of Urban Ethnic Economies', *Urban Geography* 19, 6: 489–501.
9. Kibria, N. 2006. 'South Asian Americans', in P.G. Min, ed., *Asian Americans: Contemporary Trends and Issues*, 2nd edn. Thousand Oaks, Calif.: Pine Forge Press, 206–27.
10. Le, C.N. 2007. *Asian American Assimilation: Ethnicity, Immigration, and Socioeconomic Attainment.* New York: LFB Scholarly Publishing.
11. Lee, K.Y., and S.H. Joo. 2005. 'The Portrayal of Asian Americans in Mainstream Magazine Ads', *Journalism & Mass Communication Quarterly* 82: 654–71.

12. Li, W. 1998. 'Anatomy of a New Ethnic Settlement: The Chinese Ethnoburb in Los Angeles', *Urban Studies* 35, 3: 479–501.
13. ———, ed. 2006a. *From Urban Enclave to Ethnic Suburb: New Asian Communities in Pacific Rim Countries*. Honolulu: University of Hawai'i Press.
14. ———. 2006b. 'Spatial Transformation of an Urban Ethnic Community: From Chinatown to Ethnoburb in Los Angeles', in Li (2006a: 74–94).
15. ———. 2009. *Ethnoburb: The New Ethnic Community in Urban America*. Honolulu: University of Hawai'i Press.
16. Li, W., and E. Park. 2006. 'Asian Americans in Silicon Valley: High-Technology Industry Development and Community Transformation', in Li (2006a: 119–33)
17. Lim, J.H. 2001. *Just Call Me Doctor: The (API) American Dream*. At: www.asianweek.com/2001_10_12/feature.html.
18. Lo, L. 2006. 'Suburban Housing and Indoor Shopping: The Production of the Contemporary Chinese Landscape in Toronto', in Li (2006a: 134–54).
19. Manyin, M. 2005. *CRS Issue Brief for Congress: The Vietnam–U.S. Normalization Process*. At: www.fas.org/sgp/crs/row/IB98033.pdf. (20 May 2010)
20. Min, P.G. 2006. *Asian Americans: Contemporary Trends and Issues*, 2nd edn. Thousand Oaks, Calif.: Pine Forge Press.
21. Oh, B.C. 2008. 'New Migration Motives and New Settlement Challenges: Integration Experience of Korean Immigrants in the Toronto CMA', MA research paper, Ryerson University.
22. Petersen, W. 1966. 'Success Story, Japanese-American Style', *New York Times Magazine*, 9 Jan., Section 6, 20–43.
23. Quadeer, M., and S. Kumar. 2003. 'Toronto's Residential Mosaic', *Ontario Planning Journal* 18, 5: 7–9.
24. Ray, B., G. Halseth, and B. Johnson. 1997. 'The Changing "Face" of the Suburbs: Issues of Ethnicity and Residential Changes in Suburban Vancouver', *International Journal of Urban and Regional Research* 21, 1: 75–99.
25. Saxenian, A.L. 2006. *The New Argonauts*. At: people.ischool.berkeley.edu/~anno/Papers/IMF_World_Bank_paper.pdf. (20 May 2010)
26. Singh, P., and T.V. Thomas. 2004. 'Ministering to South Asians in Canada and Beyond', *Global Missiology* (Oct.). At: ojs.globalmissiology.org/index.php/english/article/viewFile/125/362. (15 Oct. 2010)
27. Smith, C., and J. Logan. 2006. 'Flushing 2000: Geographic Exploration in Asian New York', in Li (2006a: 41–73).
28. Statistics Canada. 2009. 2006 Census Summary Table.
29. ———. 2010. 'Study: Projections of the Diversity of the Canadian Population', *The Daily*, 9 Mar. At: www.statcan.gc.ca/daily-quotidien/100309/dq100309a-eng.htm.
30. US Bureau of the Census. 2002a. *A Profile of the Nation's Foreign-Born Population from Asia* (2000 update).
31. ———. 2002b. *Survey of Business Owners*.
32. ———. 2010. American Community Survey dataset, at: www.census.gov/.
33. Wang, S. 1999. 'Chinese Commercial Activity in the Toronto CMA: New Development Patterns and Impacts', *Canadian Geographer* 43, 1: 19–35.
34. ———, R. Hii, J. Zhong, and P. Du. 2011 (forthcoming). 'Recent Trends in Ethnic Chinese Retailing in Metropolitan Toronto', *International Journal of Applied Geospatial Research*.
35. ——— and L. Lo. 2005. 'Chinese Immigrants in Canada: Their Changing Composition and Economic Performance', *International Migration* 43, 3: 35–71.
36. ——— and J. Zhong. 2008. 'Delineating Ethnoburbs in the Toronto CMA', paper presented at the Association of American Geographers meeting, Boston.
37. Zhou, M. 1992. *Chinatown: The Socioeconomic Potential of an Urban Enclave*. Philadelphia: Temple University Press.

CHAPTER 11

CONTEMPORARY PROFILES OF BLACK IMMIGRANTS IN THE UNITED STATES AND CANADA

Thomas Boswell and Brian Ray

Although immigrants from the Caribbean and Africa comprise a significant component of recent waves (since the mid-1960s) of migrants to both the United States and Canada, the presence and racialization of blacks in each country have long and somewhat distinctive histories. This chapter examines the contemporary social geography of urban black immigrants, with an emphasis on the ways in which historical processes of forced migration and more recent immigration policies have come to shape black immigrant communities in both countries, as well as the socio-economic status of black women and men who are foreign-born and US/Canadian-born. In so doing, we are particularly interested in whether there are substantial differences in the well-being of foreign-born blacks living in urban Canada as compared to the well-being of their counterparts in the US. Has the emphasis in Canadian immigration policy on selecting individuals with strong human capital created black communities in Canadian cities that differ substantially from those found in the US? In what ways does black immigrant life differ in the US and Canada, especially relative to US- and Canadian-born blacks whose cultural and social experiences have been constructed in very different histories of movement, place, and racialization?[1]

BLACK SETTLEMENT IN THE US AND CANADA

It is important to appreciate from the outset that blacks have a much more prominent presence in the US[2] than in Canada;[3] after all, approximately 40 million blacks reside in the US compared to less than 800,000 in Canada. Blacks also represent a larger percentage of the US population (13.2 per cent) than Canadian population (2.5 per cent). Nevertheless, the histories of black immigration to Canada and the US, and community formation in both countries, have been shaped by processes associated with involuntary migration. For many African Americans and a smaller number of black Canadians, forced deportation from Africa to North America beginning in the seventeenth century marked their history and continues to influence contemporary experiences in subtle and sometimes violent ways. Black immigration to Canada and the US began with the importation of **indentured labour**, transitioned into a long period of **slavery**, and much more recently has become characterized by voluntary immigration and refugee settlement. Indentured servitude and slavery, however, were much more prominent in the US, especially in terms of the number of lives affected by various forms of violence and subjugation.

The settlement geography of black immigrants in the US is complicated by secondary migrations associated with the plantation economy in the Deep South in the eighteenth and nineteenth centuries, and by subsequent moves by the southern black population to northern and western industrial cities that began in the early twentieth century. The history of black settlement in Canada is comparatively less complex. The earliest black immigrants arrived in small numbers as slaves and former slaves to Nova Scotia and parts of Ontario (Darden and Teixeira, 2009: 15), but today most black immigrants live in major cities, most notably Toronto, Montreal, and Ottawa.

Black Immigration and Settlement in the United States

Some black slaves may have accompanied Spanish settlers to Florida during the 1560s (Davis, 2006: 124), but the first recorded blacks in British North America arrived in 1619 as indentured servants in Jamestown, Virginia. Although technically not slaves, they were often treated harshly during their period of indenture, which usually lasted seven years. Once they achieved independence, they were legally treated like indentured whites who similarly had traded their labour for the cost of passage to North America. As many of the original English settlers perished under the trying conditions of the new frontier, additional blacks were imported under the indenture system as replacements. During this time the number of blacks increased until they became the second largest ethnic group after the English in many of the original Thirteen Colonies (Segal, 1995: 4). This period of black indentured immigration—from Africa to the coastal colonies— was the first major migration of blacks to the US and marks the beginning of the juxtaposed experiences of mobility and fixity that would come to shape black cultural identity (Berlin, 2010).

The institution of black slavery (the outright, permanent ownership of blacks) began to replace the system of indentured labour during the 1700s. It occurred at varying times throughout the American colonies, but took hold most firmly as more blacks were brought to the South to meet the enormous labour needs required to plant, maintain, and harvest such crops as cotton, tobacco, and sugar cane on large plantations. The migration from the east coast colonies to the Deep South is generally considered to be the first great *internal* migration of US blacks (ibid.). The first census in 1790 showed 757,208 blacks in the US, representing 19.3 per cent of the population. About 92 per cent were slaves and most were foreign-born. By the beginning of the Civil War in 1860, there were 4.4 million blacks living in the US (89 per cent being slaves) and most lived in the South (Kent, 2007: 1–16).

The slave trade was abolished in 1808 and, in turn, the immigration of blacks was greatly reduced and would not become very significant for another 150 years, until the late 1960s. The number of slaves, however, continued to grow through natural increase. Eventually, slavery, and not just the trade in slaves, did end. The Emancipation Proclamation was signed in 1863, but it applied only to freeing slaves in the states that had seceded from the Union. In 1865, the US Congress passed the Thirteenth Amendment to the Constitution, which finally outlawed the institution of slavery in the US. The first comprehensive study of black immigration to the US conducted by a black sociologist, Ira Reid (1969: 41), cited a figure of 333,000 slaves imported to the US by 1808. More recent estimates suggest a somewhat higher figure of around 400,000 (Berlin, 2010: 15). Importantly, black slaves built lives out of the wrenching disruptions of forced migration and the challenges associated with making place under conditions of subjugation. As Berlin (ibid., 18) argues, movement and place are central to understanding the black experience in the US:

> [F]luidity and fixity; or in Paul Gilroy's phrase, 'routes and roots'—ripped across some four centuries of black life . . . the alternating and often overlapping impact of massive movement and deep rootedness touched all aspects of the experience of black people, from language and theology to cuisine and music. . . . It produced, on the one hand, a malleable, flexible cultural style that became a touchstone of African American life . . . . It created, on the other, a passionate attachment to place, reflected in the earthy idioms of the rural South and the smash-mouth street jive of the modern 'hood'.

The black population experienced a second great internal migration during the 60-year period between 1910 and 1970, when about six million blacks moved from the South to the northern and western states of the US. Motives for this movement were economic and social. The collapse of the cotton economy due to boll weevil infestations and floods, as well as weak industrial development, encouraged many black migrants and the children of migrants to leave the South for industrial jobs in northern and western cities. Entrenched segregation and often violent racism continued to exist in the South, also encouraging many blacks to migrate to such cities as Chicago, Detroit, New York City, Baltimore, Philadelphia, Los Angeles, and Seattle. By 1970, the percentage of blacks living in the American South declined to 53 per cent from 90 per cent in 1910 (Shaw-Taylor, 2007: 12). This migration, and the neighbourhoods and workplaces of settlement, would again contribute in fundamental ways to shaping the social and cultural qualities of black life in the US.

Beginning in the late 1960s, a return to the South characterizes the third major shift in the internal geography of the black population. In this period, an increasing number of factories and other businesses sought to escape high labour, land, and transaction costs in the North and began to move south. As businesses migrated, blacks especially began to return to the South, but this time they sought out the region's cities. More than two million blacks had returned to the South by 2000 and now live in places like metropolitan Atlanta, Charlotte, Richmond, Orlando, Dallas, and Nashville (ibid., 34). By 2006, 59 per cent of all US-born blacks were living in the South again. Importantly, when substantial black immigration from the Caribbean and Africa recommenced in the later half of the 1960s, newcomers interacted with a larger African-American community whose culture had been shaped in part by these multiple instances of internal migration, as well as by the challenges of rural and urban place-making.

The period between the end of the slave trade and the mid-1960s marked the low point in black immigration to the US. While white America was continually renewed by new waves of European migrants, for more than 150 years between the end of the transatlantic slave trade in 1808 and the passage of the new Immigration and Nationality (Hart-Celler) Act in 1965, few foreign-born black people gained entry to the US. As Berlin (2010: 202) argues, one of the distinguishing characteristics of the black American population, even today, is that, overwhelmingly, the people were born in the US. Some black immigration from the islands in the Caribbean, particularly from Jamaica, Trinidad and Tobago, and Barbados, did occur in the first decades of the twentieth century, largely due to labour requirements associated with building the Panama Canal, but new national-origin restrictions enacted in 1924 ended this migration. Caribbean immigrants in this period settled primarily in a few northern cities, especially New York City. In 1920, about one-quarter of New York City's Harlem neighbourhood was comprised of blacks from the Caribbean. Nevertheless, by 1960 only about 125,000 black immigrants lived in the US, representing less than 1 per cent of the nation's 19 million African Americans (Kent, 2007: 4).

With passage of the Immigration and Nationality Act, the nation's immigration policy changed significantly and legal immigration for blacks was greatly facilitated. The mid-1960s marks the beginning of the second major migration of *foreign-born* blacks to the US. By 1980, the number of black immigrants had increased by almost seven times to 816,000 (ibid.). Moreover, the pace of immigration has increased: Berlin (2010: 6) estimates that more than 1.3 million blacks immigrated to the US during the 1990s, which is almost three times the number of blacks who arrived during the slave trade. By 2006, the number of foreign-born blacks had increased to slightly more than three million, repre-senting nearly 8 per cent of all blacks living in the US.[4]

Today, almost two-thirds of the black immigrants living in the US are from the Caribbean and Latin America, with most (56 per cent) coming from the West Indies[5] and Haiti. Another one-third were born in Africa. From 1960 to 2000, black immigration from the Caribbean was much greater than from Africa because of the closer proximity of the Caribbean to the US and because many Caribbean-born blacks already had a network of family and friends in the US. Gradually, however, a black African-born population base has been established in the US, serving as a pool of sponsors for future immigrants. Africa also experi-enced a number of civil wars and a substantial number of Africans migrated to the US on humanitarian grounds as government-assisted refugees. Between 2000 and 2005, Africa accounted for the greatest share of black immigrants (54 per cent), followed by the Caribbean (43 per cent). If this trend continues, African blacks eventually will replace Caribbean blacks as the major component of the nation's black immigrant population. The main Caribbean countries of origin are Jamaica, Haiti, Trinidad and Tobago, Guyana, Barbados, Bahamas, and Belize, while most Africans are from Nigeria, Ethiopia, Ghana, Liberia, Somalia, Kenya, and Sudan (Kent, 2007: 8).

Virtually all black immigrants settle in cities, especially large metropolitan areas, due to employment opportunities and social network ties (Figure 11.1). Approximately 60 per cent of black newcomers live in just five metropolitan regions: New York City, Miami–Fort Lauderdale, Washington, Boston, and Atlanta. There is some difference in the distribution of black immigrants from the Caribbean and Africa. The four leading cities for the West Indian-born blacks are New York City, Miami–Fort Lauderdale, Boston, and Atlanta, while New York, Washington, Atlanta, and Philadelphia are the principal destinations for Africans (ibid., 12).

The New York City metropolitan area is by far the most important destina-tion. Almost one-third of all black foreign-born people in the nation live in this one metropolitan area (ibid.). Figure 11.2 shows the distribution of black immigrants by census tracts in New York City. Black immigrants are highly concentrated in selected neighbourhoods, especially in central Brooklyn (Flatbush, Prospect Park, and Crown Heights), southern Queens (Richmond Hill, Jamaica, St Albans, Crown Heights, and Far Rockaway), and the northern Bronx (Williamsbridge, Baychester, and Wakefield). These neighbourhoods long have been home to African Americans and large numbers of Caribbean immigrants have also come to settle in and around these locales. Two recent studies have found that most black Caribbean immigrants choose to live in black areas nearby other co-ethnics, thus producing a triple layering effect (Boswell and Jones, 2006: 163–73; Boswell and Sheskin, 2009: 194–203). Within African-American neighbourhoods they frequently search for concentrations of West Indians and compatriots of the same islands from which they originated in the Caribbean. There are at least four possible reasons for these spatial concentra-tions. First, despite passage of the Fair Housing Act in 1968, discrimination may limit where black immigrants can successfully find housing to predominantly

Figure 11.1 Foreign-Born Blacks in Counties in the US, 2000

black neighbourhoods. Second, black immigrants, who often have higher education levels and job skills than American-born blacks, may find it easier to compete with African Americans for the better jobs in these neighbourhoods compared to the competition they might expect in white neighbourhoods. Third, the large black populations in these neighbourhoods provide a market for the goods and services produced by immigrant blacks. Fourth, many of these black neighbourhoods in New York City have had a strong Caribbean presence for decades and have established networks that assist newcomers from the Caribbean (Boswell and Jones, 2006: 174).

Black Immigration and Settlement in Canada

A history of indentured work and slavery has shaped African-American geography and culture in ways that have few immediate parallels in Canada. Although slavery did exist in British North America from the beginning of the seventeenth century until 1834, when the British Parliament passed the Slavery Abolition Act,[6] with time it became legally and politically difficult for affluent whites to purchase and 'own' black workers. In fact, some US black slaves were offered their freedom in British North America even while slavery remained legal. During the US Revolutionary period, black slaves were offered freedom and land if they fought for the Crown. When the defeated British left the newly independent US for the Canadian colonies in 1783, approximately 3,000 freed blacks also made the journey and settled primarily in Nova Scotia. Relatively few of the black Loyalist refugees ever received the land they were promised and

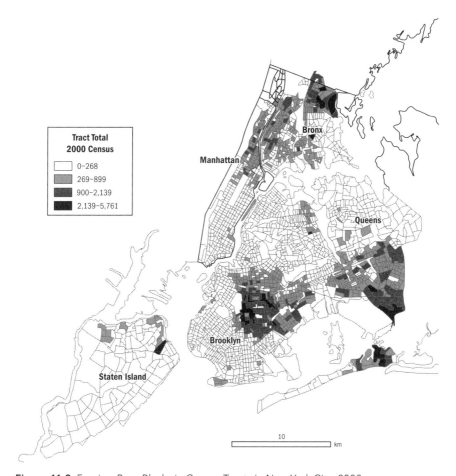

Figure 11.2 Foreign-Born Blacks in Census Tracts in New York City, 2000

eventually they lived and worked with the small number of black Africans who were held as slaves in the territory that would become Canada (Mensah, 2010).[7] While today only a small portion of Canada's black population lives in Nova Scotia (2.5 per cent), this province, and especially Halifax, holds the distinction of being one of the early locales of significant black settlement and community development. The black population that came to Canada after the Revolution was further augmented in the first half of the nineteenth century by slaves who were encouraged to desert their American masters by the promise of settler status and land during the War of 1812 (approximately 3,500 reached Canada and many settled in Nova Scotia). Another wave of US blacks reached Canada by means of the 'underground railroad', settling primarily in and around the present-day cities of St Catharines, Windsor, London, and Chatham in south-western Ontario. In 1793, Upper Canada (Ontario) passed an Abolition Act, which essentially accorded free status to US blacks, and the province became well known as a safe haven for black refugees, if not for people already held as slaves in Upper Canada.

It would be misleading to leave the impression that Canada has always enthusiastically encouraged black immigration, from the US or elsewhere. In fact, the migrations that occurred around the Revolutionary War, War of 1812, and Civil War say more about British political gamesmanship in North America than openness to US blacks and the injustices of their subjugation. From the nineteenth century until mid-1960s, immigration policy and regulations deliberately kept most non-white, non-European immigrants from entering Canada (see Chapter 2). Due to legal restrictions on black immigration, the size of the black population actually decreased by approximately 3,000 people from 1871 to 1951 (Milan and Tran, 2004), when black individuals in Canada numbered only 18,000 or 0.1 percent of the population. In fact, the ministries responsible for immigration scrupulously pursued policies that favoured white Europeans, and black applicants[8] invariably were turned down because they were deemed unlikely to assimilate to the Canadian physical and social environments. Close relatives from the Caribbean could be sponsored by Canadian citizens regardless of race, but this was somewhat of a moot point since the number of potential black sponsors was very small. Administrative procedures also kept the number of black immigrants coming to Canada in the post-war years to a minimum. One Department of Citizenship and Immigration memo from 1955 summarizes the policy, practices, and rationale as follows:

> It has long been the policy of this Department to restrict the admission to Canada of coloured or partly coloured persons. This policy has been based on unfavourable experience with respect to Negro settlements such as we have in Halifax, the general depressed circumstances of the negro [sic] in Canada and an understanding that the Canadian public . . . is not willing to accept any significant group of Negro immigrants.[9]

Arguments for curbing immigration to avoid racial tensions gained popularity during the 1960s as one major US city after another witnessed devastating periods of civil unrest. As Beauregard (2003) has demonstrated in the US, a powerful post-war discourse linking urban problems with race was led by the media, politicians, policy-makers, and educators, and the argument most certainly permeated the Canadian border and consciousness. A 1971 letter to the editor, written in response to a *Toronto Telegram* article by Aaron Einfrank—in which he argued that blacks were the cause of urban malaise and violence in the US—indicates the tenor of concern expressed by some white Toronto residents:

> We can heave a sigh of relief at being spared [Washington's] social disorder [which] derives from the influx of socially disorganized Negroes in US urban centres. . . .

Alas, the [Canadian] national government in the past four years has pursued a policy of ruthless favoritism towards the 3.5-million people of the English-speaking Caribbean nations. . . . I am not prepared to see my life disrupted, and rendered almost totally unsupportable by a mass influx of people from the . . . Caribbean. The social catastrophe of immigration from the . . . Caribbean must be immediately stopped in the interests of just social order.[10]

This man's argument could be regarded as extreme or anomalous if it had not been shared by a substantial number of Canadians. The Department of Manpower and Immigration in the late 1960s and early 1970s received a considerable volume of mail expressing similar concerns about Caribbean immigration. For example, correspondence received by the ministry about 'coloured' immigration was summarized in a memo to the assistant deputy minister (Immigration) in September 1973 in which it was noted that:

While there has always been a trickle of such correspondence it was largely of the crank variety. Now, however, the writers are deeply concerned Canadians from many walks of life . . . . The Ontario letters talk about black ghettos and 'coloureds' living on welfare and producing babies to improve their welfare . . . . Many speak of violence being on the increase, of overcrowding and ghetto-like conditions burgeoning.[11]

The Canadian public was concerned about black immigration and the urbanization of this population, but it proved impossible under Canadian law to restrict the migration of Afro-Caribbeans or Africans for reasons of race. After the clauses allowing race-based restrictions were finally removed in 1962, the number of Caribbean immigrants increased substantially during the 1970s and 1980s; by 2006, people born in the Caribbean living in Canada numbered 327,650.[12] Immigrants from Africa have also increased significantly, particularly during the past 20 years, and 398,100 people of African birth were living in Canada as of 2006. It is important to recognize, however, that the ethnoracial identities of Africans are complex and any equation of Africans with a 'black' identity is simplistic. For instance, 140,140 African immigrants in 2006 were born in Northern Africa (Algeria, Egypt, Libya, Morocco, Sudan, Tunisia, and Western Sahara), but only 7.1 per cent self-identify as black. Instead, 54.9 per cent identify as Arab and another 36.7 per cent do not consider themselves to be a racial minority. The next largest African group—East Africans—is also complex in that 49.9 per cent self-identify as black, followed by 35.4 per cent of South Asian ancestry.

Since the mid-1960s, the urban settlement of black immigrants—from the Caribbean or Africa—is undeniable. The vast majority of black immigrants live in Toronto (47.4 per cent), Montreal (22.9 per cent), and Ottawa–Gatineau (6.3 per cent). Cities such as Halifax, where historically a large proportion of blacks settled, now have a very tiny foreign-born black population (0.2 per cent). In contrast, a much larger proportion of the Canadian-born black population live in Halifax (3.4 per cent). Canada's other major gateway city, Vancouver, where the social and built environments have been transformed by Asian immigration over the last two decades, is home to only a tiny fraction of black immigrants (2.7 per cent). Some groups of black immigrants have a bias towards particular cities. For instance, the vast majority of Haitian immigrants, who speak French, live in Montreal (84.8 per cent) or Ottawa–Gatineau (7.4 per cent) and extremely few choose Canada's other major gateway cities (Toronto, 1.8 per cent; Vancouver, 0.3 per cent).

It is somewhat challenging to compare Canadian and US cities with regard to neighbourhood-level settlement and spatial concentration of foreign-born and non-immigrant black populations. The black population in Canadian cities was tiny prior to the 1960s and legally enforced spatial segregation was never a factor. Informal discrimination did keep blacks from living in particular build-ings and neighbourhoods, and in some Toronto and Montreal neighbourhoods restrictive covenants prevented property sales to black families. Racial segrega-tion and ghettoization, however, did not exist on the scale seen in the US. As a consequence, immigrants from the Caribbean and Africa who live in Toronto and Montreal overwhelmingly settle in inner-suburban neighbourhoods where relatively affordable rental and owner-occupied housing can be found (Figure 11.3). Unlike many US cities, the inner suburbs of both Toronto and Montreal are characterized by considerable housing diversity (in terms of dwelling type and tenure), and also by considerable social diversity in terms of socio-economic status (SES) and ethnocultural background. To illustrate, in Toronto and Montreal in no census tract do immigrants from the Caribbean and Africa constitute any more than 28 per cent of the resident population. Figure 11.3 shows the strongest spatial concentration of immigrants from the Caribbean and Africa are in the inner suburbs in portions of Etobicoke/West Toronto (west side) and Scarborough (east side), and not in and around the city's inner city. These two inner suburban areas, however, are quite different. Although Etobicoke/West Toronto is generally considered middle- to upper-middle-class, this particular area, where a signifi-cant number of black immigrants live, is characterized by a mixture of housing types, with a significant amount of rental housing, and average incomes are well below the metropolitan average. Moreover, while Scarborough has several neigh-bourhoods where significant portions of the population live below the poverty line, the area where black immigrants are concentrated is dominated by owned housing and the population, in general, is middle class. In both Toronto and Montreal, the neighbourhoods in which black immigrants are found are quint-essentially places of diversity in which ethnic and racial difference is an essential feature of everyday space.

Apart from the built and social complexity of locales occupied by immigrants, the black population is culturally and linguistically heterogeneous. For example, Haitian immigrants in Montreal—who constitute the majority of the black population—overwhelmingly speak French or French-based Creole. The Haitian population in Montreal also has settled in inner-suburban neighbourhoods to the

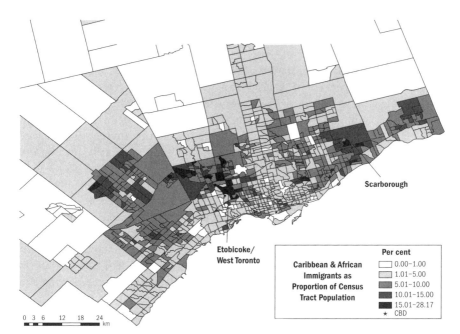

Figure 11.3 Geographic Concentration of Caribbean and African Immigrants in Metropolitan Toronto, 2006

northeast of downtown, while the only real concentration of the small English-speaking Caribbean population is found in an inner-city neighbourhood to the west of the downtown core. The African population, which is a significantly large component of Montreal's population, is also linguistically diverse and a large proportion are Muslim. While the Haitian population in Toronto is tiny, the city's black population is also heterogeneous, and the growing number of African migrants adds yet more layers of complexity. Such differences in settlement, language, culture, and religion are significant and must be kept in mind to prevent facile characterizations of the 'black' immigrant experience in Canada. In short, labels such as 'black', 'African', or 'Caribbean' camouflage considerable social complexity and the intricate ways in which people from many cultures produce place.

DEMOGRAPHIC PROFILES OF BLACK IMMIGRANTS IN THE US AND CANADA

Immigrant and Non-Immigrant Blacks

For the US component of this analysis, we use results of the 2005–7 American Community Survey (US Bureau of the Census, 2009), and the Canadian analysis relies on data from the 2006 census.[13] The data in Table 11.1 illustrate the degree to which blacks born in the US or Canada differ from their foreign-born counterparts in terms of demographic and socio-economic characteristics. It is important to emphasize from the outset that the foreign-born black population in both countries is somewhat older than the cohort of people born in Canada or the US. For example, 83.8 per cent of US black immigrants are between the ages of 15 and 64, compared to 63.4 per cent of US-born blacks, and very few immigrants are under the age of 15. Immigration requirements are selective of people in the young adult ages, permitting very few children, which tends to skew upward the average age of the foreign-born group. This is extremely important when comparing income and other measures of social status because it is almost always true that people earn less when they are younger than when they are older, and they generally attain higher levels of education as they age. Yet, even when age effects are controlled, the US and Canadian data indicate that a larger proportion of foreign-born blacks fall into higher-income categories. For example, for men and women in the 40–54 age group, foreign-born blacks in Canada had higher average total incomes than their Canadian-born counterparts ($34,556 versus $32,695). The differential was even greater in the US, where immigrant black males had an average income of $41,182 compared to $33,134 for US-born males in this age group.

Important variations exist in the geographic distribution of immigrant and US/Canadian-born blacks in both countries and these influence how differences in socio-economic status should be interpreted. In Canada, foreign-born black immigrants and Canadian-born blacks exhibit similar geographical distributions among provinces and cities (Table 11.1). Differences in distribution are more pronounced in the US, where almost half the immigrant blacks live in cities in New York and Florida, especially New York City and Miami–Fort Lauderdale. US-born blacks are more widely dispersed, but they are equally likely to be living in cities—only about 10 per cent live in rural areas.[14] In general, incomes tend to be higher in major metropolitan areas such as New York City and Toronto, although the cost of living is elevated relative to smaller cities.

Other broad similarities in demographic and social characteristics potentially influence the interpretation of socio-economic status differences in each country (Table 11.1). For example, about 54 per cent of the immigrant blacks in both countries arrived after 1990; as a consequence, a substantial proportion of newcomers are in the early stages of social and economic integration. Broad similarities also characaterize gender composition—53 per cent of foreign-born blacks in the US and 54 per cent in Canada are female. Finally, about 57 per cent of black foreign-born individuals in the US and 55 per cent in Canada came from the Caribbean (including Haiti), while approximately 30 per cent of newcomers in the US and 35 per cent in Canada came from Africa.

Affecting social status, black immigrants have achieved higher education levels and better jobs than their US-born counterparts (Table 11.1). A higher percentage of immigrant blacks 25 years and older have a college degree, and a lower percentage have not completed high school. A larger percentage of immigrant blacks also have jobs in the higher-paying management and professional occupations and a smaller percentage are working in the lower-paying construction and transportation jobs. Furthermore, proportionately fewer immigrant blacks are unemployed and live in poverty. These findings suggest immigrant blacks have higher SES[15] than US-born blacks and they are consistent with results from other studies (e.g., Logan, 2007: 49), pointing to the notion that immigrant blacks represent a 'model minority' in the US (see Chapter 10); however, more information needs to be considered before drawing a definitive conclusion.

Comparisons of immigrant and Canadian-born blacks in terms of socio-economic status are slightly more complicated (Table 11.1). The educational levels and occupations of immigrant blacks are very similar to those of Canadian-born blacks. Although the unemployment rate for immigrant blacks is substantially lower, the proportion below the poverty line matches Canadian-born blacks. We can conclude, therefore, that immigrant blacks have a somewhat higher SES than Canadian-born blacks, but the difference is not as large, nor as consistent, as in the US.

Comparing the income of blacks in Canada and the US presents some challenges because some variables in the Canadian census and the American Community Survey are not measured in the same way. In addition, the currency is not identical, further confounding simple cross-national comparisons. The data from the American Community Survey also cover three years, during which time the value of each currency fluctuated. Still, by comparing the data in Table 11.1 one can get a rough idea of the relative standards of living for black immigrants in both countries.

Immediately apparent is the convergence in mean and median incomes for immigrant blacks in both countries (Table 11.1). Average gross income for Canadian black immigrants is almost identical to that of their American counter-parts—$27,691 versus $27,407. Median incomes are also very close, although black immigrants in Canada again have a slight advantage. The similarities in income for black immigrants in both countries for the most part coincide with Jones's (2008) analysis of the SES characteristics of Jamaican immigrants in Miami and Toronto. Jones, however, found that those living in Miami had slightly higher status in terms of incomes, education levels, and types of occupations (ibid., 67–9, 158). Importantly, the study by Attewell et al. (2010) of blacks in Canada and the US also finds broad similarities in earnings, with first-generation and third-generation blacks having substantially lower incomes than white households.

Broad similarities can be seen between black immigrants in the US and Canada along two other critical SES variables, occupational structure and educational achievement (Table 11.1). If there is an edge, it is slightly in favour of the black immigrants in the US. Foreign-born blacks in the US are somewhat more likely to have a college education and to work in management and professional occupations. In contrast, black immigrants in Canada are more likely to be employed in lower-paying production and transportation jobs. The findings for both occupational status and educational achievement suggest that US black immigrants might have a slight edge over black immigrants in Canada in terms

Table 11.1 Characteristics of Foreign-Born Blacks and Blacks Born in the US and Canada

	United States		Canada	
	US-Born Blacks	Foreign-Born Blacks	Canadian-Born Blacks	Foreign-Born Blacks
Period of Arrival for Foreign-Born Population (%)				
Before 1950		0.4		0.1
1950-9		1.2		0.7
1960-9		6.5		7.5
1970-9		13.7		21.3
1980-9		24.2		16.0
1990-9		29.2		29.8
2000-7 / 2000-6*		24.8		24.6
Major States and Provinces of Residence (%)				
Georgia / Ontario	8.1	4.6	59.9	61.4
Texas / Quebec	7.7	4.3	21.7	25.5
Florida / Alberta	6.8	20.4	5.7	6.2
California / British Columbia	6.7	5.2	3.9	3.4
New York / Manitoba	6.2	27.2	1.8	2.1
North Carolina / Saskatchewan	5.9	1.6	0.6	0.6
Regions of Residence (%)				
South Atlantic / Atlantic	34.8	35.8	6.2	0.7
East North Central / Central	14.1	4.6	81.6	86.9
West South Central / Prairies	13.6	4.9	8.1	8.9
Middle Atlantic / British Columbia	11.9	35.2	3.9	3.4
East South Central / North	10.9	1.4	0.1	0.1
Gender (%)				
Female	53.7	53.2	50.1	54.0
Male	46.3	46.8	49.9	46.0
Age Groups (%)				

United States

Less than 15 years	25.8	6.3
15-64 years	63.4	83.8
65 years & older	10.8	9.9
Educational Attainment for Persons 25 Years and Older (%)		
Less than high school diploma	21.8	19.9
High school diploma but no BA	61.3	53.7
Bachelor's degree or higher	16.9	26.3
Region of Birth (%)		
Caribbean	0.1	56.6
Africa	0.1	29.6
Unemployment and Occupational Characteristics of Employed Labour Force (%)		
Unemployed	13.0	8.1
Management, professional, & related	26.2	30.3
Services	23.5	27.9
Sales & office	27.4	21.4
Farming, fishing, & forestry	0.4	0.3
Construction, extraction, maintenance, & repair	5.9	7.1
Production, transportation, & material moving	16.7	12.9
Income and Poverty Status		
Mean income	$21,799	$27,407
Median income	$13,700	$20,000
% living in families with incomes below the poverty level	24.5	15.2

Canada

Less than 15 years	54.3	7.3
15-64 years	44.2	83.2
65 years & older	0.9	6.4
Less than high school diploma	15.2	15.5
High school diploma but no BA**	63.8	66.4
Bachelor's degree or higher	21.0	18.1
Caribbean	0.0	54.8
Africa	0.0	35.4
Unemployed	12.2	9.7
Management, professional, & related	27.2	27.1
Services	10.3	13.4
Sales & office	47.3	34.2
Farming, fishing, & forestry	1.3	1.0
Construction, extraction, maintenance, & repair	5.0	5.3
Production, transportation, & material moving	8.9	19.0
Mean income (individuals)		
Total (pre-tax)	$19,643	$27,691
Wages & salaries (gross)	$15,637	$20,267
Median income (individuals)		
Total (pre-tax)	$12,000	$21,000
Wages & salaries (gross)	$6,000	$11,000
% living in families below low-income cut-off or low-income person (after tax)	26.0	26.2

*Comprises first five months of 2006 only.

**In Canada this category includes: high school certificate or equivalent; apprenticeship or trades certificate or diploma; college, CEGEP or other non-university certificate or diploma; or university certificate or diploma below the bachelor's level.

Sources: US Bureau of the Census (2009); Statistics Canada (2006a, 2006b).

of socio-economic status.

The strong similarity in total income, occupations, and educational attainment for US and Canadian black immigrants is intriguing because it suggests that Canada's policy of selecting independent immigrants based on human capital characteristics has not produced significantly stronger SES outcomes. It must be kept in mind that in Canada, as in the US, a significant share of immigrants arrive through the family reunification and refugee streams, which lessens the effect of Canada's selection system for independent applicants. Moreover, selection is based on human capital characteristics of the principal applicant, not those of his or her spouse and children, who also gain permanent residency status as dependants. While about two-thirds of US immigrants are admitted under family preferences, another 20 per cent are admitted under employment preferences, and 10 per cent arrive as refugees and asylum seekers (Martin and Midgley, 2006: 5); only about 40 per cent of the black immigrants from Africa (the source region for the second largest number of black immigrants to the US) are admitted under family preferences. Almost 30 per cent arrive as refugees and asylum seekers, and 20 per cent enter under the diversity visas program (Kent, 2007: 7). Therefore, the immigration policies of the US and Canada are more complicated than often reported and the different recruitment strategies do not appear to produce significant socio-economic differences between black immigrants in both countries.

The Socio-economic Status of Women and Men

Gender is a key factor that distinguishes Afro-Caribbean immigrants in both countries. Unlike many immigrant groups in which men represent the majority, among Afro-Caribbean immigrants women traditionally have outnumbered men and this trend continues to the present. In Canada, 56.5 per cent of all Caribbean immigrants are women, and even among the most recent cohort of newcomers (2003–6) women still constitute the majority (54.1 per cent). In the US, women constitute 53.4 per cent of all Caribbean immigrants.[16] In contrast, among immigrants from Africa to Canada there is almost parity between women and men (51.5 per cent men) while a higher percentage of African immigrants to the US are men (55 per cent). As the African immigrant population is small relative to the numbers coming from the Caribbean, the larger number of African men seems muted compared to the Caribbean population. As a consequence, among black foreign-born individuals living in the US and Canada, women are the numerical majority (53 and 54 per cent, respectively).

The reasons for the predominance of women among Afro-Caribbean migrants in the US and Canada can be traced to labour market conditions in many Caribbean countries and the nature of immigration regulations and employment opportunities in North American cities. As Foner (2005: 158) notes, since the days of plantation slavery Caribbean women have worked outside of the home. Even after emancipation, many women were involved in various forms of paid employment; however, in countries like Jamaica, female unemployment rates have usually been more than twice those of men, a fact that has been a significant push factor encouraging migration. Researchers also have pointed to socio-cultural factors as encouraging migration in that economic independence is regarded as a fundamental dimension of self-image and in most Caribbean countries no social barriers discouraged women from migrating without a male partner (ibid., 159).

In the US, Caribbean women have outnumbered men in the legal migration stream in most years since the mid-1960s. For people not migrating to join existing family members in the US, the 1965 legislation favours individuals with particular occupational experiences and skills, and women typically have been more likely to attain labour certification than men. Many Caribbean women qualify to enter the US because they respond to the demand for domestic labour in American cities. Women also easily obtain entry visas as nurses. Given the gender stratification of the labour market, women applicants have been more willing to take jobs for which there is strong demand and their applications have been looked on more favourably by immigration officials. Similarly, in Canada women domestics led the first movement of Afro-Caribbean immigrants to Canada in the early post-war years. The largest flow of Afro-Caribbeans to Canada began in 1955 when a Household Servant Scheme launched by the federal government permitted 100 young women to migrate. The program was a success, at least for officials and employers, and by 1959 the quota was raised to 280 women per year. As in the US, the domestic recruitment program was a response to a shortage of household labour for affluent households in large Canadian cities. Women who came to Canada under this program entered as 'temporary' labour migrants but could convert to permanent residence status after their initial two-year contract had finished. Once they attained permanent residence status, they could then sponsor close family members and in essence they became the nuclei around which the Afro-Caribbean community grew in cities like Toronto and Montreal. With the elimination of some of the most racist dimensions of immigration policy and regulations in the 1960s, more Afro-Caribbeans migrated to Canada as independent economic applicants, not just through domestic labour recruitment schemes. Women still outnumbered men in the migration flow, a trend that continues to the present day. Even among Haitians, who often left their country of birth for reasons of political persecution as well as poverty and an absence of opportunity, in every year since 1980 women have predominated (Mensah, 2010: 113).

The larger number of women than men in the black population, especially among the Afro-Caribbeans, is integral to understanding the socio-economic status of the communities in both countries. As Ray and Rose argue in Chapter 7, gender inequalities endemic to North American society most certainly structure the nature of opportunities for women. While in both countries few major differences distinguish black immigrant women from men with regard to demographic characteristics, immigration timing, region of origin, and regions of settlement, they are much more dissimilar in socio-economic characteristics. For instance, black immigrant women in both the US and Canada are less likely than their male counterparts to hold a university degree (Table 11.2). Women are also more likely than men to be employed in low-paying service, sales, and office occupations. Although defined slightly differently in the US and Canada, 'service' occupations truly highlight the gender gap. In Canada, only 7.6 per cent of employed black immigrant men work in these occupations compared to 19 per cent of all women workers. Conversely, a much larger proportion of men work in construction, maintenance and repair, production/manufacturing, and transportation occupations—in the US, 20.8 per cent of men work in production and transportation occupations compared to only 5 per cent of women.

These gender differences in educational and occupational characteristics are clearly reflected in lower incomes for women (Table 11.2). In both countries, total income (employment, government transfers, and interest/investment) for women is substantially less than that for men. In the US, a gap of $8,620 in mean gross income occurs between men and women; in Canada the gap is $5,389. Canadian data indicate that the income gap—$3,292—is marginally less severe when after-tax income is considered, suggesting that the redistributive qualities of the tax and social welfare system have some effect in reducing the inequality between women and men. The Canadian data, however, make clear that the incomes of employed black immigrant women are substantially lower than those of men—a gross wage and salary difference of $4,982 exists. Gender inequalities are a persistent characteristic of the Canadian labour market, especially among older workers and those without post-secondary education qualifications, and as Pendakur and Pendakur (2002) have demonstrated, the incomes of foreign-born visible minority women are particularly low relative to other groups when age, gender, and minority status are controlled. These data emphasize Foner's (2005: 166) point that black immigrant women are somewhat unique in that they experience a triple oppression in the labour market: lower-class social status, being black in white-dominated societies, and gender discrimination.

When a sizable proportion of a social group has relatively low income, it is not surprising that poverty is a problem. Among black immigrants, a higher proportion of women than men live below the poverty line either in families or in lone-person households (Table 11.2). Although the measurement of poverty is different in both countries,[17] the gender difference does persist. For instance, in the US, 16.8 and 13.3 per cent of black immigrant women and men, respectively, live in families with incomes below the poverty level, while the same is true of 28.3 and 23.8 per cent of black immigrant women and men in Canada. The poverty of black immigrant women is compounded by the fact that they are much more likely to lead lone-parent families, whereas a larger proportion of men are married or living on their own. In Canada, 32.2 per cent of black immigrant women live in lone-parent families and less than half live in married or common-law households. The proportion of foreign-born black women living in lone-parent households in the US is slightly lower (24.6 per cent) and more individuals live in married-couple households (61.9 per cent); still, a significant portion of the population live in households with only one adult household

Table 11.2 Characteristics of Foreign-Born Black Women and Men, US and Canada

	United States			Canada	
	Foreign-Born Men	Foreign-Born Women		Foreign-Born Men	Foreign-Born Women
Region of Birth (%)					
Caribbean	52.9	59.9	Caribbean	51.8	57.4
Africa	33.3	26.4	Africa	37.9	33.2
All other	13.8	13.7	All other	7.7	7.8
Educational Attainment for Persons 25 Years and Older (%)					
Less than high school diploma	18.9	20.8	Less than high school diploma	13.8	16.9
High school diploma but no BA	52.1	55.1	High school diploma but no BA*	64.2	68.2
Bachelor's degree or higher	29.0	24.1	Bachelor's degree or higher	22.0	14.9
Household Type for Individual Respondents (%)					
One family, married or common law	62.7	61.9	One family, married or common law	59.1	45.7
One family, lone parent	19.9	24.6	One family, lone parent	14.4	32.2
Multiple family	n.a.	n.a.	Multiple family	4.7	6.2
Non-family (single & multiple person)	15.1	12.1	Non-family (single & multiple person)	21.8	15.9
Unemployment and Occupational Characteristics for Employed Labour Force (%)					
Unemployed	8.1	8.1	Unemployed	8.9	10.5
Management, professional, & related	26.9	33.8	Management, professional, & related	25.5	28.7
Services	19.7	36.0	Services	7.6	19.0
Sales & office	18.2	24.6	Sales & office	26.0	42.3
Farming, fishing, & forestry	0.5	0.1	Farming, fishing, & forestry	1.7	0.3
Construction, extraction, maintenance, & repair	13.8	0.5	Construction, extraction, maintenance, & repair	9.8	0.8
Production, transportation, & material moving	20.8	5.0	Production, transportation, & material moving	29.4	8.9
Income and Poverty Status					
Mean income	$32,015	$23,395	Mean income (individuals) Total (pre-tax)	$30,572	$25,183
			Wages & salaries (gross)	$22,930	$17,948
Median income	$24,000	$16,900	Median income (individuals) Total (pre-tax)	$23,000	$20,000
% living in families with incomes below the poverty level poverty level	13.3	16.8	% living in families below low-income cut-off or low-income person (after tax)	23.8	28.3

*In Canada this category includes: high school certificate or equivalent; apprenticeship or trades certificate or diploma; college, CEGEP, or other non-university certificate or diploma; or university certificate or diploma below the bachelor's level.

n.a. = not available

Sources: US Bureau of the Census (2009); Statistics Canada (2006a, 2006b).

head. Given the prevalence of this family type, it is not surprising that poverty is such a significant factor in women's lives in both countries.

There is reason for optimism, especially regarding the social mobility of the children of immigrants (i.e., the second generation). A landmark study by Kasinitz et al. (2008: 176–7) of the second generation living in New York City points to considerable convergence in the earnings of several ethnic and racial groups, including West Indians, when educational attainment is controlled in the analysis. Second-generation women overall are less likely to be in the labour force and working full-time than men, although the gender gap is narrowest for West Indians. Similarly, comparative US–Canada research by Attewell et al. (2010) finds that the second generation does substantially better than their immigrant parents in both countries. No strong evidence in either country suggests downward mobility into an impoverished underclass, and in Canada second-generation blacks approach socio-economic parity with Canadian-born whites (see also Reitz and Zhang, 2011). Attewell et al. caution, however, that gender should be incorporated into an analysis of these broad trends. While second-generation black women in Canada have wages similar to their white counterparts, the wages of black men are still 15 per cent lower than those of white men, even after controlling for other socio-demographic characteristics. 'This finding points to the need for more finely grained research on the inter-action of gender and race in the Canadian labour market, and indeed in many aspects of Canadian life' (Attewell et al., 2010: 490).

CONCLUSION

Among a number of important similarities between immigrant blacks living in the US and Canada, most notable is that they are overwhelmingly (more than 90 per cent) concentrated in major metropolitan areas. New York City, Miami–Fort Lauderdale, Washington, and Atlanta in the US and Toronto, Montreal, and Ottawa in Canada are the places where immigrant blacks negotiate settlement and encounter complicated histories of discrimination that can be a challenge to making a new life. The black immigrant communities in both countries are dominated by individuals from the Caribbean, although in recent years numbers of migrants from sub-Saharan Africa have been growing. The black immigrant population in both countries is an amalgamation of cultural and linguistic communities. In many ways, categories such as 'black' and 'black community' reveal more about processes of labelling in North America than about an under-lying commonality that implicitly unites people from very different parts of the world.

Our research indicates that immigrant blacks in both countries have a higher SES than US- and Canadian-born blacks. Despite differences in immigration laws and policies, immigrant blacks in both countries have similar socio-economic characteristics, although US blacks appear to be slightly better off in terms of education achievement levels and types of jobs. Our research also emphasizes that gender must be considered when interpreting the socio-economic status of black immigrants, especially relative to other immigrant groups, given that women are the numerical majority. Black immigrant women, like women generally in the US and Canada, struggle in lower-paying jobs, have lower education levels than Canadian- and US-born women, and often lead lone-parent households. The income inequality experienced by black immigrant women is especially striking when compared to that of men.

In terms of settlement geography, immigrant blacks are highly concentrated in African-American neighbourhoods in US cities, but in Canada blacks do not dominate any neighbourhoods and they are highly likely to live with many other ethnic and racial groups. It has been argued that this geography of settlement often creates double challenges for black immigrants in the US: they can experience discrimination from whites and African-American neighbours may not regard them as being culturally 'black' given their foreign-born status (Waters, 1999: 7, 65, 192, 199, 287; Shaw-Taylor, 2007: 22; Jones, 2008: 125–6; Berlin, 2010: 217–19). Recent research, however, suggests that second-generation blacks seem to triumph in the face of place-based inequities. In both countries, the second generation achieves significantly better incomes and social mobility than their immigrant parents. The same, however, is not true of third and subsequent generations; their SES pales in relation to that of the second generation (Kasinitz et al., 2008; Attewell et al., 2010).

The fact that immigrant blacks in the United States are funnelled into predominantly African-American neighbourhoods can have a profound influence on their lives because racial segregation for blacks is unlike segregation for other racial groups (Crowder and Tedrow, 2001: 88–104; Boswell and Jones, 2006: 155–74; Logan, 2007: 56–64; Boswell and Sheskin, 2009: 188–203). Inner-city poor black neighbourhoods, when compared to other neighbourhoods, tend to be characterized by numerous social problems including higher crime rates, elevated rates of unemployment, less effective schools, and greater school dropout rates. By the same token, middle-class suburban black neighbourhoods also share many of these same problems (Pattillo-McCoy, 1999).

The conditions of segregation for immigrant blacks are very different in the United States and Canada, suggesting that the future for Canadian black immigrants, their children, and grandchildren may be brighter than for black immigrants in the US. The history of segregation in US cities is deep and meaningful, and it continues to have consequences for US- and foreign-born blacks. The finding of Attewell et al.—that third-generation blacks in the US, who grow up immersed in the spatial segregation of American cities, do less well than the second generation with strong ties to an immigrant experience—raises concern about the long-run prospects for the descendants of today's black immigrants. In contrast, the absence in Canada of a history in which space was used to create geographies of exclusion and deprivation means that substantially fewer Canadian black immigrants are trapped in poor neighbourhoods. They are not likely to be as disadvantaged by associations with a culture of poverty to the same degree that many, especially working-class, black immigrants are in the US.

QUESTIONS FOR CRITICAL THOUGHT

1. Why is the spatial segregation of US- and foreign-born blacks more than a neutral fact? How does such segregation influence life experiences and opportunities for social mobility among immigrants and their children? Is there evidence of growing or diminishing spatial concentration among black immigrants in Canadian cities?

2. Canada has an official policy of multiculturalism whereas the US does not. Based on what you have read in this chapter regarding the experiences of black immigrants in the US and Canada, do you favour a government-sponsored multiculturalism policy like Canada's, or do you think such a policy slows the integration of black immigrants into the mainstream of Canadian and American society?

3. If the decision were up to you, what policies would you put in place to address some of the socio-economic challenges experienced by blacks in Canada and the US? For instance, would you direct your attention to the quality of housing and neighbourhoods, or to anti-racism programs in schools and workplaces? What kinds of policies would make a difference?

SUGGESTED READINGS

1. Berlin, Ira. 2010. *The Making of African America: The Four Great Migrations.* New York: Viking Press. This history of the African-American population begins with the black slave trade that brought 400,000 Africans to the eastern tidewater of the United States and ends with broader voluntary immigration from Africa, Latin America, and Europe that has focused on a few of the larger cities of the eastern US.

2. Foner, Nancy. 2005. *In a New Land: A Comparative View of Immigration.* New York: New York University Press. Foner compares immigrants living in New York City with those in other cities in the US and Europe. The book also examines differences and similarities between New York's two great immigrant waves: those who arrived a century ago and those arriving since the late 1960s.

3. Henry, F. 1994. *The Caribbean Diaspora in Toronto.* Toronto: University of Toronto Press. This study of post-war Caribbean immigrants in Toronto emphasizes processes of community development, racialization, and experiences of discrimination.

4. Jones, Terry-Ann. 2008. *Jamaican Immigrants in the United States and Canada: Race, Transnationalism and Social Capital.* New York: LFB Scholarly Publishing. Basing her work on questionnaire surveys of immigrant Jamaicans and on the immigration literature dealing with race, transnationalism, and social capital, Jones compares Jamaican immigrants living in Miami and Toronto.

NOTES

1. This is similar to a question asked by Mary Waters in her study of black West Indian immigrants living in Brooklyn, New York (Waters, 1999: 3).
2. In accordance with the US Bureau of the Census definition, here 'black' refers to individuals who consider themselves to be black or African American alone or in combination with one or more other races (Tables B01003 and B02009, US Bureau of the Census, American Community Survey, 2008).
3. The black population in Canada includes only individuals who self-identify as 'black' in response to a census question about visible minority status. In 2006, there were 783,795 blacks in Canada, making blacks the third largest visible minority group after South Asians and Chinese. Individuals who identify as members of more than one racial group (2.6 per cent of the total) are not included in the overall 'black' count.
4. This figure is derived from calculations based on data in the US Bureau of the Census Public Use Microdata files of the 2005–7 American Community Surveys.
5. For the purposes of this chapter, we define the West Indies to include the islands in the Caribbean where English is the official language. It thus excludes Spanish-speaking islands such as Cuba, the Dominican Republic, and Puerto Rico. It also includes Guyana in South America and Belize in Central America.
6. Slavery's abolition has a complicated legislative history in Canada. The Legislative Assembly of Upper Canada passed a law in 1793 that effectively made it illegal to buy new slaves and began a process of phasing slavery out of existence, and over time court rulings in the other Canadian provinces made it much more difficult for individuals to hold slaves (usually by requiring proof of ownership).
7. For example, the black Loyalists reportedly joined approximately 500 black slaves in Nova Scotia (Mensah, 2010: 47).
8. 'Black' is used here as it was by the Department of Citizenship and Immigration to refer to people who phenotypically did not appear to be 'white'. This included Afro-Caribbeans, as well as Asians, Asian Africans, and anyone else who did not look white. In this regard, judgements regarding an applicant's race were subjective.
9. Memo from the Director, Immigration Branch, to Deputy Minister, Department of Citizenship and Immigration, 14 Jan. 1955, Library and Archives Canada (LAC), RG 76, Vol. 830, File # 552–1–644, Part 2.
10. E. Carrigan, 'Limit Immigration', *Toronto Telegram*, 21 July 1971, p. 17.
11. Memo from Director, Programs and Procedures Branch, to Assistant Deputy Minister (Immigration), 20 Sept. 1973, LAC, RG 76, Vol. 83–84/349, Box 107, File IM 5750–5.
12. An additional 88,720 people born in Guyana were also living in Canada in 2006. For historical and cultural reasons, Guyanese immigrants are frequently considered part of the Caribbean diaspora in Canada. All of the data presented in this chapter, however, only consider immigrants from the Caribbean proper and Bermuda.
13. The analysis is based primarily on data from DLI (Data Liberation Initiative) special tabulations. When special tabulation data are not available, analysis is based on public use micro-data files for individuals from the 2006 Canadian census.
14. US Bureau of the Census, 2000 Census of Population, Table PCT2, 'Urban and Rural', Summary File 4.
15. SES is a multiple variable concept. Normally it is measured taking into consideration educational achievement, the occupations people have, and measures of income and sometimes poverty levels. Measures of these variables are found in Tables 11.1 and 11.2 in this chapter.
16. The US data are from the 2000 census (*American FactFinder*).
17. The measure of poverty in Canada is based on the concept of a low-income cut-off (LICO). The LICO is an income threshold at which families or individuals would spend 20 percentage points more than a comparable average family or individual on food, shelter, and clothing. In the US, poverty is based on family income relative to a pre-defined poverty threshold for particular family configurations. The poverty thresholds vary depending on: family size, number of children, and location of residence.

REFERENCES

1. Attewell, P., P. Kasinitz, and K. Dunn. 2010. 'Black Canadians and Black Americans: Racial Income Inequality in Comparative Perspective', *Ethnic and Racial Studies* 33, 3: 473–95.

2. Beauregard, R.A. 2003. *Voices of Decline: The Postwar Fate of U.S. Cities*. New York: Routledge.

3. Berlin, Ira. 2010. *The Making of African America: The Four Great Migrations*. New York: Viking Press.

4. Boswell, Thomas D., and Terry-Ann Jones. 2006. 'The Distribution and Socio-economic Status of West Indians Living in the United States', in John W. Frazier and Eugene L. Tettey-Fio, eds, *Race, Ethnicity, and Place in a Changing America*. Binghamton, NY: Global Academic Publishing, 155–80.

5. ———— and Ira M. Sheskin. 2009. 'Deconstructing the Black Populations of New York City and Miami-Dade County', in Frazier et al. (2009: 185–211).

6. Crowder, Kyle D., and Lucky M. Tedrow. 2001. 'West Indians and the Residential Landscape of New York', in Nancy Foner, ed., *Islands in the Sun: West Indian Migration to New York*. Berkeley: University of California Press, 81–114.

7. Darden, Joe T., and Carlos Teixeira. 2009. 'The African Diaspora in Canada', in Frazier et al. (2009: 13–34).

8. Davis, David Brion. 2006. *Inhuman Bondage: The Rise and Fall of Slavery in the New World*. London: Oxford University Press.

9. Foner, Nancy. 2005. *In a New Land: A Comparative View of Immigration*. New York: New York University Press.

10. Frazier, John W., Joe T. Darden, and Norah F. Henry, eds. 2009. *The African Diaspora in the United States and Canada at the Dawn of the 21st Century*. Binghamton, NY: Global Academic Publishing.

11. Jones, Terry-Ann. 2008. *Jamaican Immigrants in the United States and Canada: Race, Transnationalism and Social Capital*. New York: LFB Scholarly Publishing.

12. Kasinitz, P., J.H. Mollenkopf, M.C. Waters, and J. Holdaway. 2008. *Inheriting the City*. New York and Cambridge, Mass.: Russell Sage Foundation and Harvard University Press.

13. Kent, Mary Mederios. 2007. 'Immigration and America's Black Population', *Population Bulletin* 62, 4: 1–16.

14. Logan, John R. 2007. 'Who Are the Other African Americans? Contemporary African and Caribbean Immigrants in the United States?', in Yoku Shaw-Taylor and Steven A. Tuch, eds, *The Other African Americans: Contemporary African and Caribbean Immigrants in the United States*. New York: Rowman & Littlefield, 49–67.

15. Martin, Philip, and Elizabeth Midgley. 2006. 'Immigration: Shaping and Reshaping America', *Population Bulletin* 61, 4: 1–28.

16. Mensah, Joseph. 2010. *Black Canadians: History, Experience, Social Conditions*. Halifax: Fernwood.

17. Milan, A., and K. Tran. 2004. 'Blacks in Canada: A Long History', *Canadian Social Trends* (Spring): 2–7.

18. Pattillo-McCoy, M. 1999. *Black Picket Fences*. Chicago: University of Chicago Press.

19. Pendakur, K., and R. Pendakur. 2002. 'Colour My World: Have Earnings Gaps for Canadian-born Ethnic Minorities Changed over Time?', *Canadian Public Policy* 28, 4: 489–512.

20. Reid, Ira De Augustine. 1969. *The Negro Immigrant: His Background, Characteristics and Social Adjustment, 1899–1937*. New York: Arno Press and the New York Times.

21. Reitz, Jeffrey G., and Ye Zhang. 2011. 'National and Urban Contexts for the Integration of the Immigrant Second Generation in the United States and Canada', in Richard Alba and Mary Waters, eds, *New Dimensions of Diversity: The Children of Immigrants in North America and Western Europe*. New York: New York University Press.

22. Segal, Ronald. 1995. *The Black Diaspora: Five Centuries of the Black Experience outside Africa*. New York: Farrar, Straus, and Giroux.

23. Shaw-Taylor, Yoku. 2007. 'The Intersection of Assimilation, Race, Presentation of Self, and Transnationalism in America', in Yoku Shaw-Taylor and Steven A. Tuch, eds, *The Other African Americans: Contemporary African and Caribbean Immigrants in the United States*. New York: Rowman & Littlefield, 1–47.

24. Statistics Canada. 2006a. *Census of Canada, 2006*, special tabulations made available through Data Liberation Initiative.

25. ———. 2006b. *Census of Canada, 2006*, Public Use Microdata File.

26. US Bureau of the Census. 2003. *2000 Census of Population*, Summary File 4. Washington.

27. ———. 2009. 2005–7 American Community Survey. Public Use Microdata Samples (PUMS). Washington.

28. Waters, Mary C. 1999. *Black Identities: West Indian Immigrant Dreams and American Realities*. Cambridge, Mass.: Harvard University Press.

CHAPTER 12

LATIN AMERICAN IMMIGRANTS: PARALLEL AND DIVERGING GEOGRAPHIES

Luisa Veronis and Heather Smith

INTRODUCTION

This chapter provides a comparative geography of Latin American immigration to cities in the United States and Canada. Our aim is to examine and to reflect on how broader trends and patterns intersect with local and micro-scale forces to yield different settlement outcomes and experiences at both the national and urban scales. To help readers understand the degree of difference between Latin American immigrant geographies in each country, the chapter begins with an overview of the contemporary Latin American immigrant population in the US and Canada, focusing on spatial distribution and selected markers of demography and economic performance. This section is followed by a historical overview of Latin American immigration on each side of the border, paying particular attention to the explanatory influence of macro- and micro-scale factors. Because it provides fruitful ground for comparative work with the potential to contribute to broader theories of (im)migration and North American urban geography, the chapter concludes with an examination of the new geographies of settlement outside main gateway cities—a shared focus of recent work in both countries.

It is important to clarify what we mean by 'Latin Americans'. Throughout this chapter, we use the term to refer to individuals who are resident in either the US or Canada but were born in Mexico or Central or South America.[1] For the most part, this foreign-born population claims heritage from one of the countries colonized by Spain. Notable exceptions are Brazil, Guyana, and Suriname, whose colonial legacies differed and yielded Portuguese, English, and Dutch, respectively, as dominant languages rather than Spanish. While some of our comments refer to the Hispanic/Latino ethnic group to which the majority of Central and South American immigrants to Canada and the US belong, when we refer to Latin Americans in this chapter, we are referring exclusively to the foreign-born.

LATIN AMERICAN IMMIGRANTS IN THE US AND CANADA

Regions and Countries of Origin

A useful starting point to illustrate differences and parallels is an examination of the most recent census data on Latin American immigrants in the US and Canada.[2] Specific variables—immigrant status, citizenship, educational achievement, employment, occupation, and household income—highlight key differences in the US versus Canadian Latin American immigration experience.

A reflection of the long-standing history of Latin American immigration to the US compared to the recent arrival of this group in Canada, the starkest difference between the Latin American populations in the two countries is their comparative size. Whereas there are only 409,415 Latin American immigrants in Canada, there are 16,734,115 immigrants from Latin America in the US.[3] Representing about 5.6 per cent of the US population, and 44.4 per cent of the total US immigrant population, the foreign-born from Latin America are an influential group with significant implications for social policy and politics. On occasion, they have played influential roles in shaping domestic policy (e.g., immigration) as well as foreign policy (e.g., US relations with Latin America), and politicians increasingly are compelled to address the interests and expectations of this group at both local and national scales.

In contrast, Latin-American immigrants make up about 1.3 per cent of the Canadian population and 6.3 per cent of the foreign-born population. Thus, the group's social and political presence is significantly weaker than its counterpart in the US. However, Latin Americans are one of the fastest-growing immigrant groups in Canada; growing at a rate of 13.2 per cent[4] in the 2001–6 period. The result of new immigration rather than natural increase, as compared to 3.3 per cent natural increase for the Canadian-born population, this trend suggests that the Latin American presence will become more significant in the near future.

As Table 12.1 shows, the national origins of Latin Americans in each country also display clear differences. In the US, those born in Mexico are by far the largest foreign-born Latin American group, representing 68.6 per cent of the total. Those born in El Salvador and Guatemala are the next most dominant, representing only 6.4 per cent and 4.3 per cent, respectively. Immigrants from South America represent only 15.3 per cent of the Latin American total, with Colombia contributing the highest number.

In Canada, South Americans represent the largest share of the total at 64.6 per cent—a proportion far higher than in the US. Also, in Canada no single national group dominates over others, at least not as dramatically as Mexican immigrants do in the US. Even though Mexicans are the largest group in Canada, they represent only 15 per cent of the Latin American population, closely followed by Salvadorans and Colombians. The next cluster of dominant groups includes

Table 12.1   Countries of Origin of the Latin American Foreign-Born Populations, US and Canada

Latin America and Caribbean		US (2006–8) Population Count	%of Total	Canada (2006)* Population Count	% of Total
	Caribbean**	3,373,906	16.8	327,650	44.5
	Central America and Mexico	14,177,884	70.5	144,800	19.6
	South America	2,556,231	12.7	264,615	35.9
Latin American total		16,734,115		409,415	
Central America and Mexico		14,177,884	84.7	144,800	35.4
	Mexico	11,478,376	68.6	61,470	15.0
	Belize	n.a.	n.a.	2,145	0.5
	Costa Rica	82,167	0.5	3,700	0.9
	El Salvador	1,078,643	6.4	43,805	10.7
	Guatemala	720,852	4.3	16,150	3.9
	Honduras	428,543	2.6	5,365	1.3
	Nicaragua	233,592	1.4	9,245	2.3
	Panama	98,609	0.6	2,920	0.7
	Central America n.e.c.	57,082	0.3	n.a.	n.a.
South America		2,556,231	15.3	264,615	64.6
	Argentina	167,208	1.0	19,910	4.8
	Bolivia	70,679	0.4	4,050	1.0
	Brazil	339,771	2.0	17,850	4.3
	Chile	89,013	0.5	27,315	6.6
	Colombia	596,511	3.6	42,550	10.4
	Ecuador	400,946	2.4	13,895	3.4
	Falkland Islands	n.a.	n.a.	15	0.0
	French Guiana	n.a.	n.a.	70	0.0
	Guyana	251,651	1.5	88,720	21.6
	Paraguay	n.a.	n.a.	7,680	1.9
	Peru	384,959	2.3	23,200	5.6
	Suriname	n.a.	n.a.	820	0.2
	Uruguay	49,698	0.3	7,150	1.7
	Venezuela	160,750	1.0	11,385	2.8
	South America n.e.c.	45,045	0.3	n.a.	n.a.

*Total immigrant status population, including non-permanent residents.

**For Canada, Caribbean includes Bermuda.

n.a. = not available.

n.e.c.= not elsewhere classified.

Sources: Statistics Canada, *Census of Canada, 2006*; US Bureau of the Census, 2006–8 American Community Survey, three-year estimates.

Chileans, Peruvians, and Argentines. In other words, when compared to the US, Canada has a more even representation of immigrants from Latin American countries, which has significant implications for differential settlement experiences and internal group dynamics. Before we examine the macro- and micro-scale factors that have contributed to these differences, we now turn to a detailed contemporary portrait of Latin American immigrants in Canada and the US.

Geographical Distribution: National Scale

Reflecting proximity to place of origin, comparative degrees of economic opportunity, and historically developed social networks, about 74 per cent of the Latin American foreign-born reside in seven US states, with the highest populations by far in California (5,386,259) and then Texas (2,771,027), Florida (1,196,771), and New York (1,051,705) (Figure 12.1). As for country of birth concentration, settlement of the Mexican-born is focused on Texas and California, which together account for 58.3 per cent of all Mexican-born immigrants in the country. Colombians, on the other hand, concentrate in Florida, New York, and New Jersey, where 63 per cent of all the Colombian-born immigrants in the US reside.

Latin Americans in Canada concentrate in just four provinces (Figure 12.2). About 57 per cent of all Latin Americans live in Ontario; 20.6 per cent live in Quebec; and 16.8 per cent are distributed almost equally between British Columbia and Alberta.

Geographical Distribution: Metropolitan/Urban Scale

In both countries, Latin American immigrants gravitate to traditional gateway and large urban centres (Tables 12.2 and 12.3). Data for 2006–8 show that the Los Angeles, New York, Houston, Chicago, Dallas–Fort Worth, Miami, and Riverside–San Bernadino **metropolitan statistical areas** (MSAs) had the highest total number of Latin American foreign-born among their populations. They were also the only metros with numbers in excess of 500,000. With over 2.5 million foreign-born Latin Americans among its population, Los Angeles is home to over 15 per cent of the nation's total Latin American foreign-born population. In El Paso and Miami, over 90 and 85 per cent of the foreign-born come from Latin America, representing over a quarter of these cities' total populations. Indeed, over half of the other US metropolitan areas listed in Table 12.2 also show Latin Americans as representing more than 50 per cent of their total foreign-born populations. Not reflected in this table is the recent settlement shift of immigrants away from traditional gateway states and cities towards second- and third-tier cities, particularly in the US South and Mountain West.

Table 12.3 illustrates significant discrepancies between the national and urban geographies of Canada and the US. Canada's smaller urban system (both city size and number of cities) and its relatively smaller Latin American foreign-born population mean that the size and share of Latin Americans in Canadian cities pales in comparison to their counterparts in US cities. Nevertheless, Latin

American immigrants are much more concentrated in Canada's two largest census metropolitan areas (CMAs) (and immigrant gateways): almost 60 per cent of the total population born in Latin America reside in Toronto (40 per cent) and Montreal (17.2 per cent). The percentages of the population born in Latin America

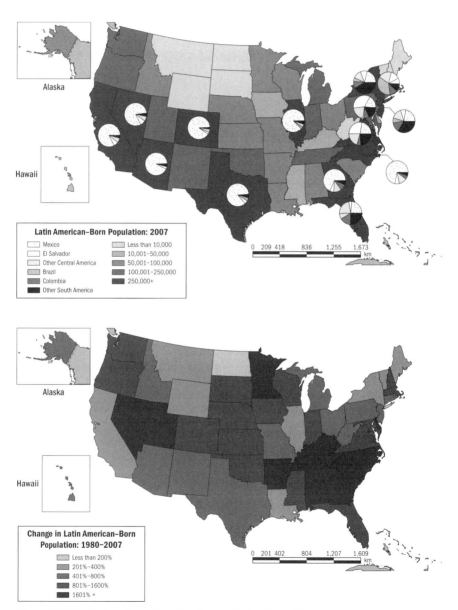

Figure 12.1 Latin American–Born Immigrants in the United States

Source: Data from US Census Bureau. Cartography by Laura Simmons and Thomas Ludden, University of North Carolina, Charlotte.

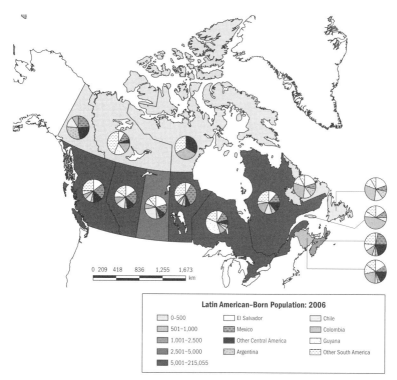

Figure 12.2 Latin American–Born Immigrants in Canada, 2006

Source: Data from Statistics Canada, *Census of Canada, 2006*. Cartography by Thomas Ludden, University of North Carolina, Charlotte.

(among the foreign-born and the total population) are relatively low compared to US cities. Latin Americans, however, represent a larger share of the foreign-born population in some second-tier cities (Quebec City and London and Kitchener–Waterloo in Ontario) and third-tier cities (Gatineau and Sherbrooke in Quebec and Abbotsford, BC); these are non-traditional gateway cities and belong to the 'new geographies of settlement', discussed below. Finally, Leamington, Ontario, and Lethbridge, Alberta, are smaller agricultural towns with populations under 100,000 that rely extensively on Mexican migrants (both permanent residents and temporary seasonal agricultural workers) to support their economies. The over-representation of Mexican migrants is exceptional.

SELECTED DEMOGRAPHIC AND SOCIO-ECONOMIC DETAILS

Immigrant Status

Three components of immigrant status differentiate the US Latin American population from that of Canada: time of arrival, naturalization/citizenship rates, and authorized versus undocumented status. Table 12.4 shows how immigrants

Table 12.2 Top 20 Metropolitan Settlement Areas, Latin American Foreign-Born in the US

Metropolitan Area	Total Population	Total Foreign-Born	Foreign-Born from L. America	% of Foreign-Born from L. America	% of Total City Population	% of Total US L. American Population
Los Angeles–Long Beach–Santa Ana, Calif.	12,818,132	4,394,068	2,528,562	57.5	19.7	15.1
New York–Northern New Jersey–Long Island (NY/NJ/Penn.)	18,925,869	5,317,616	1,446,105	27.2	7.6	8.6
Houston–Sugar Land–Baytown, Tex.	5,603,882	1,198,503	808,761	67.5	14.4	4.8
Chicago–Naperville–Joliet (Ill./Ind./Wis.)	9,502,094	1,675,949	789,453	47.1	8.3	4.7
Dallas–Fort Worth–Arlington, Tex.	6,150,828	1,090,385	726,455	66.6	11.8	4.3
Miami–Fort Lauderdale–Pompano Beach, Fla	5,403,075	1,994,677	687,278	34.5	12.7	4.1
Riverside–San Bernardino–Ontario, Calif.	4,054,985	893,600	652,853	73.1	16.1	3.9
Phoenix–Mesa–Scottsdale, Ariz.	4,160,999	692,038	482,073	69.7	11.6	2.9
San Francisco–Oakland–Fremont, Calif.	4,222,756	1,245,521	393,741	31.6	9.3	2.4
Washington–Arlington–Alexandria (DC/Va/Md/WV)	5,306,742	1,073,591	365,083	34.0	6.9	2.2
San Diego–Carlsbad–San Marcos, Calif.	2,965,943	671,608	352,790	52.5	11.9	2.1
Atlanta–Sandy Springs–Marietta, Ga	5,251,899	674,422	285,168	42.3	5.4	1.7
Las Vegas–Paradise, Nev.	1,821,359	397,205	226,270	57.0	12.4	1.4
San Jose–Sunnyvale–Santa Clara, Calif.	1,789,271	649,751	188,140	29.0	10.5	1.1
El Paso, Tex.	731,496	198,834	184,694	92.9	25.3	1.1
Denver–Aurora, Colo.	2,454,378	309,255	175,463	56.7	7.2	1.1
San Antonio, Tex.	1,982,788	213,491	163,434	76.6	8.2	1.0
Boston–Cambridge–Quincy (Mass./NH)	4,494,144	716,264	149,703	20.9	3.3	0.9
Austin–Round Rock, Tex.	1,590,744	228,455	146,012	63.9	9.2	0.9
Fresno, Calif.	895,357	189,757	132,379	69.8	14.8	0.8

Source: US Bureau of the Census, 2006–8 American Community Survey, three-year estimates.

Table 12.3 Top 20 Metropolitan Settlement Areas, Latin American Foreign-Born in Canada*

Metropolitan Area	Total Population	Total Foreign-Born	Foreign-Born from Latin America	% of Foreign-Born from L. America	% of Total City Population	% of Total Canadian L. American Population
Toronto	5,113,150	2,396,485	164,025**	6.8	3.2	40.0
Montreal	3,635,570	782,290	70,370	8.9	1.9	17.2
Vancouver	2,116,580	870,470	25,935	2.9	1.2	6.3
Calgary	1,079,310	264,650	14,235	5.3	1.3	3.4
Ottawa–Gatineau (Ont. and Que.)	1,130,760	211,380	11,340	5.3	0.9	2.7
Ottawa	846,805	187,950	8,935	4.8	1.1	2.2
Gatineau	283,960	23,430	2,400	10.2	0.8	0.6
Edmonton	1,034,945	198,635	9,615	4.8	0.9	2.3
Winnipeg	694,698	127,175	9,220	0.0	1.3	2.3
London, Ont.	457,720	91,800	8,035	8.7	8.7	2.0
Kitchener-Waterloo, Ont.	451,235	107,105	8,325	7.8	1.8	2.0
Hamilton, Ont.	692,911	172,015	7,685	4.5	1.1	1.9
Quebec City	715,515	28,440	3,420	12.0	0.5	0.8
Oshawa, Ont.	330,594	54,900	3,555	6.5	1.1	0.9
Windsor, Ont.	323,342	78,280	3,070	4.0	0.9	0.8
Victoria, BC	330,088	65,555	2,225	3.4	0.7	0.5
Abbotsford, BC	159,020	38,470	2,365	6.1	1.5	0.6
Sherbrooke, Que.	186,952	11,420	2,000	17.5	1.1	0.5
Guelph, Ont.	127,009	26,610	1,225	4.6	1.0	0.3
Lethbridge, Alta	95,196	11,620	1,215†	10.5	1.3	0.3
Leamington, Ont.	49,741	12,540	5,615‡	44.8	11.3	1.4

*Total immigrant status population,, including non-permanent residents.

**This includes a significant number of immigrants from Brazil (7,545) and Guyana (70,220).

†Among these, 465 are Mexican-born.

‡Among these, 5,005 are Mexicans, including 1,440 non permanent residents.

Source: Statistics Canada, Census of Canada, 2006.

from different countries and regions of birth compare with regard to the first two of these variables. The most striking difference between the two countries is the share of Latin American immigrants who are naturalized: 30.9 per cent in the US versus 73.3 per cent in Canada. Bloemraad (2006) suggests that the higher number of undocumented immigrants is a contributing factor in the US's lower naturalization rates. In contrast, public policy and legal structures in Canada favour naturalization that embraces multiculturalism and allows dual citizenship.[5] Despite these differences, in both countries time of arrival shapes naturalization and citizenship rates. For those who have been in either country longer, higher rates of citizenship tend to follow. Rates also vary among groups and sub-regions. In the US, Mexicans show the lowest overall rates of naturalization, even after extended time in the country (roughly tied with other Central Americans), while South Americans present higher rates. In Canada, Central Americans show somewhat higher percentages. Lower percentages among South Americans probably are related to the large share of recent arrivals from Colombia.

Major differences exist between the two countries in regard to undocumented migration. Within the US the extent of undocumented status among Latin Americans is a significant issue of public and government concern. While the flow of undocumented migrants to the country is thought to have slowed in recent years, steady growth over the 1990–2006 period brings the total current estimate to 11.9 million (10.4 million adults and 1.5 million children). A 2009 report by the Pew Hispanic Center indicates that about 76 per cent of this population is of Hispanic origin, with the majority (59 per cent) coming from Mexico (Passel and Cohn, 2009). At this estimated level, undocumented immigrants represent 4 per cent of the total American population and 5.4 per cent of its workforce. Undocumented immigrants are also the parents of an increasing number of US-born citizens. The same report revealed that almost half of the undocumented migrant households are couples with children and that of those children, 73 per cent are US-born. Such statistics are a complicating factor in national-scale efforts to revise federal immigration policy.

In contrast, undocumented migration is not as significant in Canada; the estimated count is about 200,000 or 0.65 per cent of the population. Nevertheless, since the early 2000s there has been a relative increase of undocumented migrants, especially from Latin American countries experiencing economic, social, and political unrest such as Argentina, Colombia, Ecuador, and Venezuela.

Education

Education, employment, and occupational status are important markers of economic well-being and degree of integration for any immigrant or ethnic group. Although a number of differences exist in the census categories of the two countries, three broad patterns can be discerned in the group's educational attainment. First, Latin Americans in Canada tend to have more education than those in the US; second, Latin Americans in Canada tend to be closer to the

Table 12.4 Selected Demograpic Characteristics of the Foreign-Born Latin American Population by Region of Birth, US and Canada

USA (2006–2008)

Immigration Status	Total Latin American FB	Mexico	Other Central America	Caribbean	South America
	20,108,021	11,478,376	2,699,508	3,373,906	2,556,231
Naturalized Citizen	30.9%	21.9%	30.0%	54.0%	42.0%
Entered 2000 or later	1.8%	1.3%	2.0%	2.7%	2.8%
Entered 1990 to 1999	6.2%	3.9%	5.5%	11.4%	10.4%
Entered before 1990	22.9%	16.7%	22.5%	39.9%	28.7%
Not a Citizen	69.1%	78.1%	70.0%	46.0%	58.0%
Entered 2000 or later	27.1%	28.6%	28.3%	17.9%	31.1%
Entered 1990 to 1999	24.5%	28.8%	23.5%	15.5%	18.0%
Entered before 1990	17.5%	20.7%	18.2%	12.6%	8.9%

	Total	Latin America	Mexico	Central America	Caribbean	South America
Educational Attainment						
Population 25 years and over	31,356,641	16,289,221	9,005,626	2,233,020	2,926,098	2,124,477
Less than high school graduate	32.2%	47.5%	60.6%	48.8%	27.4%	18.0%
High school graduate (includes equivalency)	23.2%	25.8%	23.8%	25.1%	30.0%	29.6%
Some college or associate's degree	17.7%	15.2%	10.4%	16.0%	23.1%	23.8%
Bachelor's degree	16.0%	7.8%	3.7%	7.4%	12.9%	18.5%
Graduate or professional degree	10.9%	3.7%	1.4%	2.7%	6.6%	10.2%
Employment Status						
Population 16 years and over	35,330,144	18,777,689	10,597,660	2,562,367	3,222,265	2,395,397
In labour force	67.6%	71.0%	70.5%	76.3%	66.9%	72.5%
Civilian labour force	67.5%	70.8%	70.5%	76.1%	66.7%	72.3%
Employed	63.6%	66.4%	66.2%	71.5%	61.8%	68.4%
Unemployed	3.8%	4.4%	4.3%	4.7%	4.9%	3.9%
Per cent of civilian labour force	5.6%	6.2%	6.1%	6.1%	7.3%	5.3%
Armed Forces	0.2%	0.1%	0.1%	0.2%	0.3%	0.2%
Not in Labour	32.4%	29.0%	29.5%	23.7%	33.1%	27.5%

	Total	Latin America	Mexico	Central America	Caribbean	South America
Occupation						
Management, professional, and related occupations	27.5%	13.4%	7.6%	11.3%	25.2%	26.0%
Service occupations	23.0%	28.2%	28.1%	31.4%	28.0%	25.2%
Sales and office occupations	18.1%	15.2%	11.7%	14.7%	22.6%	22.0%
Farming, fishing, and forestry occupations	1.9%	3.3%	5.5%	1.2%	0.3%	0.2%
Construction, extraction, maintenance, and repair occupations	13.1%	19.5%	23.5%	21.1%	9.1%	12.8%
Production, transportation, and material moving occupations	16.4%	20.4%	23.5%	20.4%	14.8%	13.9%

Earnings in the past 12 months (In 2008 Inflation adjusted dollars for full-time, year-round workers)

	Total	Latin America	Mexico	Central America	Caribbean	South America
Population 16 years and over with earnings	15,859,410	8,913,032	4,964,842	1,330,965	1,471,382	1,145,843
$1 to $9,999 or less	2.5%	3.0%	3.3%	3.2%	2.4%	2.1%
$10,000 to $14,999	7.8%	10.9%	12.8%	11.6%	7.0%	6.7%
$15,000 to $24,999	23.8%	31.9%	36.6%	33.0%	22.2%	22.8%
$25,000 to $34,999	19.3%	22.2%	22.4%	22.1%	21.6%	22.2%
$35,000 to $49,999	17.1%	16.3%	14.2%	16.4%	20.8%	19.6%
$50,000 to $74,999	14.3%	10.0%	7.5%	9.2%	15.4%	14.8%
$74,000 or more	15.1%	5.7%	3.2%	4.4%	10.6%	11.8%
Median earnings, (dollars) for full-time, year-round workers						
Male	34,364	27,151	25,485	26,952	35,728	35,865
Female	30,405	23,666	20,051	22,239	30,823	29,280

Poverty Status in the Past 12 months

	Total	Latin America	Mexico	Central America	Caribbean	South America
Population for whom poverty status is determined	37,128,875	19,830,94	11,326,289	2,666,846	3,311,719	2,526,100
Below 100 per cent of the poverty level	15.9%	19.8%	23.5%	16.7%	16.4%	11.1%
100 to 199 per cent of the poverty level	23.9%	31.2%	35.8%	30.2%	23.3%	21.8%

	Total	Latin America	Mexico	Central America	Caribbean	South America
At or above 200 per cent of the poverty level	60.3%	49.0%	40.6%	53.1%	60.3%	67.1%

Poverty Rates for Families for Whom Poverty Status is Determined

	Total	Latin America	Mexico	Central America	Caribbean	South America
All families	14.6%	19.8%	24.1%	16.6%	14.9%	9.7%
With related children under 18 years	19.1%	24.4%	28.2%	20.6%	19.5%	12.5%
With related children under 5 years only	17.1%	223.3%	27.0%	21.4%	17.9%	13.3%
Married-couple family	10.9%	15.2%	19.7%	10.8%	8.3%	6.5%
With related children under 18 years	13.5%	17.9%	22.2%	12.6%	8.7%	7.4%
With related children under 5 years only	11.5%	16.7%	21.0%	14.2%	7.4%	7.8%
Female householder, no husband present, family	30.8%	36.3%	46.3%	32.5%	27.3%	20.9%
With related children under 18 years	39.9%	44.4%	53.4%	39.9%	34.8%	27.7%
With related children under 5 years only	42.8%	47.0%	56.5%	42.5%	34.8%	33.6%

Canada (2006)

Immigration Status	Total Latin American FB	Central America	Caribbean (and Bermuda)	South America
	74,420	147,235	330,085	267,100
Naturalized citizen	73.3%	70.5%	79.1%	67.7%
Entered 2001 to 2006	2.8%	3.8%	2.0%	3.4%
Entered 2003 to 2006	0.0%	0.0%	0.0%	0.0%
Entered 2001 to 2002	2.8%	3.8%	2.0%	3.4%
Entered 1991 to 2000	20.7%	25.8%	19.9%	18.9%
Entered before 1991	48.8%	39.2%	56.5%	44.5%
Not a citizen	26.7%	29.5%	20.9%	32.3%
Entered 2001 to 2006	13.2%	12.0%	8.6%	19.6%
Entered 2003 to 2006	10.3%	9.3%	6.2%	15.8%
Entered 2001 to 2002	2.9%	2.7%	2.4%	3.8%
Entered 1991 to 2000	4.7%	4.9%	5.0%	4.2%
Entered before 1991	3.7%	3.0%	4.3%	3.3%
Non-permanent residents	5.1%	9.7%	3.0%	5.2%

	Total	Latin America	Central America	Caribbean (and Bermuda)	South America
Educational Attainment					
Total population 15 years and over	25,664,225	660,825	120,685	306,700	233,440
No certificate, diploma or degree	23.8%	21.1%	32.7%	18.3%	19.0%
Certificate, diploma or degree	76.2%	78.9%	67.3%	81.7%	81.0%
High school certificate or equivalent	25.5%	24.0%	23.2%	23.1%	25.8%
Apprenticeship or trades certificate or diploma	10.9%	12.6%	10.5%	15.2%	10.4%
College, CEGEP or other non-university certificate or diploma	17.3%	19.5%	15.2%	22.5%	17.9%
University certificate or diploma below bachelor level	4.4%	6.3%	4.3%	6.8%	6.6%
University certificate, diploma or degree at bachelor's level or above	18.1%	16.4%	14.2%	14.1%	20.4%
Labour Force Activity					
Total population 15 years and over	25,664,220	660,830	120,690	306,695	233,445
In the labour force	66.8%	72.2%	74.1%	71.2%	72.6%
Not in the labour force	33.2%	27.8%	25.9%	28.9%	27.4%
Participation rate	66.8%	na	74.1%	71.2%	72.6%
Employment rate	62.4%	na	68.5%	65.5%	67.2%
Unemployment rate	6.6%	na	7.5%	7.9%	7.4%
Occupation (National Occupational Classification for Statistics 2006)					
Total labour force 15 years and over	17,146,135	476,985	89,375	218,210	169,400
Occupation – Not applicable	1.7%	2.7%	2.5%	2.8%	2.6%
All occupations	98.3%	97.3%	97.5%	97.2%	97.4%
Management occupations	9.5%	6.6%	5.1%	6.2%	7.8%
Business, finance and administrative occupations	17.6%	18.6%	13.5%	19.9%	19.5%
Natural and applied sciences and related occupations	6.5%	5.8%	5.1%	4.7%	7.6%
Health occupations	5.5%	7.5%	4.0%	11.1%	4.5%
Occupations in social science, education, government service and religion	8.2%	6.6%	5.7%	7.1%	6.6%

	Total	Latin America	Central America	Caribbean (and Bermuda)	South America
Occupations in art, culture, recreation and sport	2.9%	2.1%	2.0%	1.9%	2.5%
Sales and service occupations	23.5%	23.8%	27.7%	22.5%	23.4%
Trades, transport and equipment operators and related occupations	14.9%	14.7%	17.7%	14.0%	14.1%
Occupations unique to primary industry	3.8%	1.6%	5.2%	0.7%	1.0%
Occupations unique to processing, manufacturing and utilities	5.8%	10.1%	11.6%	9.2%	10.5%
Total Income in 2005					
Total population 15 years and over	25,664,225	660,830	120,685	306,700	233,445
Without income	4.8%	4.5%	5.2%	3.6%	5.2%
With income	95.2%	95.5%	94.8%	96.4%	94.8%
Under $5,000	10.0%	10.1%	11.4%	9.2%	10.7%
$5,000 to $9,999	9.4%	9.5%	11.7%	8.5%	9.7%
$10,000 to $19,999	19.7%	21.2%	23.6%	20.4%	21.0%
$20,000 to $29,999	14.3%	15.9%	17.4%	16.0%	14.9%
$30,000 to $39,999	12.4%	14.2%	13.5%	15.2%	13.2%
$40,000 to $49,999	8.9%	9.1%	7.6%	10.0%	8.8%
$50,000 to $79,999	13.7%	11.4%	7.6%	12.9%	11.5%
$80,000 and over	6.7%	4.1%	2.0%	4.3%	5.0%
Average income $	35,498	na	25,362	31,914	31,294
Median income $	25,615	na	20,289	26,160	23,815
Income Status in 2005					
Total persons in private households	30,628,940	696,825	129,880	316,945	250,000
Total persons in economic families	26,358,390	605,625	117,430	266,055	222,140
Persons in economic families below low income cut-off before tax	3,144,530	119,765	23,350	50565	45850
Prevalence of low income before tax in 2005 for economic family members %	11.9	19.8%	19.9	19.0	20.6
Persons in economic families below low income cut-off after tax	2,274,755	86,985	16,945	36555	33485

	Total	Latin America	Central America	Caribbean (and Bermuda)	South America
Prevalence of low income after tax in 2005 for economic family members %	8.6	14.4%	14.4	13.7	15.1
Total persons 15 years and over not in economic families	4,270,545	91,195	12,450	50885	27860
Persons not in economic families below before-tax low income cut-off	1,556,490	42,510	6,370	22630	13510
Prevalence of low income before tax in 2005 for persons not in economic families %	36.4	46.6%	51.1	44.5	48.5
Persons not in economic families below after-tax low income cut-off	1,209,865	37,095	5,455	19690	11950
Prevalence of low income after tax in 2005 for persons not in economic families %	28.3	40.7%	43.8	38.7	42.9

national average in terms of educational attainment, whereas Latin Americans in the US, on average, have less education than the nation as a whole; and third, South Americans in both countries have more education than other Latin American groups.

Employment and Occupational Status

Comparatively low levels of educational attainment among some of the subgroups do not seem to be hindering Latin American immigrant participation in the labour force. In both countries, Latin Americans demonstrate participation and employment rates above the national average. In terms of differences among various subgroups, however, as shown in Table 12.4 the Mexican-born in the US lag behind the Central and South American-born (70.5 per cent as opposed to 76.3 per cent and 72.5 per cent, respectively). While Latin American immigrants in the US show higher unemployment rates than the nation as a whole (4.4 per cent vs 3.8 per cent), these rates are overall higher in Canada (over 7.5 per cent vs 6.6 per cent). Canada, historically, has somewhat higher unemployment rates than the US. Of note, too, in this regard is the fact that newcomers often have higher rates of unemployment than more established immigrants (see Chapter 6), especially in Canada where authorized entry is less contingent on an offer of work.

Employment and unemployment rates reflect people's participation in the formal labour market, but do not capture what is undoubtedly a large segment of the Latin American-born population that is working informally, especially in the US. Research on day labourers suggests that on any given day over 100,000 people across the US are working as, or looking for, day labour work. Gathering at informal hiring sites like gas stations, home improvement store parking lots, or busy street corners, day labourers primarily find low-paid hourly work in the landscaping and construction sectors (Valenzuela et al., 2006). In contrast, the Canadian informal sector is somewhat less significant and the incidence of day labourers almost nil.

Again, despite differences in the US and Canadian data, Central Americans (including Mexicans in the US) concentrate in occupations requiring lower levels of education (sales, service, and construction) compared to the national average, whereas South Americans' occupations (administration, management, and professional occupations) reflect their higher educational achievement and thus their overall occupational distribution is closer to that of the nation as a whole.

How do these education and occupation concentrations translate into economic standing? Latin American immigrants in both countries have incomes lower than the national average; moreover, Central Americans (including Mexicans in the US) tend to have lower incomes than South Americans, probably a reflection of differences in education levels. Specifically, Latin Americans tend to be over-represented among those earning less than US$35,000 in the US and Cdn$50,000 in Canada. As is the case with employment and unemployment rates, variations exist within place-of-birth subgroups (see Table 12.4) and by immigration period; migrants who are more established tend to do better and are closer to the national rates.

As a result of their employment, occupation, and income trends, Latin American immigrants in both countries experience a higher incidence of poverty and low income compared to the population at large. In the US, poverty rates indicate Mexicans and Central Americans to be trailing behind not just their South American counterparts but also the total US and foreign-born populations. In Canada, Latin Americans experience a much higher prevalence of low income before tax for economic family members (family members that participate in the labour force) compared to the national average, but Central Americans and Mexicans fare somewhat better than South Americans. This trend may be explained by the large numbers of Colombians (including refugees) who have recently arrived in Canada, and by the fact that poverty rates tend to be higher for newcomers.

What, then, underlies the differences in size, geography, and character of Latin American communities on each side of the border? And what might be the implications of these differences for experiences of settlement and integration? We answer these questions by examining the groups' respective histories of immigration, paying particular attention to selected countries of origin and conditions of exit, immigrant status upon arrival, and immigration policy. We also consider geographic proximity and geopolitical relations between the sending and receiving countries. Each of these factors intersects, shaping international migration flows, settlement processes, and the distinctive formation of immigrant communities and identities between and within Canada and the US.

HISTORICAL GEOGRAPHIES OF LATIN AMERICAN IMMIGRATION IN THE US AND CANADA

Latin American Immigration and Urban Settlement in the US

Latin American immigration in the United States has a much longer and varied history than is the case in Canada (Table 12.5). The shared border between the US and Mexico has long provided opportunity for Latin American immigrant crossings driven by goals of economic gain, family reunification, and access to employment and educational opportunity. These pull factors have been augmented by general and nation-specific push factors of civil war and political unrest, extreme poverty, and flight from persecution and oppression.

In terms of contemporary Latin American immigration to the US, it is important to understand the role played by macro-level policy in framing the character of immigrant flows; alignment with US economic needs; commitment to humanitarian values; and post-arrival settlement geographies. The establishment of the **Bracero Program (Mexican Farm Labour Program)** in 1942 is an example of alignment between US labour demand and diminished opportunity in Mexico. Developed initially to address wartime labour shortages in the agricultural sectors of California, the Bracero program issued a limited number of temporary visas each year to Mexicans willing to work primarily on US farms (Cerrutti and Massey, 2004). For many Mexicans, the program provided a critical opportunity since agricultural productivity in their own country remained hampered by limited investment and rural out-migration, and scarce work prospects existed in urban labour markets. Before its closure in 1964, Bracero facilitated the border crossing of more than 4.5 million guest workers, with as many as 300,000 to 450,000 arriving each year throughout the mid-1950s. As is the case with more recent programs, Bracero was both a response to, and a facilitator of, undocumented immigration into the country. With its quota system and agricultural focus, the program did not meet America's full demand for unskilled labour. Over the Bracero years, undocumented workers continued to flow into the US looking for and finding work. Even when informal-sector wages were lower than the mandatory minimums set by Bracero, they were still markedly higher than those in Mexico or in the other countries of Central America. When Bracero ended, many who would have entered the US through the program now arrived independently and without authorization, further swelling the number of undocumented Mexicans in the US.

The Hart-Celler Act (1965) and the **Immigration Reform and Control Act** (IRCA, 1986) also changed the policy context of Latin American immigration to the US. Prior to 1965, while quota systems prohibited or limited migration of particular national groups to the United States, such quotas never applied to

Latin Americans. The Hart-Celler Act for the first time placed numeric restrictions on migrants coming from what was termed the 'western hemisphere' (Cerrutti and Massey, 2004). Collectively, a 120,000 persons per annum cap was set for migrants coming from North America, Latin America, and the Caribbean, with preference given to those with immediate family in the United States or with needed skill sets (Miyares, 2004). Those entering the country as refugees or asylum seekers, at least initially, fell outside these caps. Over the next decade the Hart-Celler Act was refined and by the late 1970s a global policy was established into which refugees/asylum seekers were folded, thereby reducing further the number of migrants who could come officially from the countries of Latin America (Cerrutti and Massey, 2004).

The IRCA of 1986 has had perhaps the greatest impact on contemporary patterns of Latin American immigrant settlement in the US. As the number of undocumented immigrants began to rise and the US economy faltered, the IRCA addressed the combined effect of these issues on three fronts. As Massey and Capoferro (2008: 28) explain:

> First, to eliminate the attraction of American jobs for immigrants, it imposed sanctions on employers who knowingly hired undocumented workers. Second, to deter people from trying to enter the United States illegally in the first place, it allocated new resources to expand the Border Patrol. Finally to wipe the slate clean and begin afresh, it authorized an amnesty for undocumented migrants who could prove continuous residence in the US after January 2, 1982, which was combined with a special legalization program for undocumented farm-workers that was added to appease agricultural growers.

Collectively, these policy contexts laid a foundation for the evolution of a new immigrant geography across the US (Cerrutti and Massey, 2004; Massey, 2008; Iceland, 2009). Beginning in the late 1980s, Latin American immigrants began streaming into states such as Arizona, Colorado, Georgia, Michigan, Nevada, North Carolina, Oregon, Pennsylvania, and Utah (Gozdziak and Martin, 2005; Smith and Furuseth, 2006; Jones, 2008; Massey and Capoferro, 2008; Ansley and Schefner, 2009). While Mexican migrants dominated many of these streams, an increasing number were from the other countries of Latin America and were undocumented. In many cases, cities and communities within these new destination states were unaccustomed to previous waves of immigrant arrival and unprepared for this new settlement. In the 'New South', for example, where immigrant growth rates exceeded 1,000 per cent between 1980 and 2007, this translated into the large-scale insertion of culturally distinct Spanish-speaking immigrants into long-standing nativist and biracial, bicultural communities (see Box 12.1 for the example of Charlotte, North Carolina). While traditional immigrant settlement states and cities continue to retain their appeal, labour surpluses, wage competition, rising costs of living, and immigrant backlash as conveyed through the passing of local policy ordinances encourage a growing

BOX 12.1	New Settlement Destinations: Charlotte, North Carolina

When Roberto Suro, executive director of the Pew Hispanic Center, visited Charlotte, North Carolina, in 2006 to speak at a conference about the changing demographics of the region, he suggested that the city would be remembered as a vanguard of the nation's contemporary processes of Latinization. Charlotte, like many other cities across the US, has experienced tremendous and unexpected growth of its Latin American immigrant population over the last 30 years. While Mecklenburg County (the county home of the city) had only 17,875 foreign-born among its population in 1990, by 2000 that figure had skyrocketed to 68,349. Almost 45 per cent of these immigrants came from Latin America, with the largest subgroup by far being those from Mexico. Indeed, the total number of immigrants from Mexico residing in the county in 2000 (18,239) outnumbered the total number of all immigrants in the county 10 years earlier. By 2007 Mecklenburg County had 25,556 Mexican-born residents—50.5 per cent of all Latin American immigrants. The rapid growth of Charlotte's immigrant population (more than twice the national average between 1980 and 2000) and its skew towards Mexicans (and to a lesser extent Central Americans) are critical elements in the city's designation as a pre-emerging immigrant gateway (Singer, 2004).

In pre-emerging gateways, immigrant newcomers tend to be more recently arrived with lower rates of citizenship than migrants in more established gateways; tend to be poorer with lower levels of English literacy, and show a marked tendency for suburban rather than central-city residence. This certainly holds true for Charlotte, where Smith and Furuseth (2004) have documented the dispersed settlement of the city's Latin American immigrant population across three aging middle-ring suburban districts characterized by anonymity, affordability, and accessibility of rental units in multi-family apartment complexes near arterial roadways.

But what brings these migrants to Charlotte? Economic opportunity is the primary draw. While agricultural and food-processing opportunities attract Latin American immigrants to the rural communities of the state, opportunities in the growing and diversifying service-based economy bring them to the cities. Coincident with Charlotte's rise as the nation's second largest banking centre (Bank of America and Wachovia were headquartered in the city in the 1990s) came a need for workers across the service spectrum and in both the formal and informal labour markets. Newcomers found a land of opportunity with year-round employment in construction and landscaping, as well as a pioneering and entrepreneurial spirit in the local economy. At least initially, they also found a 'context of welcome' in which newcomers rarely encountered anyone asking about documentation or disdainful of their inability to speak or read English. For those **secondary migrants** who had arrived by way of earlier stops in California and Arizona where job competition and conflict about immigrant rights were on the rise, this reception was a refreshing change.

It was, however, short-lived. The continued pace and scale of immigrant growth in Charlotte in the first decade of the twenty-first century, combined with the maturation of the Latin American community from pioneering to more family-oriented, translated into their growing visibility in a city traditionally defined both from within and without by its distinctively southern biracial and bicultural profile. The arrival of a foreign-born, 'brown-skinned', linguistically and culturally distinct group has challenged the city's progressive 'New South' identity, tested its tradition of southern hospitality, and challenged its cash-strapped but critical social services. Between 2000 and 2005, enrolment of Hispanic children in the Charlotte–Mecklenburg school system more than doubled, from 4.5 per cent to 10.1 per cent. As the presence of Latin American immigrants becomes more commonplace across the suburban neighbourhoods and public spaces of the city, nervousness grows around issues of property value, quality of life, allocation of jobs, access to services, and the impacts of cultural change. In this local context, which is fuelled by the macro-scale influence of national debates around immigration reform and policy, myth-making and misinformation about Latin American immigrants and broader processes of migration and settlement abound. As a consequence, despite good intentions, service provision missteps and planning errors result. These in turn lead to growing confusion, fear, mistrust, and discrimination, which ultimately affect the daily existence of Latin American immigrants coming to or already settled in the city and restrict their ability to craft and control their own identities and ways of integrating and contributing to their new communities. Their identities and futures are being structured *for* them, not *by* them, and there is an overall failure to engage this group for broader community development and progress.

number of Latin American immigrants to seek alternative destinations in which economic opportunity is more plentiful, their labour more valued, and their presence comparatively welcome.

Whereas Mexican immigration into the US is largely framed by the historic interplay of economic and legislative factors, Central and South American immigration is tied strongly to the push factors of political and civil instability in the 1980s. Two groups, Colombians and Salvadorans, illustrate different regional migration paths. Colombian migration to the US dates back to just after World War I, when migrants from Bogotá came to New York City, but most Colombian migrants arrived post-1960 with the greatest flows coinciding with the country's political instability and violence in the 1980s. Unlike Salvadorans, many Colombian migrants were members of their country's middle class and arrived with greater educational and professional skills. As a consequence, many—but not all—have greater and earlier social mobility than their counterparts from Central America whose rural backgrounds often correlate with poverty and low levels of education and literacy (Arreola, 2004; Dixon and Gelatt, 2006). Among South American immigrants to the US, Colombians are the largest and fastest-growing subgroup. Between 1960 and 2000, the Colombian population nearly doubled, growing by 497,288 persons, significantly outpacing Ecuadorans (by 290,995), Peruvians (by 271,083) and Brazilians (by 198,442), the next four highest growth groups (Arreloa, 2004; Dixon and Gelatt, 2006). With concentrations in New York and Florida reflecting particular homeland origins,

Colombians have shown a clear tendency towards urban residence with settlement in particular neighbourhoods such as New York City's Jackson Heights, where they acted as vanguard for future groups of Andean (Ecuadorian and Peruvian) immigrants (Miyares, 2004). Colombians even informally named this neighbourhood 'Chaperino' after a middle-class suburb in their capital (Arreola, 2004).

Civil war in the 1980s also drove hundreds of thousands of migrants from El Salvador northward seeking freedom from violence and instability (Terrazas, 2010). Between 1980 and 1990, the number of Salvadoran immigrants coming to the US rose from 94,000 to 465,000. Family reunification and new arrivals fleeing both economic and environmental hardship extended and expanded this migration stream such that by 2008 there were more than 1.1 million across the country—many of whom (about 57,000 or 5 per cent of the Salvadoran total) were undocumented (Arreola, 2004; Terrazas, 2010). The increasing number of Salvadoran immigrants since the late 1990s is partially tied to this group's receiving **Temporary Protected Status (TPS)** from the US government following Hurricane Mitch (nationals of Guatemala, Honduras, and Nicaragua were also granted TPS). Granted on humanitarian grounds, TPS suspends deportation orders for undocumented immigrants already in the county and permits recipients to request work authorization from the US Citizenship and Immigration Service (Terrazas, 2010).

Reflecting the long-standing overland migration paths through their own country, Mexico, and into the US, and the draw of their diversified immigrant economies, more than half of the nation's Salvadoran immigrants live in California and Texas, with smaller settlement in New York (Arreola, 2004; Terrazas, 2010). Concentration is also evident at the metropolitan scale, with 65.9 per cent of all Salvadorans living in the Los Angeles, San Francisco, Houston, Dallas–Fort Worth, New York, or Washington, DC, MSAs.

Despite their continued concentration in gateway states, Salvadorans are part of the changing geography of Latin American settlement at both the national and local scales. Nationally, they display significant and growing concentrations in new destination states such as Maryland and Virginia. Within Washington, DC, Salvadorans represent 12.3 per cent of all foreign-born and are the largest single immigrant group in the MSA (Price and Singer, 2008; Terrazas, 2010). They also have a marked preference for suburban settlement, a hallmark feature of Singer's (2004) typology of emerging and pre-emerging immigrant gateways—urban areas unfamiliar with large-scale or historically consistent immigrant settlement but that have experienced significant immigrant growth since 1980, particularly from Latin America (see Box 12.1).

Recent scholarship focuses on how new destination cities and suburbs have become repositories for the effects of macro-scale forces (Ansley and Shefner, 2009; Varsanyi, 2010). Inexperienced with foreign-born or culturally distinct populations, new destinations have struggled with the extent of their receptivity and the need to adjust public service structures to accommodate newcomers. Local school systems, emergency rooms, hospitals, and health units, parks and recreation, and myriad other public services have had to reconsider how best and most equitably to provide for rapidly growing immigrant populations whose arrival and settlement in their communities is driven by the forces of global migration, economic globalization, and federal immigration policy. Policing is perhaps the most contentious example of the tension that exists between the macro- and micro-scale factors shaping the contemporary Latin American immigrant experience in the US.

The desire to stem undocumented migration has led to what many view as a militarization of the border with targeted crackdowns at some of the busiest crossings (e.g., El Paso, Texas: Operation Hold-the-Line/Operation Blockade; and San Diego, California: Operation Gatekeeper), sharp increases in the number of officers hired to patrol the border, and the building of fences and walls over large stretches of the desert borderlands between the US and Mexico. At a local level, this border militarization has extended into cities and communities many miles away through mechanisms such as the 287(g) program, which facilitates the deputization of county sheriffs and municipal police officers to enforce aspects of federal immigration policy (e.g., deportation) if people are picked up and/or charged with other (e.g., traffic) offences and found to be in the country 'illegally' (Cerrutti and Massey, 2004; Massey and Capoferro, 2008; Nevins, 2010; Rodriguez et al., 2010; Varsanyi, 2008, 2010; also see Chapter 1).

As the contemporary history of Latin American immigration into the US continues to unfold, it is clear that new geographies are emerging. While the traditional balance of macro-scale push and pull factors continues to hold, micro-scale factors have grown in importance as the ones that both frame and define the immigrant experience of those coming from Mexico, Central America, and South America.

Latin American Migration and Urban Settlement in Canada

In contrast to the US, Latin American immigration to Canada is much more recent (Table 12.5) and, thus, interest in the group's experiences is both relatively recent and limited. In spite of its novelty, Latin American immigration to Canada is demographically diverse in terms of countries of origin, immigrant status, and class. These features, along with important differences in the sending and receiving contexts, have resulted in distinct experiences of settlement, integration, and community formation among Latin Americans in Canada compared to the US. To start, macro-scale factors such as geopolitics, geographical location, and national immigration policies played a key role in shaping distinct experiences of Latin American immigration. Historically, Canadian geopolitical interests in Latin America have been less significant, which combined with geographical distance has meant that immigration to Canada was somewhat less natural than to the US. Furthermore, Canadian immigration policy was restrictive until 1967, favouring migrants from Europe over other world regions.

The first two groups from Latin America, the Lead and Andean waves in the late1950s to early 1970s, consisted of a small number of migrants. Changes

Table 12.5 Waves of Latin American Immigration to the United States and Canada

| Immigration Waves | United States | | Canada | | |
	Time Period	Characteristics	Immigration Waves	Time Period	Characteristics
'Lead' Wave	Pre-1848	Portions of western US part of Mexican Republic Mexican-American, Chicano heritage established Continuous flows and interaction between two countries Legacy today of majority Hispanic counties across borderland states	n.a.	n.a.	n.a.
Open Wave	1920s	Latin Americans exempt from regional or country-specific quotas Relatively open borders	n.a.	n.a.	n.a.
Bracero Wave	1942–64	Temporary agricultural guest worker program for Mexicans Particular focus on needs of California Over 4.5 million Bracero visas issued	'Lead' Wave	1956–65	South Americans of European descent: urban intelligentsia from major Latin American cities
				1965–9	Venezuelans, Argentines, and Peruvians: mostly blue-collar groups
Andean Wave	1980s–present	Colombians seeking refuge from political instability and violence Many Colombians and some Ecuadorians middle-class, well-educated, and skilled Migration to US has translated for many into downward socio-economic mobility	Andean Wave	1973–5	Colombians, Ecuadorians, and Peruvians (economic refugees) admitted in higher proportions High estimates of 'undocumented' migrants from Peru and Guatemala Mostly blue-collar groups; skilled and unskilled labourers

Coup Wave	1970s–80s	Politically motivated immigration, steady increase in numbers from Chile, Uruguay, and Argentina in response to authoritarian/military rule in Southern Cone. Many highly educated with professional and technical skills. Exile perspective leads to some return migration with re-establishment of democratic rule	Coup Wave	1973/4–1978/9	Significant numbers of Chilean refugees, as well as refugees from Argentina and Uruguay (often sponsored by the Canadian government) and exiles fleeing the military coups and dictatorships in their countries. Chilean intelligentsia with left-leaning views, urban professionals, and skilled labourers. Some blue-collar workers (especially those involved in unions) among Argentines and Uruguayans
Central American Wave	1980s–present	Salvadorans and Guatemalans largest Central American groups. Fleeing civil conflict, seeking economic opportunity. Humble origins, limited education, many undocumented and employed in US in low-wage service and manufacturing	Central American	1983–present	Salvadorans and Guatemalans: mostly urban poor, rural middle classes, and peasantry. Lower average educational levels than the Lead or Coup Waves, and perhaps less skill specialization than the Andean Wave. 1990s–present: family reunification
IRCA Wave	Post 1986–present	Amnesty to those with demonstrated continuous residence since 1982. Growing anti-immigrant sentiment in traditional gateway states and cities. Emergence of new destinations through processes of primary and secondary migration. Numbers of undocumented continue to rise (especially Mexicans and Central Americans). Localization of national immigration policy. Post-1970 skilled, professional, and technical migration continues, playing its part in overall polarized profile of group	Skilled Worker Wave	1990s–present	Highly educated professionals from throughout Latin America arriving under the 'skilled worker' and 'business class' categories. Since 2000, arrival of significant numbers of Colombian refugees and some undocumented migrants from countries facing social, economic, and political unrest such as Argentina, Ecuador, Peru, and Venezuela

Sources: Adapted from Mata (1985) with data from Roniger and Sznajder (1999); Pellegrino (2001); Pellegrino and Vigorito (2005); Arreola (2004); Zuniga and Hernandez-Leon (2005); Massey (2008).

to Canadian immigration policy in 1967 (the removal of restrictions on place of birth) facilitated the arrival of the Coup wave in the 1970s and the Central American wave in the 1980s. Both waves consisted of a majority of refugees, most of them sponsored by the Canadian government through special programs. Again, geopolitics and differences in its relationship to Latin America led Canada to recognize asylum claimants from the region whereas the US did not, or did to a much lesser extent. The two refugee waves mark the emergence of a Latin American community in Canada, especially in cities such as Montreal, Toronto, and to a lesser extent Vancouver. But the forced nature of their migration meant that they had very particular experiences of migration and settlement marked by what Nolin (2006) identifies as 'transnational ruptures' (see also Ginieniewicz and Schugurensky, 2006). While the trauma of displacement and the violence of exile add to the challenges refugees face in the integration process, important class differences distinguish the two refugee waves and have implications for their ability to participate in Canadian society. The Coup wave consisted of urban professionals fleeing the military dictatorships in their countries, while Central American refugees were mostly peasants displaced by the civil unrest stirring the region. These micro-scale differences played an important role in the groups' respective socio-economic achievement in Canada; South American refugees generally did better than Central Americans.

The most recent wave of Latin American immigration reflects changes to Canadian immigration policy, particularly efforts to attract skilled workers to meet the labour needs of Canada's post-industrial economy (see Chapter 1). Since the 1990s, most migrants from Latin America are professionals with higher levels of education; nevertheless, significant numbers still arrive under the family reunification program (especially from Central America) and as refugees (especially from Colombia and, more recently, Mexico). While most Latin American newcomers settle in Canada's three main gateway cities— where most employment opportunities are located—since the 2000s a growing number of both skilled workers and refugees have been settling in cities such as Calgary, Ottawa–Gatineau, Edmonton, and London, Ontario (see Table 12.3 and Box 12.2). Thus, in the 1990s, Latin American immigrants became much more diverse in terms of national origin, socio-economic class, political inclinations, and so on. This growing internal diversity has posed a challenge for building a united community and shared identity as Latin Americans in Canada and in its cities (Veronis, 2010).

The nature and form of Latin American immigration to Canada have influenced the empirical focus and theoretical approaches to studying the group. Goldring (2006) identifies two main research focuses: transnationalism and integration. We would add a third: temporary and precarious migration. Canadian research done under the umbrella of transnationalism is concerned with the unique experiences of trauma among Latin American refugees (Nolin, 2006) and forms of transnational solidarity (Landolt, 2007, 2008; Nolin, 2006). Given the significant share of Latin American refugees in Canada, more attention has been paid to their experiences than in the US, including those of Chileans (Ginieniewicz and Schugurensky, 2006), Salvadorans (Landolt, 2007, 2008), Guatemalans (Nolin, 2006), and, more recently, Colombians.

Studies dealing with questions of integration examine different aspects of Latin American collective organizing and participation in Canadian cities. The focus has been on the nature and form of Latin American community organizing in Toronto, including both grassroots (Landolt, 2008) and more formal political and social organizing (Landolt and Goldring, 2009), as well as Latin Americans' involvement in the non-profit sector (Veronis, 2010). These studies generally offer a multi-scalar analysis of immigrant participation by taking into account the role of both the local and transnational contexts in shaping Latin Americans' collective organizing.

Finally, broader changes in international migration have fuelled interest in the experiences of temporary migrants, especially seasonal agricultural workers, mostly Mexicans working in southwestern Ontario and British Columbia (Basok, 2004; Becerril, 2007), and other migrants in precarious conditions, such as exotic dancers (Diaz Barrero, 2007). Cast in the broader context of globalization (especially in its neo-liberal form), these studies take a human rights perspective and critically demonstrate the emergence of unequal forms of citizenship and belonging. This research also contributes to the critical examination of how nations in the global North increasingly use immigration policy instrumentally to create temporary citizens who contribute to their economies without constituting a social burden (see Chapter 1).

The emergence of Latin American immigrant settlement in regions and cities outside traditional gateways is a shared focus of recent scholarship in both the US and Canada (*Canadian Ethnic Studies*, 2005). In Canada, while the three largest CMAs still receive the most significant share of Latin American newcomers, growing numbers settle in second- and third-tier cities, particularly in southwestern Ontario, southern Quebec, and Alberta (see Box 12.2 for the example of Ottawa–Gatineau). Nevertheless, compared to the US, this shift has occurred much more recently, since the early 2000s. In 2001–6, mid- and small-sized cities in southwestern Ontario and southern Quebec experienced important growth rates (albeit not as dramatic as those in the US) of their Latin American populations: 145 per cent growth in London, Ont.; 193 per cent in Quebec City; and 204 per cent in Sherbrooke, Quebec.

Moreover, the nature and broader context of Latin American immigrant flows to new settlement destinations in Canada differ significantly from the US experience. In Calgary, the main pull factor has been a growth in employment opportunities related to Alberta's booming energy sector, attracting both immigrants and migrants with temporary work permits. In Ontario and Quebec, most of this growth is led primarily by the arrival of large numbers of Colombians—both skilled workers and refugees—who have fled (voluntarily or involuntarily) the socio-political unrest and violence that have stirred the country for over two decades. Colombian immigrants settle outside Canada's main gateway cities for a number of reasons. Many come from smaller cities and thus prefer living in smaller urban centres; this settlement pattern sometimes is the result of **chain migration**, as in the case of London, Ontario, where 88 per cent of the 2,290 Colombians arrived in 2001–6. Finally, a significant number of Colombian refugees are government-sponsored, in which case their destination is assigned to them, particularly in the province of Quebec where the government has

BOX 12.2   New Settlement Destinations: Ottawa–Gatineau, Ontario–Quebec

Ottawa–Gatineau is Canada's fourth largest CMA; it features the fifth largest foreign-born population as well as the fifth largest population born in Central and South America (Table 12.3). The case of Ottawa–Gatineau illustrates contemporary trends of immigration from Latin America, including settlement in cities other than the three major gateways and the experiences of skilled worker immigrants and those of Colombian refugees. Ottawa and Gatineau form Canada's National Capital Region and comprise the only CMA located on an interprovincial border, between the provinces of Ontario and Quebec—a border that is politically and symbolically significant, between English and French Canada. One important consequence is that different provincial and municipal policies and laws regulate everyday life on each side of the border, including in the area of immigration and settlement, with implications for access to resources such as employment, housing, and social services (health care, public transit, education, and childcare). Nevertheless, the border does not represent a major barrier for the population's daily practices; it is common for individuals to live on one side and to work, shop, and entertain themselves on the other. A second significant consequence of the transborder position of Ottawa–Gatineau relates to the cultural and linguistic characteristics of each side: Quebec culture and French dominate in Gatineau; Ottawa is predominantly English-speaking and reflects Canada's anglophone culture. Nevertheless, the National Capital Region is generally bilingual, a characteristic reinforced by the presence of the federal government, the largest employer in the region.

In 2006, there were 11,340 Latin American foreign-born individuals in Ottawa–Gatineau. The three largest groups are the same as those on a national scale, albeit in a different ranking: El Salvador (2,150), Colombia (1,780), and Mexico (1,200). The majority arrived before 1996 (7,030 or 62 per cent) and most were migrants associated with the refugee waves from South and Central America. In the past decade, however, the number of Latin American newcomers has grown, especially skilled workers and refugees from Colombia. In the 1996–2006 period, 3,930 Latin Americans settled in Ottawa–Gatineau, representing 34.7 per cent of the group; most of the growth took place between 2001 and 2006 with the arrival of 2,500 Latin Americans. Since 1996, the largest numbers of arrivals were from Colombia (1,525), Peru (465), Mexico (460), and Venezuela (240). These data help to illustrate the diversity of Latin Americans in Canada. Furthermore, different factors and processes shape the migration and settlement experiences of different migrant groups, in this case Colombian refugees and skilled workers.

Colombians are the most numerous and fastest-growing group: 85.7 per cent of Colombians arrived in Ottawa–Gatineau after 1996. Whereas all Latin American nationalities have a majority of individuals residing in Ottawa, Colombians are the only group almost equally distributed between Ottawa (975) and Gatineau (800). Between 2001 and 2006, a majority of Colombians settled in Gatineau (620 versus 530 in Ottawa). These settlement patterns need to be understood in light of Quebec's immigration policy: most Colombians in Gatineau are government-sponsored refugees and thus their experiences are markedly different from those of skilled workers (who migrate voluntarily). Skilled workers originate from throughout Latin America and are attracted to the area primarily by employment opportunities in the high-tech sector and in the federal government. The quality of life (shorter commute times, less congestion, a greener environment) is also an attractive factor, especially for families. Skilled workers generally prefer living in Ottawa, probably because most jobs are located there and because of linguistic and cultural preferences.

Clearly, Latin American newcomers in Ottawa–Gatineau arrive with different needs, interests, and abilities for integration (e.g., resources, knowledge of Canada's official languages, education levels, etc.). In turn, these differences pose two main challenges: (1) for service provision—to meet the needs of refugees who often arrive with trauma and those of skilled workers looking for adequate employment; and (2) for community formation. With regard to the latter, the group's diversity represents a significant challenge to mobilize and organize along common interests. Latin Americans tend to organize along national lines at best; but conflicts arise even within national groups due to differences in class, time, and conditions of arrival, or political inclinations. For example, Colombians in Ottawa–Gatineau are divided into different factions based on homeland politics.

The transborder context of Ottawa–Gatineau presents additional challenges for immigrants' settlement, especially skilled workers—e.g., access to jobs, recognition of foreign credentials (Reitz, 2001; Bauder, 2003)—compared to other Canadian regions. Newcomers need to grapple with different settlement policies and programs on each side of the border, as well as the region's bilingualism (including Canada's language and culture politics). In Gatineau, services to immigrants are provided exclusively in French as per Quebec's language policies and immigration agenda, i.e., to strengthen French language and culture. Services in Ontario are offered primarily in English with limited programs in French. But newcomers are eligible for settlement programs only in their province of residence, i.e., immigrants in Gatineau can take language courses in French only, whereas those in Ottawa can choose between either English- or French-language training. These policies and programs are at odds with the region's bilingualism and employers' preference (especially the federal government) for workers' facility in both official languages.

The case of Latin American immigrants in Ottawa–Gatineau helps to illustrate how macro-scale factors (the broader context) and micro-scale factors (migrants' individual characteristics) combine and lead to different experiences of settlement and integration. Furthermore, this immigrant profile highlights the power of local characteristics and place-based processes in shaping these experiences.

identified smaller urban centres as ideal destinations for refugee settlement in an effort to regionalize immigration (Box 12.2).

Finally, the significant presence of Mexican migrants in southwestern Ontario needs to be highlighted. This is a special case because the group includes not only immigrants and permanent residents, but also a large number of temporary migrants with work permits. Over 7,000 Mexicans are admitted to Canada annually under the Seasonal Agricultural Workers Program, of which 90 per cent work in Ontario (Basok, 2004). While these workers stay in Canada for only about eight months in a given year, they have a significant impact on the rural communities and small towns where they are employed because these regions have had relatively little exposure to diversity. As a result, as Basok has documented, the local reaction to the presence of Mexican temporary migrants has led to tensions and conflicts within these smaller and predominantly 'whiter' communities.

CONCLUSION

Important differences distinguish the experiences of Latin American immigrants in Canada and the US. While contemporary literature in Canada focuses on issues of transnationalism, integration, and temporary and precarious migration, in the United States the tension between federal immigration policy and local impact and the multi-scalar challenges posed by undocumented immigrants are subjects of recent scholarly focus. In both countries, the emergence of new geographies of settlement and adjustment is an important area of attention with implications for public policy and everyday lived experience. It also provides fruitful terrain for cross-national comparisons and an opportunity for new theoretical developments on immigration in North American cities. New geographies of Latin American immigrant settlement raise important questions about issues of receptivity and identity formation by disrupting traditional ethnoracial relations and notions of citizenship and belonging. It is our hope that this chapter serves as a springboard for further research in these areas.

QUESTIONS FOR CRITICAL THOUGHT ───────────────

1. What are the main differences between the Latin American immigrant population in Canada and in the US and what factors account for those differences?

2. What factors help explain why Latin American immigrants in Canada and the US are gravitating towards 'new destination' communities? How are these new destinations different from the primary settlement areas of the past?

3. What are the macro factors that help to explain the different experiences of Latin American immigrants in Canada and the US? What are the micro factors? Is one scale more important than the other?

SUGGESTED READINGS

1. Goldring, Luin. 2006. 'Latin American Transnationalism in Canada: Does It Exist, What Forms Does It Take, and Where Is It Going?', in Vic Satzewich and Lloyd Wong, eds, *Transnational Identities and Practices in Canada*. Vancouver: University of British Columbia Press, 180–201. This chapter provides an overview of Latin American immigration to Canada by examining the experiences of various national subgroups over time. The overall experience is characterized by strong transnational ties to countries of origin combined with efforts to integrate in Canadian society.

2. Massey, Douglas S., ed. 2009. *New Faces in New Places: The Changing Geography of American Immigration*. New York: Russell Sage Foundation. This interdisciplinary volume explores emerging patterns of immigrant settlement in urban and rural US communities. Several chapters also address community reaction to new immigrant groups and illuminate the complex terrain of receptivity and welcome in these new immigrant destinations.

3. Singer, Audrey, Susan W. Hardwick, and Caroline B. Brettell, eds. 2008. *Twenty-First Century Gateways: Immigrant Incorporation in Suburban America*. Washington: Brookings Institution Press. Focusing on the reception, impact, and adaptation of immigrants across a range of new gateway cities, the volume focuses particular attention on the dynamics of suburban lure and transformation and their intersection with immigrant arrival and settlement.

4. Smith, Heather A., and Owen J. Furuseth, eds. 2006. *Latinos in the New South: Transformations of Place*. Aldershot, UK: Ashgate. One of the first volumes to spotlight the unexpected and large-scale settlement of Latino immigrants in the contemporary US South. Across different states and localities, authors focus on the challenges and opportunities immigrants bring to their new communities and on the ways in which the distinctive culture and historical legacy of the region impact immigrant experience and identity.

5. Veronis, Luisa. 2010. 'Immigrant Participation in the Transnational Era: Latin Americans' Experiences with Collective Organizing in Toronto', *Journal of International Migration and Integration* 11, 2: 173–92. This article examines Latin American immigrants' efforts to participate and build community in Toronto since the mid-1970s by focusing on the group's collective forms of organizing and the challenges of intra-group diversity.

NOTES

1. Although the island countries of the Caribbean sometimes are considered part of Latin America, the Caribbean immigrant experience in Canada and the US is not a focus of this chapter, except where otherwise noted. Caribbean immigration to the US and Canada is partly covered in Chapter 11. Thus, the focus of our analysis is countries of origin, not mother tongue per se.

2. Unless otherwise noted, data come from US Bureau of the Census's 2006–2008 American Community Survey, three-year estimates, and from Statistics Canada, *Census of Canada, 2006.*

3. 'Latin Americans' includes immigrants to the US and Canada from non-Spanish-speaking South American countries (Brazil, Guyana, and Suriname) but does not include those from the Caribbean.
4. Latin Americans grew from 209,410 in 2001 to 277,330 in 2006 (excluding Guyanese-born and Brazilian-born immigrants).
5. Of the growing numbers of migrants admitted under Canada's temporary foreign worker program (see Chapter 1), domestic workers are eligible for permanent status and naturalization while agricultural workers are not.

REFERENCES

1. Ansley, Fran, and Jon Shefner, eds. 2009. *Global Connections, Local Receptions: New Latino Immigration to the Southeastern United States*. Knoxville: University of Tennessee Press.
2. Arreola, Daniel D., ed. 2004. *Hispanic Spaces, Latino Places: Community and Cultural Diversity in Contemporary America*. Austin: University of Texas Press.
3. Basok, Tanya. 2004. 'Post-national Citizenship, Social Exclusion and Migrants Rights: Mexican Seasonal Workers in Canada', *Citizenship Studies* 8, 1: 47–64.
4. Bauder, Harald. 2003. '"Brain Abuse", or the Devaluation of Immigrant Labour in Canada', *Antipode* 35, 4: 139–53.
5. Becerril, Ofelia. 2007. 'Transnational Work and the Gendered Politics of Labour: A Study of Male and Female Mexican Migrant Farm Workers in Canada', in Goldring and Krishnamurti (2007: 157–72).
6. Bloemraad, Irene. 2006. *Becoming a Citizen: Incorporating Immigrants and Refugees in the United States and Canada*. Berkeley: University of California Press.
7. *Canadian Ethnic Studies*. 2005. Special issue: 'Thinking about Immigration outside Canada's Metropolitan Centres', 37, 3.
8. Cerrutti, Marcela, and Douglas S. Massey. 2004. 'Trends in Mexican Migration to the United States, 1965–1995', in Jorge Durand and Douglas S. Massey, eds, *Crossing the Border: Research from the Mexican Migration Project*. New York: Russell Sage Foundation, 17–44.
9. del Pozo Artigas, José, ed. 2006. *Exiliados, emigrados y retornados. Chilenos en América y Europa, 1973–2004*. Santiago: RIL editores.
10. Diaz Barrero, Gloria Patricia. 2007. '"Living" There: Transnational Lives and Working Conditions of Latina Migrant Exotic Dancers', in Goldring and Krishnamurti (2007: 145–56).
11. Dixon, David, and Julia Gelatt. 2006. *Detailed Characteristics of the South American Born in the United States*. Washington: Migration Policy Institute.
12. Ginieniewicz, Jorge, and Daniel Schugurensky, eds. 2006. *Ruptures, Continuities and Re-learning: The Political Participation of Latin Americans in Canada*. Toronto: Transformative Learning Centre, OISE/University of Toronto.
13. Goldring, Luin. 2006. 'Latin American Transnationalism in Canada: Does It Exist, What Forms Does It Take, and Where Is It Going?', in Vic Satzewich and Lloyd Wong, eds, *Transnational Identities and Practices in Canada*. Vancouver: University of British Columbia Press, 180–201.
14. ——— and Sailaja Krishnamurti, eds. 2007. *Organizing the Transnational: Labour, Politics and Social Change*. Vancouver: University of British Columbia Press.
15. Gozdziak, Elzbieta M., and Susan F. Martin. 2005. *Beyond the Gateway: Immigrants in a Changing America*. Lanham Md: Lexington Books.
16. Iceland, John. 2009. *Where We Live Now: Immigration and Race in the United States*. Berkeley: University of California Press.
17. Jones, Richard, ed. 2008. *Immigrants outside Megalopolis*. Lanham, Md: Lexington Books.
18. Landolt, Patricia. 2007. 'The Institutional Landscapes of Salvadoran Refugee Migration: Transnational and Local Views from Los Angeles and Toronto', in Goldring and Krishnamurti (2007: 191–205).
19. ———. 2008. 'The Transnational Geographies of Immigrant Politics: Insights from a Comparative Study of Migrant Grassroots Organizing', *Sociological Quarterly* 49, 1: 53–77.
20. ———and Luin Goldring. 2009. 'Immigrant Political Socialization as Bridging and Boundary Work: Mapping the Multi-

layered Incorporation of Latin American Immigrants in Toronto', *Ethnic and Racial Studies* 32, 7: 1226–47.

21. Massey, Douglas S., ed. 2008. *New Faces in New Places: The Changing Geography of American Immigration*. New York: Russell Sage Foundation.

22. ―― and Chiara Capoferro. 2008. 'The Geographic Diversification of American Immigration', in Massey (2008: 25–50).

23. Mata, Fernando. 1985. 'The Four Immigrant Waves from Latin America to Canada: Historical, Demographic and Social Profiles', unpublished manuscript, York University.

24. Miyares, Ines M. 2004. 'Changing Latinization of New York City', in Arreola (2004: 145–66).

25. Nevins, J. 2010. *Operation Gatekeeper and Beyond: The War on "Illegals" and the Remaking of the U.S.–Mexico Boundary*, 2nd edn. New York: Routledge.

25. Nolin, Catherine. 2006. *Transnational Ruptures: Gender and Forced Migration*. Aldershot, UK: Ashgate.

27. Passel, Jeffrey S., and D'Vera Cohn. 2009. *A Portrait of Unauthorized Immigrants in the United States*. Washington: Pew Hispanic Center. At: pewhispanic.org/reports/report.php?ReportID=107.

28. Pellegrino, Adela. 2001. 'Trends in Latin American Skilled Migration: "Brain Drain" or "Brain Exchange"?', *International Migration* 39, 5: 111–32.

29. ―― and Andrea Vigorito. 2005. 'Emigration and Economic Crisis: Recent Evidence from Uruguay', *Migraciones Internacionales* 3, 1: 57–81.

30. Price, Marie, and Audrey Singer. 2008. 'Edge Gateways: Immigrants, Suburbs, and the Politics of Reception in Metropolitan Washington', in Audrey Singer, Susan W. Hardwick, and Caroline B. Brettell, eds, *Twenty-First Century Gateways: Immigrant Incorporation in Suburban America*. Washington: Brookings Institution Press, 137–70.

31. Reitz, Jeffrey. 2001. 'Immigrant Skill Utilization in the Canadian Labour Market: Implications of Human Capital Research', *Journal of International Migration and Integration* 2, 3: 347–78.

32. Rodriguez, Cristina, Muzaffar Chishti, Randy Capps, and Laura St John. 2010. *A Program in Flux: New Priorities and Implementation Challenges for 287(g)*. Washington: Migration Policy Institute.

33. Roniger, Luis, and Mario Sznajder. 1999. *The Legacy of Human Rights Violations in the Southern Cone: Argentina, Chile, and Uruguay*. New York: Oxford University Press.

34. Singer, Audrey. 2004. *The Rise of New Immigrant Gateways*. Washington: Brookings Institution, Center on Urban and Metropolitan Policy.

35. Smith, Heather A., and Owen Furuseth. 2004. 'Housing, Hispanics, and Transitioning Geographies in Charlotte, North Carolina', *Southeastern Geographer* 44, 2: 216–35.

36. ―― and ――, eds. 2006. *Latinos in the New South: Transformations of Place*. Aldershot, UK: Ashgate.

37. Suro, Roberto. 2006. 'The New Latino South: Understanding Immigration in Context', remarks made at UNC Charlotte Urban Institute First Annual Regional Conference, 'The Changing Face of the New South: Latinos in the Greater Charlotte Region', 24 Apr.

38. Terrazas, Aaron. 2010. *Salvadoran Immigrants in the United States*. Washington: Migration Policy Institute.

39. US Bureau of the Census. 2008. 2006–8 American Community Survey, three-year estimates. Washington.

40. Valenzuela Jr, Abel, Nik Theodore, Edwin Meléndez, and Ana Luz Gonzalez. 2006. *On the Corner: Day Labor in the United States*. Los Angeles: UCLA Center for the Study of Urban Poverty.

41. Varsanyi, Monica W. 2008. 'Rescaling the "Alien", Rescaling Personhood: Neoliberalism, Immigration, and the State', *Annals, Association of American Geographers* 98, 4: 877–96.

42. ――, ed. 2010. *Taking Local Control: Immigration Policy Activism in U.S. Cities and States*. Stanford, Calif.: Stanford University Press.

43. Veronis, Luisa. 2010. 'Immigrant Participation in the Transnational Era: Latin Americans' Experiences with Collective Organizing in Toronto', *Journal of International Migration and Integration* 11, 2: 173–92.

44. Zuniga, Victor, and Ruben Hernandez-Leon, eds. 2005. *New Destinations: Mexican Immigration in the United States*. New York: Russell Sage Foundation.

CHAPTER 13

CROSSING THE 49TH PARALLEL: AMERICAN IMMIGRANTS IN CANADA AND CANADIANS IN THE US

Susan Hardwick and Heather Smith

INTRODUCTION: INVISIBLE IMMIGRANTS AT THE BORDERLAND

As earlier chapters in this book have illustrated, Canada and the United States are among the world's largest immigrant receiving nations. To date, however, very few scholars have examined Canadian immigrants in the US or Americans who have migrated to Canada, and no research has yet compared these groups' migration and settlement experiences.[1] This lack is surprising not only because of the large numbers involved in these dual migration flows but also because immigration between the two countries has a long and fascinating history. Estimates for 2007 indicated about 840,000 Canadian-born immigrants living in the US, representing 0.28 per cent of the overall population and about 2.2 per cent of the total foreign-born. Comparatively, there were about 316,350 US-born immigrants residing in Canada in 2006, representing approximately 1 per cent of the total population and about 5 per cent of the total foreign-born population.[2] To help fill the gap in studies of cross-border migration across the **49th parallel**, this chapter comparatively maps, documents, and analyzes contemporary Canadian immigration to the US and American immigration to Canada.

The changing politics of the Canadian–US border over the last decade make this chapter especially timely and relevant. A marked tightening of the international boundary occurred after terrorist attacks on the World Trade Center in September 2001. One manifestation of this tightening has been a series of increasingly restrictive policies related to crossing what has long been known as the 'world's longest undefended border'. Passports, **NEXUS cards**, or enhanced driver's licences are now required of all Canadian citizens wishing to cross the US border, with the equivalent Western Hemisphere Travel Initiative compliant documents also required for all Americans who cross into Canada. Electronic passport scanning facilitates the tracking of cross-border movement with greater frequency and precision than ever before. While these changes may seem minor (or even common sense, given current geopolitical realities), they are consequential particularly in the context of growing anti-American sentiment in Canada and growing anti-immigrant sentiment in the US. At the scale of the individual immigrant, such changes make Canadian- and American-born immigrants more visible and more aware of their foreign and alien status. Such awareness can also translate into altered behaviour patterns, a shifting sense of identity, and a re-evaluation of migration and settlement decisions.

RESEARCH QUESTIONS, METHODS, DATA SOURCES

This chapter focuses on three questions about Canadians in the US and Americans in Canada. (1) What are the statistical and spatial patterns of the Canadian-born population in the US between 1970 and 2007 and US-born population in Canada from 1971 to 2006 at three scales of analysis: national, provincial/state, and metropolitan area? (2) What interrelated political, socioeconomic, and cultural factors shaped the migration decision-making of these two groups of immigrants in time and place? And, in keeping with the urban focus of this volume: (3) What does Canadian settlement in selected Canadian-rich US cities look like and how does this pattern compare with that of US-born residents in selected American-rich Canadian cities? While the different scales and geographies of the cities profiled in the chapter (Toronto and Atlanta) make direct comparison impossible, this assessment sheds further light on the urban impacts of the affluence and 'whiteness' that define both migrant groups in the contemporary era.

We launched our search for answers to these questions by gathering and analyzing statistical data from the censuses of the two countries and then completing a cartographic analysis at the provincial and state scale showing the total number of Canadians living in the US in 1970, 1980, 1990, 2000, and 2007 and the US-born population in Canada in 1971, 1981, 1991, 2001, and 2006. The maps shown in Figures 13.1 and 13.2 provide additional information on the distribution patterns of Canadians in the US and Americans in Canada after 1970 by state, province, and territory.

To learn more about the push–pull factors that shaped these migration flows at various periods of time in each place, we then consulted secondary sources

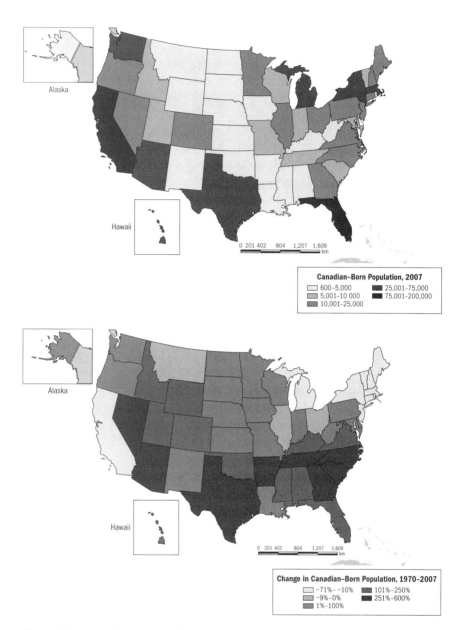

Figure 13.1 Canadian-Born Population Distribution in the US, 2007, and Percentage Change in Canadian-Born Population in the US, 1970–2007

Source: *US Census of Population*, various years. Cartography by Laura Simmons and Thomas Ludden, University of North Carolina, Charlotte.

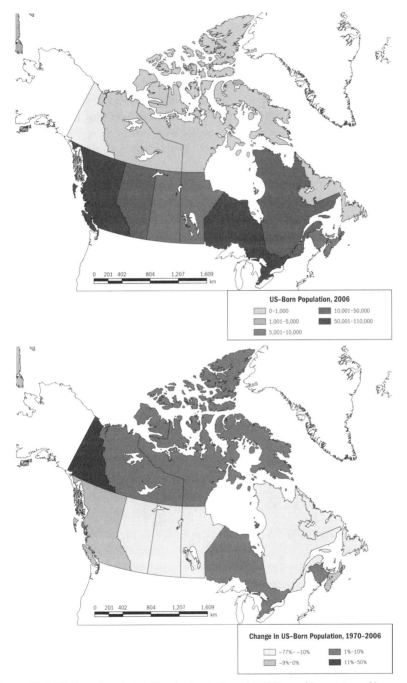

Figure 13.2 US-Born Population Distribution in Canada, 2006, and Percentage Change in US-Born Population in Canada, 1971–2006

Source: Statistics Canada, *Canadian Census of Population*, various years. Cartography by Laura Simmons and Thomas Ludden, University of North Carolina, Charlotte.

on North American immigration patterns and issues. In addition, we conducted preliminary unstructured interviews with Canadians in the US and distributed survey questionnaires to American immigrants in Canada based on a snowball method of sampling for the selection of interviewees on both sides of the border. While our American-based research is in its pilot stages, in the case of our more advanced Canadian work, careful attention was paid during the selection process to choose individuals who represented as wide a range of ages, genders, ethnicities, races, socio-economic classes, and location patterns in different parts of Canada as possible. Exploratory fieldwork and participant observation in Canada were also conducted as a precursor to work planned for a more comprehensive follow-up study to be conducted following the release of new census data in 2010 (in the US) and 2011 (in Canada).

NORTHWARD FLOWS: AMERICAN SETTLEMENT IN CANADA

The earliest major out-migration of Americans to Canada occurred prior to, during, and after the Revolutionary War. Politically and economically motivated **United Empire Loyalists** and other subsequent groups of disenfranchised Americans headed north primarily because of acute dissatisfaction with government policies in the new American Republic. Canada's image as a place of refuge has encouraged ongoing American migration north over the years, with another dramatic example occurring in the decades prior to the US Civil War when runaway slaves found safety in Upper Canada (Ontario) and the Maritime colonies with the help of the **Underground Railroad** (Winks, 1997). Slavery had been abolished in British North America by the judiciary in the 1820s and, especially after 1833, when Britain enacted abolitionist legislation, the question of slavery in what would become Canada was moot. Unlike the US, British North America did not have fugitive slave laws at this time. Canadian law was based on the common-law doctrine mandating that 'a slave became free upon touching free soil' (Dickerson, 1999: 2). This more open and welcoming approach became especially important after 1850 when the US Congress passed a strict fugitive slave law because this new legislation encouraged even more African Americans to seek the relative safety and security of life in Canada.

A larger wave of mostly white American migrants relocated to Canada during the early years of the twentieth century. The rich prairie soils, abundant and affordable land, and well-organized immigrant recruitment efforts by the Canadian government were magnets for many Midwest and Great Plains farmers in search of more affordable land north of the border. Some arrived with religious groups such as the Latter Day Saints, Doukhobors, and Mennonites. This migration stream from farms and small towns in the central US added to the agricultural development of Canada's Prairie provinces and southeast British Columbia, and was followed by subsequent groups of predominantly white Americans migrating to Canada throughout the middle and later decades of the last century. Along with these groups came others who left the US to attend school or seek academic jobs in Canadian colleges and universities in the

1970s and 1980s. A great deal of anti-American sentiment was expressed towards many of these American academics due to the shortage of academic positions in Canada in certain disciplines during this time period as well as the large number of faculty positions held by Americans (see Mathews and Steele, 1969). Still others came to Toronto and other large Canadian cities as economic elites to open newly expanding branch offices of US insurance companies, banks, and other transnational business operations.

An even larger part of the mid-twentieth-century migration wave included the largest out-migration of political refugees in the history of the US during the Vietnam War years. Canada accepted 1,700 American **war resisters** in 1963 alone as **landed immigrants**. As American involvement in Vietnam intensified, Prime Minister Trudeau's administration made the contested decision in 1969 that 'henceforth American Vietnam draft and military resisters, that is, both "dodgers" and "deserters" would be admitted to Canada without regard to their draft or military status' (Hagan, 2000: 609). Thereafter, the number of draft-age US migrants to Canada increased exponentially as news of this more open and formalized Canadian policy reached the US (Kasinsky, 1976). As a result, it is estimated that there were as many as 100,000 American war resisters and draft dodgers in Canada at the end of the Vietnam War (Jones, 2005). By the end of the 1970s with the Vietnam War over, the number of Americans migrating to Canada stabilized to approximately 5,000 per year (Kobayashi and Ray, 2005: 2). Beginning in 1999 and continuing up to the present, the population of documented immigrants in Canada who were born in the United States has once again been on the rise, from a total of 5,533 per year in 1999 to 10,943 in 2006. Corresponding with the years following former US President George W. Bush's election in 2000 and re-election in 2004, this out-migration was due in part to opposition to the ultra-conservative political climate in the US during these eight years (Markels, 2004; Hardwick and Mansfield, 2009). The majority of these new immigrants settled in Canada's largest cities. Others moved to smaller cities and towns located in close proximity to the US border. Towns such as Nelson and other American-rich settlement nodes in the Kootenay Mountains in southeastern British Columbia, for example, have long been home to Americans who migrated to British Columbia during the Vietnam War era (see Box 13.1). A significant number of US immigrants also reside on Canada's east coast. Halifax, Nova Scotia, for example, is currently home to two diverse groups of American origin: the descendants of African-American slaves who first came to the area more than a century ago, and a group of American Buddhists who followed their spiritual leader to Halifax from Colorado in the 1970s.

SOUTHWARD FLOWS: CANADIAN MIGRATION AND SETTLEMENT IN THE UNITED STATES

The story of Canadian migration into the US has a long history dating back to the 1600s and the early days of exploration and fur trading, which led small groups of migrants westward from the St Lawrence River, across the Great Lakes, and south

| BOX 13.1 | Resisterville: A Vietnam-Era American Enclave in British Columbia |

The largest period of out-migration of Americans in the history of the nation happened during the Vietnam War era. Most left their homeland for Canada as war resisters or to escape the draft. Others migrated to places like Sweden and Australia. Of the 100,000 or so who settled in Canada, about half remain today. It is estimated that at least 40,000 of the people who left the US during this period from the early 1960s to the early 1970s ended up in British Columbia, with the majority settling in small towns in the remote Kootenay Mountains or Slocan Valley in southeastern British Columbia.

Nelson has a long history of welcoming military protestors. More than a century ago, some Doukhobors, a pacifist sect originally from Russia, came to the area after persecution in their homeland for refusing to bear arms and after their communal lifestyle and land-use patterns in Saskatchewan, where they initially settled, caused resentment among other newcomers to the Canadian prairies and became unacceptable to government authorities. Later, a group of American Quakers, another pacifist Christian group, settled nearby. Today, Nelson is known as a peaceful community with a special affinity for protecting political refugees (including, recently, a number of American soldiers who refused deployment to the wars in Iraq and Afghanistan). 'We are a peace-loving community and proud of it . . . they are a part of this community's fabric', according to John Dooley, Nelson's mayor and chairperson of its police board (as cited in Hutchinson, 2007).

The urban landscape of Nelson provides evidence of the large number of counterculture residents from the US who reside in the area. Vegetarian restaurants, candle and 'head' shops, and alternative bookstores abound. Another recent reminder of the many Americans who live in this part of British Columbia occurred in the summers of 2006 and 2007 when the Doukhobor community in the neighbouring town of Castlegar helped host the first-ever reunion for Vietnam War-era migrants in Canada. These two 'Our Way Home' events drew more than 1,000 American-born Canadians who left the United States in the 1960s and 1970s. Almost all of the more than 1,000 people who attended these reunions had never met before this event. These American immigrants are largely indistinguishable from 20 million other inhabitants of Canada due to their predominately white skin colour and their desire to blend in as quickly as possible after arrival. Even after US President Jimmy Carter granted this group amnesty in 1977, many chose to remain in Canada and integrated rapidly into the mainstream Canadian population.

Despite its unique role in the story of Vietnam-era US migration to Canada, the urban social geography of Nelson has a great deal in common with other American-rich settlements in Canada. In Nelson, as in much larger cities like Vancouver and Toronto, Americans most often reside in white urban or suburban enclaves or in smaller towns with predominately white populations, which suggests that many are drawn to places of privilege. As we discuss later in this chapter, although nothing has yet been said by other scholars about these preferred residential patterns—or the complex social, cultural, and/or economic processes that have shaped them—our preliminary findings indicate that Can–Am migration, 'whiteness', and privilege seem to overlap in Canadian cities at all scales of analysis. If such proves to be the case in other parts of Canada and also among Canadians who reside in metropolitan areas in the United States, this Can–Am enclaving may have important implications for North American urban social geographies today and in the future.

into the regions surrounding the Mississippi River. The British colonization of francophone Acadia (present-day Nova Scotia) and the French and Indian Wars prompted the 1755 expulsion and forced migration of thousands of **Acadians** southward down the eastern seaboard to new but often difficult lives in New York, Pennsylvania, Maryland, Virginia, and the Carolinas (Brasseaux, 1987; Hamilton, 2006). Following France's 1763 handover of the Louisiana Territory to Spain, many Acadians migrated even farther, disembarking in the port of New Orleans and contributing indelibly to the region's **Cajun** culture and mythology (Brasseaux, 1987; Simpson, 2000).

Throughout the 1800s, the expanding US industrial economy provided opportunities in infrastructure development (i.e., the Erie Canal opened in 1825), manufacturing, mining, textiles, and logging, along with better pay and job security (Belanger and Belanger, 1999; Hamilton, 2006). Following the British adjustment of lumber tariffs, many Maritimers relocated to the Boston area, and those from Ontario and Quebec resettled into communities around the Great Lakes and in the Midwest (Simpson, 2000). There was also available farmland, especially in the expanding American West. In 1850, the census documented at least 148,000 Canadian-born residents in the US (Cooper and Grieco, 2004). By 1900, Canadians numbered over 1.1 million and comprised over 11 per cent of all US foreign-born (Vedder and Gallaway, 1970). Only Germans and Irish had higher numbers (Simpson, 2000: 17). During this same period, about 10 per cent of Detroit's population, 9 per cent of Boston's, and 5 per cent of the population of both Buffalo and Seattle were thought to be Canadian-born (Simpson, 2000: 17).

In addition to Canadian-born migrants directly resettling in the United States, a flow of migrants from Europe migrated to the US via Canada—some immediately, others after longer periods of time (some having Canadian-born children) before moving south. In an effort to oversee this form of **through-migration**, in the late 1890s the US required steamship and railway carriers to ensure that all passengers crossing the border were in fact eligible to enter the country, and an increasingly large number of American border inspectors were stationed at Canadian ports and land crossings (Ramirez, 2001; Hamilton, 2006).

Although the impacts of the Great Depression dampened Canadian emigration and prompted some return migration, the enduring trend was that southward migration exceeded northward. From the 1930s until the mid-1960s Canadians headed south in steady numbers (an average of about 10,000 per year until 1946, rising to over 50,000 in the early 1960s). The continued escalation, coupled with the increasingly educated and skilled profile of the migrant group through the 1950s and 1960s, saw simmering Canadian concerns about the **brain drain**, first voiced in the 1920s, come to a full boil. As Simpson (2000: 30–1) details, in addition to the considerable media attention this issue received, concerns about cross-border moves (especially by Canadian-educated individuals or highly skilled and educated immigrants who had already relocated to Canada from elsewhere in the world) were expressed broadly across the political and professional spectrum. Between 1955 and 1968, the largest professional groups leaving Canada for the US were those trained and working in the fields of health care, engineering, and education. In studies that asked migrants why they left, responses consistently included comments about higher salaries, broader opportunities for career advancement, and a search for better and more stimulating jobs.

The flow of Canadian brain-drain immigrants southward decreased after the passage of the 1965 **Hart-Celler Immigration and Naturalization Act** in the United States, which introduced annual immigration quotas and limited the numbers coming to the US from any one country. Canadians who had been accustomed to moving across the border with relative ease and few restrictions were now included among the less-favoured immigrant masses. Three other factors also contributed to the slowing of the Canadian brain drain in the post-1965 era: the involvement of the US in the Vietnam War; economic stagflation; and unrest around issues of race-based civil rights in the US. Thus, while about 50,035 Canadians left for the US in 1952, only 18,592 left in 1972 (ibid., 32). As discussed earlier, this same period coincided with an increasing number of Americans coming north as draft resisters or seeking new lives in a more diverse and liberal context.

Not until the mid-1990s and the passage of the **North American Free Trade Agreement (NAFTA)** did the scale and substance of Canadian emigration to the United States once again capture political and public attention. While the annual quotas remained in place, NAFTA launched a program to provide a subset of Canadians with renewable temporary working status in the US (and the reverse for Americans interested in working in Canada). Trade NAFTA (TN) status and TN visas allowed college-educated Canadian professionals to seek and secure work in the US with the only requirements for crossing the border being proper documentation and a letter of offer from a US employer. TN visa holders also could bring their families with them as non-working dependants. With an increasing number of Canadians taking advantage of this program either to live and work in the US permanently (by converting to H-1B status as temporary workers with specialized knowledge) or to move back and forth across the border as contract work opportunities dictated, concerns grew about another brain drain from Canada to the US. Throughout the 1990s, employment opportunities lured about 28,000 Canadians south on an annual basis. While we do not know definitively how many of those carried TN visas (a range of other programs, such as student visas, temporary work permits, and intra-company transfers, could be used to enter the US), this figure includes about 20,000 Canadian-born and 8,000 people who migrated to the US through Canada, emphasizing the enduring role of Canada as an immigrant way station to its southern neighbour (Ramirez, 2001; Hamilton, 2006: 37; Michalowski and Tran, 2008).

At the turn of the twenty-first century, Canadians accounted for 2.64 per cent of the total foreign-born population in the US and represented the eighth largest immigrant group following those born in Mexico, the Philippines, India, China (including Hong Kong and Taiwan), Vietnam, Cuba, and Korea (North and South) (Cooper and Greico, 2004). The most recent US census data for 2007 suggest that while the overall number of Canadian immigrants in the US is once again increasing, with 840,197 as of 2007, the pace of their migration may be slowing and their representation in the total and foreign-born populations declining due to a combination of economic and political factors and ever larger numbers of new immigrants from Latin America and other parts of the world.

The 2007 statistics show Canadians as 0.28 per cent of the total US population and 2.26 per cent of the total foreign-born population in the US today. This may be a small change from 2000, but it is potentially significant given the tradition of US-to-Canada flows northward mirroring those of Canadian flows southward to the United States. The same factors that might dissuade Canadians from moving south might also work to encourage Americans to migrate to Canada. These slight changes over 2000 to 2007 also hold significance when placed in the broader context of overall contemporary patterns.

As in the past, two major concerns framed 1990s-era discourse about Canadian emigration: the highly skilled and educated profile of Canadian emigrants and the use of Canada as a way station for highly skilled and well-educated immigrants coming from other nations of the world. That Canada continued to lose her sons and daughters, as well as significant numbers of her own international immigrants, precipitated a flurry of concern and study (DeVoretz and Layrea, 1998; Helliwell, 1999; Iqbal, 1999, 2000; Stewart-Patterson, 1999; Watson, 1999; DeVoretz, 1999; Frank and Belair, 1999; Shillington, 2000; Cervantes and Guellec, 2002; Finnie, 2006; Zarifa and Walters, 2008). In more recent years, these trends have led to the development of targeted programs designed to encourage some of these migrants to return home to Canada.

As has been the case now for almost a century, Canadians who choose to settle south of the border are on average more affluent, more highly educated, and more professionally skilled than the overall population of the US (Michalowski and Tran, 2008). They are also 'whiter' than both the US and Canadian populations overall. Indeed, between 1980 and 2007, the percentage of Canadian-born migrants in the US who are white never dropped below 92 per cent. As we will see later in the chapter, this frequently overlooked characteristic of Canadian immigrants in the US may well have implications at the inter-urban and intra-urban scales.

MIGRATION IN THE POST-9/11 ERA

What have been some of the impacts of the changing political and economic scene on Canadian–American migration since 9/11? Data analyzed from our preliminary surveys and interviews indicate that this convergence of economic and political change at the Canadian–US border has had different results among many American immigrants in Canada compared to Canadians living in the US. Despite increasingly restrictive policies governing border crossings post-9/11, more Americans have migrated to Canada since 2000 than at any time in more than 40 years. According to our interviewees, Americans continued to leave for Canada in the years following the election of George W. Bush as President in 2000 and his re-election in 2004 because of their desire to escape the ultra-conservative climate of their homeland during this period. According to a 36-year-old former Californian who now resides in Vancouver, for example:

> I moved up here about eight years ago after getting fed up with the whole conservative problem in the US and I will honestly never ever go back. It will take a lot more than electing Obama to get me to return. Canada is safe, clean and a caring place. I'm staying and that's all there is to it. (Personal interview conducted by S. Hardwick, June 2005, Vancouver)

As in earlier decades, Americans also continue to migrate northward in search of universal health care in Canada, more open policies in support of gays and lesbians, gun control, stronger support for multiculturalism, and feelings of greater safety and security north of the Canada–US border. This former American who now lives in Moncton, New Brunswick, reported that universal health care was a primary pull factor in her decision to leave the US:

> I had been thinking about leaving Boston for a very long time because of the lack of support for multicultural policies. But I realized this past year (even after Obama was elected President) that politicians were *never* going to be able get any kind of health-care legislation passed in the US and so I just put my house on the market, packed up, and headed north. (Personal interview conducted by S. Hardwick, 11 Sept. 2009, Moncton, NB)

Conversely, Canadian emigration rates have been on the decline in the first decade of the new century. It is not surprising that the events of 9/11 and the election and re-election of conservative politicians may have dampened many Canadians' interest in moving to the US and *perhaps* even precipitated a return home for some. Borderland legislation approved during the Bush years may well be discouraging others from their earlier interest in migrating south. In addition, new technological advances now allow for consistent tracking of cross-border flows and document lengths of stay for Canadian migrants to the US. These changes are especially significant for Canadian **snowbirds** whose overstays in the US can result in the loss of Canadian health insurance and other social service benefits provided by their homeland (see Box 13.2).

With regard to how the post-9/11 era has affected them, one former Toronto resident who moved to North Carolina in 1997 explains:

BOX 13.2   Canadian Snowbirds in the Sunshine State

In a report titled 'Winter . . . by the numbers,' Statistics Canada (2007) detailed that in 2006 the state of Florida received 2.1 million visits by Canadians. While Florida ranked second to New York in the number of visits, it far outpaced it in terms of the number of nights spent in the state by Canadians and in terms of the Canadian dollars spent there. In 2006, 37.8 million nights were spent in Florida by Canadians (more than five times the number of nights spent in New York) and $2.4 billion was spent by Canadians visiting the state (more than three times that spent in New York). Underlying these figures is the distinctive phenomenon of snowbird seasonal migration.

According to the government of Canada, snowbirds 'are mainly retired or semi-retired people who live away from their northern homes during significant portions of each winter.' The Canadian Snowbird Association[3] is more specific, defining Snowbirds as 'those spending 31 nights or more in a southern destination' (Canada, 2009: 1). Florida is the destination of choice for about 70 per cent of all Canadian snowbirds spending over one month in the US Sunbelt. While there, the leisure and consumer culture of Canadian snowbirds is integral, if not critical, to local economies. Indeed, snowbirds have spending patterns similar to native Floridians and sometimes make big-ticket purchases while away from Canada. As would be expected, they tend to spend more when the Canadian dollar hovers at par with its American counterpart (Jackovics, 2009; Sasso, 2010). Many Canadian snowbirds own or rent second homes in the state, with rates of homeownership rising since the post-2008 economic downturn in the US (Jackovics, 2009).

Local newspaper articles, commentaries, and advertising flyers indicate an embracing of Canadian snowbird residents across many Florida communities. '(They) are not tourists', notes a Florida Visitors' Bureau representative, 'They are part-time residents' (quoted ibid., 1). In Hollywood, Florida, where a considerable number of Canadian snowbirds hail from Quebec, two of the local newspapers are published in French and many of the local bank branches have ATMs with French-language capacity (Nadeau, 2008). In Zephyrhills, local leadership has been courting Canadian snowbirds since the 1930s, recognizing the economic advantages they bring to the community through their daily, leisure, and real estate spending as well as through their support of local utilities. In the winter months, city managers see considerable spikes as snowbird customers sign up for utility service (Sasso, 2010). And, in Lakeland, an annual mature lifestyle celebration and trade show, the 'Snowbird Extravaganza', attracts Canadians; its 2010 advertising slogan to potential vendors was 'Snowbirds: They've got time, money and an interest in finding new ways to spend both of them' (Medipac and the Canadian Snowbird Association, 2010).

The difference between Canadian and Floridian winters is clearly a driving explanatory factor in the seasonal snowbird migration. But other factors are closely tied to this group's demographic position as **baby boomers** who have retired or who are semi-retirees. As the Snowbird Extravaganza slogan alludes, Canadian snowbirds are a comparatively affluent and privileged group. In a study of seasonal migration of elderly adults in Florida that included both American and Canadian snowbirds, it was found that they were overwhelmingly white (94 per cent) and non-Hispanic with higher levels of education; income, and overall health than permanent Florida residents (Smith and House, 2006). For many snowbirds the allure of an active, outdoor, and/or creative lifestyle shared convivially among a cohort of other people of similar age and background draws them to places where large numbers of other seasonal migrants cluster. Indeed, over the years French-Canadian snowbirds have gravitated to communities along the Atlantic coast of Florida (especially around Hollywood) while Anglo-Canadian snowbirds have congregated on the Gulf coast and in the Palm Beach and Orlando areas (Jarvis, 2002). In this way, the location and richness of social networks have played a critical role in enticing snowbirds to leave their permanent residences for long periods of time and to gravitate to particular communities. The Canadian Snowbird Association estimates that the average length of stay in the US for snowbird migrants is more than four months.

Despite their affluence and its positive impact on the Florida economy, the long-term but impermanent presence of Canadian snowbirds is not without conflict and controversy. As detailed in a paper in the *Florida Historical Quarterly*, over the years there have been repeated episodes of irritation about perceptions that Canadian snowbirds are under-taxed given their high rates of property ownership and residence in the state. As well, resentment has arisen over their use of state-supported services despite their status as foreigners; their overt display of their Canadian heritage (speaking French, flying the Maple Leaf flag); and their tendency to enclave and patronize only Canadian-owned businesses (ibid.). Such points of tension are a stark reminder for many snowbirds that they are neither permanent residents nor Florida citizens but rather visitors—and foreign-born ones at that.

As with any other immigrant group, then, there are issues of conflicted and complicated identity. On the migratory continuum, Canadian snowbirds fall closer to immigrants than to tourists and that status has implications for how they are made welcome into community life. Still, despite those who display too obviously their Canadian heritage, there is the perspective that 'Canadian ethnicity melts in the Sunshine state' and that most Canadian snowbirds simply blend in with mainstream Floridian culture (Goldfield, 1989). Another view is presented by Desrosiers-Lauzon (2009), who suggests that Canadian snowbirds adopt an exaggerated form of Canadian identity, one that in some cases hearkens back to French and English colonizers.

As is the case with Canadian immigrants to the US generally, little research explores the phenomenon of Canadian seasonal migration to the American Sunbelt states. What literature exists is found mostly in journals on aging or gerontology and rarely examines the distinctive and myriad geographies of this group. The snowbird tendency to enclave as francophone or anglophone Canadians, their conflicted identities as both Canadian and American residents, as both residents and visitors, and their role as economic anchors in many Florida communities have significant geographic implications that warrant attention. For immigration scholars, snowbird migration is an untapped and fruitful area of future research.

> When you think about 9/11 itself, you do question your choice about where you are living. Should you be living close to or back with family? Also, you do think, 'Should I just move back to Canada?' because I don't think that Canada is the same kind of target for terrorist acts, at least not then . . . I mean at the point of 9/11 . . . I didn't agree at that point with US foreign policy and . . . you are torn. You are living in a country and you want to support it and it was obviously a horrible incident that happened but then when you think of the aftermath and the way that the government handled it . . . it moved them away from world support and they went at it alone. (Personal interview conducted by H. Smith, 2009, Wilmington, NC)

More succinctly and practically, another respondent, originally from Manitoba and now resident in the US for over 17 years, says that the post-9/11 era for him has translated into 'more bothersome screening at airports'.

There are also the less tangible affects. An immigrant who migrated to the US through Canada (originally from the United Kingdom) and has been in the US for over a decade provides this response when asked about the affects of 9/11 and its aftermath.

> I think it has created a level of anxiety in issues surrounding documentation that would not have happened otherwise. I have not agreed politically with the United States response to 9/11; however, I don't think that that affected me personally. (Personal interview conducted by H. Smith, 2009, Charlotte, NC)

It is particularly interesting that while respondents commented about their dislike of post-9/11 US foreign policy and government action, they did not say they had considered leaving the US for this reason. Clearly, future research will continue to investigate this perspective.

URBAN SOCIAL GEOGRAPHIES OF MIGRATION, SETTLEMENT, PRIVILEGE, AND WHITENESS

The story of Canadian immigrants in the US and American immigration to Canada embodies a long and complex history. In this concluding section, we consider the comparative stories of Can–Am migrants by examining their urban residential patterns. This is a remarkably underexplored aspect of the immigrant experience, especially given the degree of spatial concentration among these groups at the provincial and state levels, and a demonstrated tendency for urban settlement. US cities that were home to the largest number of Canadians in 2000 and the most American-rich Canadian cities for the census year 2006 are listed in Table 13.1. This broad overview is followed by tentative analyses of the residential patterns of Americans in Toronto and of Canadians in Atlanta, Georgia. The residential distribution of Canadians and Americans within these cities

Table 13.1   Top 20 American-Rich Canadian Cities (2006) and Canadian-Rich US Cities (2000)

Rank	Canadian CMA	US-Born 2006	% Population US-Born 2006	American MSA	Canadian-Born 2000	% Population Canadian-Born 2000
1	Toronto	41,280	0.81	Los Angeles–Riverside–Orange County, Calif.	68,919	0.42
2	Vancouver	24,775	1.17	New York–Northern New Jersey–Long Island (NY/NJ/Conn./Penn.)	48,443	0.23
3	Montreal	16,665	0.46	Boston–Worcester–Lawrence (Mass./NH/Maine/Conn.)	41,963	0.72
4	Calgary	11,035	1.02	Detroit–Ann Arbor–Flint, Mich.	36,791	0.67
5	Ottawa–Gatineau	8,920	0.79	San Francisco–Oakland–San Jose, Calif.	35,055	0.50
6	Edmonton	7,470	0.72	Seattle–Tacoma–Bremerton, Wash.	29,502	0.83
7	Windsor, Ont.	6,630	2.05	Miami–Fort Lauderdale, Fla	25,937	0.67
8	St Catharines–Niagara, Ont.	6,170	1.58	Tampa–St Petersburg–Clearwater, Fla	19,226	0.80
9	Victoria, BC	6,120	1.85	Phoenix–Mesa, Ariz.	17,876	0.55
10	Hamilton, Ont.	5,990	0.86	Chicago–Gary–Kenosha (Ill./Ind./Wisc.)	16,531	0.18
11	Winnipeg	4,590	0.66	Washington–Baltimore (DC/Md/Va/WV)	15,775	0.21
12	London, Ont.	4,275	0.93	San Diego, Calif.	13,968	0.50
13	Halifax	3,370	0.90	Portland–Salem (Oreg./Wash.)	12,400	0.55
14	Oshawa, Ont.	2,030	0.61	Dallas–Fort Worth, Tex.	11,717	0.22
15	Kelowna, BC	1,995	1.23	Houston–Galveston–Brazoria, Tex.	10,768	0.23
16	Abbotsford, BC	1,980	1.25	West Palm Beach–Boca Raton, Fla	10,172	0.90
17	Saskatoon	1,740	0.74	Atlanta, Ga	10,045	0.24
18	Kingston, Ont.	1,525	1.00	Hartford, Conn.	9,360	0.79
19	Quebec City	1,395	0.19	Philadelphia–Wilmington–Atlantic City (Penn./NJ/Del./Md)	9,292	0.15
20	Moncton, NB	1,250	0.99	Denver–Boulder–Greeley, Colo.	9,023	0.35

Sources: Statistics Canada, *Canadian Census of Population*, 2006; US Census Bureau, *US Census of Population*, 2000.

might provide some clues into the ways in which the distinctive demographic characteristics of each group (overwhelmingly white, generally affluent, highly educated, and broadly skilled) may be contributing—albeit in a very small way—to the shaping of intra-urban geographies of privilege in each place.

Reflecting the extent to which urban fortunes are tied to historic legacy and economic restructuring, the US cities in which Canadian immigrants have primarily settled have changed over the last three decades. In 1970, the metropolitan statistical areas (MSAs) of Detroit, Los Angeles, Boston, New York, and San Francisco had the highest numbers of Canadian-born residents, ranging from just under 70,000 for Detroit to just over 20,000 for San Francisco. Thirty years later, Detroit's Canadian-born population had dropped by more than 30,000 and the city had slipped to fourth place behind Los Angeles, New York, and Boston. Even more noticeable over this period is the significant rise in Canadian-born residents in cities across the Sunbelt South. A function of both seasonal and permanent snowbird migration among Canadian retirees and the economically motivated migration of younger groups, the steady growth of Canadians in cities like Miami, Tampa, Phoenix, West Palm Beach, and Atlanta is a significant trend. So, too, is the growing Canadian-born presence in hallmark post-industrial cities such as Portland, Oregon, Denver, and Washington, DC.

Shifting to the intra-urban scale, the 2000 settlement pattern is of special interest. Canadian immigrants in the Atlanta MSA have made this metro area the only non-Florida southern city to newly appear on the 'Top 20 Canadian-born' list of US metros. Atlanta, like many other large cities across the US South, has experienced considerable population and economic growth in the post-industrial era. One of the South's flagship cities, Atlanta had a total population of 4,112,198 in 2000 with 10,045 Canadian-born, representing 0.24 per cent of the total population but 2.37 per cent of the total foreign-born. Like many other immigrants, Canadians are likely drawn to the Atlanta area because of its growing service-based economy and its warmer climate relative to Canada. The residential location of Canadians living in this city shows a notable pattern of concentration (Figure 13.3). A distinctive suburban settlement pattern stretches northward a significant distance from the central business district. A clear concentration of Canadian-born immigrants lives in and immediately surrounding Alpharetta—an affluent, overwhelmingly white peripheral suburb. The city's website (www.alpharetta.ga.us) details Alpharetta's higher-than-average median household income ($92,718), education level (57 per cent with college degrees or higher compared to 34 per cent for the MSA), and ethnoracial profile (16 per cent 'visible minority', compared to an MSA split of 63 per cent white and 29 per cent black). Several questions emerge. To what extent does this pattern connect to the suburban settlement trends of other new immigrant groups in other new destination cities? To what extent does this pattern constitute a form of enclaving or self segregation based on race and privilege? Given the importance of economic success and lifestyle as a motivating factor to leave Canada for the US, it is important to ask questions about how those motivations translate into decision-making choices about where to live at the intra-urban level. In places like Atlanta, where historical legacies of race and racism have created segregated cities, it would be most interesting to know more about how Canadian immigrants make residential decisions and reconcile their multicultural heritage.

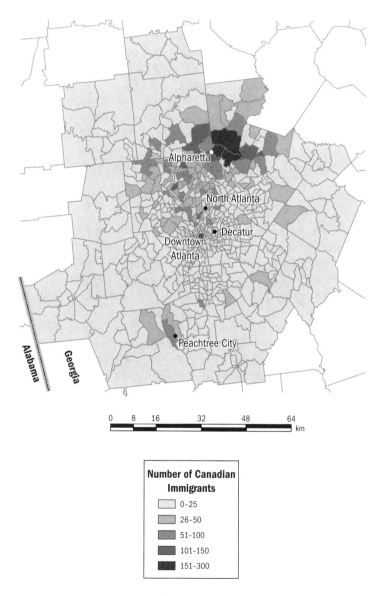

Figure 13.3 Canadian Population in Atlanta MSA by Census Tract, 2000

Source: *US Census of Population*, 2007. Cartography by Laura Simmons and Thomas Ludden, University of North Carolina, Charlotte.

In Canada, the cosmopolitan cities of Toronto, Vancouver, and Montreal are home to the largest number of US-born migrants—with Calgary not far behind. Other important clusters of American residents live in smaller urban places such as Saint John, New Brunswick, and Halifax, in smaller southern Ontario cities, and in Nelson and other small towns in southeastern British Columbia. Factors influencing American settlement vary in each of these places, with descendants of the early Loyalists and former American slaves more important in the Maritimes and southern Ontario while business-class migrants, Vietnam War draft dodgers, and more recent retirees, academics, and midlife mavericks are more common in Toronto, Vancouver, and Nelson. The surprisingly large population of Americans in Calgary over the past two decades or so is largely the result of Alberta's connection with the Texas oil industry.

A closer examination of the spatial distribution of Americans in Canada's largest metropolitan area, Toronto, reveals some interesting patterns that may also exist in other Canadian cities and that reflect the contemporary white and privileged profile of the American migrant to Canada. The Toronto census tract analysis shown in Figure 13.4 reveals a densely populated zone of American settlement in two distinctive parts of the metropolitan area: in the urban core in a broad wedge along Yonge Street, and in a more peripheral location to the west of downtown in the waterfront suburbs such as Oakville. Although a significant number of affluent visible minorities reside in the Yonge Street wedge, both of these Toronto areas are populated largely by white, wealthy, and highly educated Canadian-born residents, with the same traits as many of the American-born immigrants who reside there.

Toronto's 'corridors of wealth' have long attracted American migrants to neighbourhoods in this part of downtown Toronto and its outer western suburbs, and these patterns linger today. Although some of the earlier arrivals from the US who settled in Toronto (e.g., during the Vietnam War years) initially settled in less expensive parts of the city, their preference for residing in areas of urban whiteness and privilege seems to have endured. It is predicted from the findings for Toronto that the distribution patterns of Americans in other Canadian cities may parallel these trends. If so, a closer look at the ways in which American-born immigrants in Canada are counterparts to processes of gentrification and white flight suburbanization is warranted.

CONCLUSION

Documenting, analyzing, and predicting the nuances of whiteness and privilege and the residential preferences and patterns of predominately white Can–Am migrants in the US and Canada is beyond the scope of this chapter. We raise it as an issue largely to bring it into the light. Based on our preliminary findings, we speculate that except for rare exceptions such as African Americans who initially settled in parts of Nova Scotia and southern Ontario upon arrival in Canada as beneficiaries of the Underground Railroad during the decades prior to the American Civil War, the majority of American immigrants in Canada

Figure 13.4 American Population in Toronto by Census Tract, 2006

Source: Statistics Canada. *Canadian Census of Population*, 2006. Cartography by Laura Simmons and Thomas Ludden, University of North Carolina, Charlotte.

have long been (and will continue to be) white, well-educated, and economically secure. Likewise, most of the Canadians who have settled in the US over the years have been white and economically secure. And while it is true that the proportional representation of white Canadian immigrants to the US is on a downward slide, a reflection of Canada's growing multicultural non-immigrant population, the fact remains that as recently as 2007, over 90 per cent of all Canadian immigrants to the US categorized themselves as white. That this fact is rarely, if ever, mentioned in the literature (even that addressing the post-1990s brain drain) is remarkable.

Although this trend has not been documented or analyzed in prior published literature, the predominately 'white-on-white' socio-spatial pattern identified in this chapter in both an established (Toronto) American immigrant destination and a newer (Atlanta) Canadian immigrant destination is an important discovery that has implications for cities located on both sides of the border and thus bears further investigation. How does the predominant whiteness of Americans in Canada and Canadians in the US affect their adjustment to new places of residence? What impact does the whiteness of an immigrant group have on their post-resettlement residential choices, economic opportunities, and attachment to place? How does whiteness help or hinder immigrants in the construction of their national or transnational identities and/or affect their sense of belonging to new lives and landscapes? How are their residential choices implicated in broader patterns of urban and social change? These and other related questions about the geographies of whiteness, privilege, immigration, and identity have not yet been adequately theorized or analyzed in the context of Canadian and US immigration or, for that matter, in other parts of the world. These related themes and theories, therefore, remain important avenues for future research by urban social geographers and other scholars.

Our research has resulted in the discovery of some unexpected and interrelated gaps in what is known about the geography of Canadian migration to the US and US migration to Canada. First and foremost, very little has been written about American immigrants in Canada and Canadians in the US. Perhaps because of their linguistic, religious, and ethnic invisibility on both sides of the border, these two groups have been largely ignored in prior work on North American immigration, rendering them almost invisible. There is also a dearth of work on inter- and intra-urban geographies of these groups—a surprising gap given the strong relationship between all immigrant groups and urban dynamics. In addition, there is the equally overlooked finding that most Can–Am migrants are white and privileged. At this preliminary stage of our work, findings suggest that today, as in the past, white migrants form the vast majority of those who cross the Canada–US border each day. This fact is particularly important during this restrictive **Homeland Security** era in the United States, when racial profiling and other similarly abusive approaches to controlling and managing border crossings are an all-too-frequent event.

As discussed in our introduction, one of our overarching goals has been to lay a foundation for a larger and more nuanced future study of Americans in Canada and Canadians in the US. It is clear from this preliminary study that a great deal remains to be accomplished to document and analyze the spatial patterns, migration flows, and shifting identities of these comparative North American immigrant groups. Of particular importance is the need for a more in-depth analysis of the pre- and post-migration journeys, decision-making, and residential patterns of these groups at a variety of different scales at differing time periods. Newly released census data that will provide evidence of the total number of Canadians in all states and metropolitan areas in the US in 2010, followed by the release of similar Canadian census data for Americans in Canadian provinces, territories, and metropolitan areas in 2011, will make possible a more up-to-date analysis of the spatial patterns of Can–Am migrants in the US and Canada. Meanwhile, the foundational data and ideas in this chapter provide an exciting beginning point for a more intense study of the patterns and related social, political, and economic processes shaping the lives of these two groups of immigrants now and in the years ahead.

QUESTIONS FOR CRITICAL THOUGHT

1. List and discuss at least three of the primary push–pull factors that encouraged Canadians to migrate to the United States and Americans to move to Canada.

2. Why might theories related to 'whiteness' be particularly useful in analyzing and understanding the spatial patterns of American immigrants residing in Canadian cities and/or Canadian immigrants who live in US cities?

3. Who are Acadians? Where did this group originate in Canada, and why did many thousands of Acadians 'migrate' to Louisiana?

SUGGESTED READINGS

1. Hagan, John. 2001. *Northern Passage: American Vietnam War Resistors in Canada*. Cambridge, Mass.: Harvard University Press. Hagan provides a comprehensive and carefully researched analysis of American war resisters and draft dodgers in Canada from the late 1960s through the 1970s. Of particular note are sections that provide a larger historical and political context of the experiences of US migrants in Canada during this time period.

2. Iqbal, Mahmood. 2000. 'Brain Drain: Empirical Evidence of Emigration of Canadian Professionals to the United States', *Canadian Tax Journal* 48, 3: 674–88. This article documents the significant increase in Canadian professional immigration to the US over the late 1980s and 1990s. It places this growth in the context of NAFTA and addresses why the 'brain drain' continues to be an issue worthy of policy and scholarly attention.

3. Konrad, Victor, and Heather Nicol. 2008. *Beyond Walls: Re-inventing the Canada–United States Borderlands*. Aldershot, UK: Ashgate. Geographers Konrad and Nicol analyze the shifting politics of the US–Canada border in the post-9/11 era and provide a comprehensive study of borderlands theory, globalization, and borderlands culture and environment in the North American context.

4. Simpson, Jeffrey. 2000. *Star-Spangled Canadians: Canadians Living the American Dream*. Toronto: HarperCollins. Simpson provides the most comprehensive assessment to date of the processes and motivations shaping historical and contemporary Canadian migration to the US. In addition to exhaustive archival statistical research, Simpson also draws from hundreds of interviews with Canadian immigrants in the US.

NOTES

1. A few notable exceptions to this lack of published literature on Americans in Canada include geographer Randy Widdis's seminal publications on historical patterns and flows at the Canada–US borderland (e.g., Widdis, 1997), political scientist John Hagan's comprehensive work on Vietnam War draft dodgers and war resisters in Canada (Hagan, 2001), and the paper prepared for the Migration Policy Institute by geographers Audrey Kobayashi and Brian Ray (2005) in the months following former US President George W. Bush's re-election in 2004. Work on Canadians in the United States is even more limited with the important exception of Jeffrey Simpson's book, *Star Spangled Canadians: Living the American Dream* (2000), and Jack Jedwab's large-scale analyses of census and national poll data on this topic published by the Association for Canadian Studies (Jedwab, 2010). During the past two years, several new books and articles have been published on Canada–US borderland issues that document some of the impacts of US post-9/11 legislation. Of most use in framing this chapter were Victor Konrad and Heather Nicol's *Beyond Walls: Re-Inventing the Canada–United States Borderlands* (2008) and *Canada–US Border Securitization: Implications for Binational Cooperation* by Donald K. Alper and James Loucky (2009). See also Heather Nicol's 2005 article in *Geopolitics*: 'Resiliency or Change? The Contemporary Canada–US Border'.
2. Unless otherwise noted, contemporary data cited in this chapter were drawn from US Census Bureau, 2006–2008 American Community Survey (ACS), and Statistics Canada, *Census of Population*, 2006.
3. The Canadian Snowbird Association (www.snowbirds.org) 'is a national not-for-profit advocacy organization dedicated to actively defending and improving the rights and privileges of travelling Canadians.'

REFERENCES

1. Alper, Donald K., and James Loucky. 2009. *Canada–US Border Securitization: Implications for Binational Cooperation*. Orono: University of Maine, Canadian–American Center.
2. Belanger, Damien-Claude, and Claude Belanger. 1999. 'French Canadian Emigration to the United States 1840–1930', *Readings in Quebec History*. Montreal: Marianopolis College. At: faculty.marianopolis.edu/c. belanger/quebechistory/readings/leaving. htm. (31 Oct. 2010)
3. Brasseaux, Carl A. 1987. *The Founding of New Acadia: The Beginnings of Acadian Life in Louisiana, 1765–1803*. Baton Rouge: Louisiana State University Press.
4. Canada, Government of. 2009. *Types of Canadian Travelers*. At: canadian

international.gc.cn/miami/commerce.can/ economic.economiq. (17 June 2010)
4. Cervantes, Mario, and Dominique Guellec. 2002. 'The Brian Drain: Old Myths, New Realities', *OECD Observer* 230 (Jan.). At: oecdobserver.org/news/fullstory.php/ aid/673/The   _brain_drain:_Old_Myths_ new_realities.html. (6 Nov. 2009)
6. Cooper, Betsy, and Elizabeth Grieco. 2004. *The Foreign-Born from Canada in the United States*. Washington: Migration Policy Institute. At: www.migrationinformation. org/USFocus/display.cfm?ID=244. (6 Nov. 2009)
7. Desrosiers-Lauzon, Godefroy. 2009. 'Canadian Snowbirds as Migrants', *Canadian Issues*/Themes Canadiens 7, 2: 27–32.

8. DeVoretz, Don J. 1999. 'The Brain Drain Is Real and It Costs Us', *Policy Options* (Sept.): 18–24.

9. —— and Samuel A. Laryea. 1998. *Canadian Human Capital Transfers: The United States and Beyond.* C.D. Howe Institute Commentary 115. Toronto: C.D. Howe Institute.

10. Dickerson, James. 1999. *North to Canada: Men and Women against the War.* Westport, Conn.: Praeger.

11. Finnie, Ross. 2006. *International Mobility: Patterns of Exit and Return of Canadians, 1982 to 2003.* Statistics Canada Analytical Studies Branch Research Paper Series, Catalogue no. 11F0019MIE—No. 288. Ottawa: Minister of Industry.

12. Frank, Jeff, and Eric Belair. 1999. *South of the Border: Graduates from the Class of '95 Who Moved to the United States.* Hull, Que.: Minister of Public Works and Government Services Canada, Catalogue no. MP43–366/2–1999.

13. Goldfield, David. 1989. 'Book review of *Shades of the Sunbelt: Essays on Ethnicity, Race, and the Urban South', Journal of Southern History* 55, 4: 747–9.

14. Hagan, John. 2001. *Northern Passage: American Vietnam War Resisters in Canada.* Cambridge, Mass.: Harvard University Press.

15. Hamilton, Janice. 2006. *Canadians in America.* Minneapolis: Lerner.

16. Hardwick, Susan, and Ginger Mansfield. 2009. 'Discourse, Identity, and "Homeland as Other" at the Borderlands', *Annals, Association of American Geographers* 99, 2: 383–405.

17. Helliwell, John F. 1999. 'Checking the Brain Drain: Evidence and Implications', *Policy Options* (Sept.): 6–17.

18. Hutchinson, Brian. 2007. 'Resistorville, B.C.', *National Post*, 10 May 2007. At: www. canada.com/nationalpost/news/story.html ?id=eb7eb231-b6b6-4537-8525-f04. (10 May 2007)

19. Iqbal, Mahmood. 1999. 'Are We Losing Our Minds?', *Policy Options* (Sept.): 34–43.

20. ——. 2000. 'Brain Drain: Empirical Evidence of Emigration of Canadian Professionals to the United States', *Canadian Tax Journal* 48, 3: 674–88.

21. Jackovics, Ted. 2009. 'Snowbirds Swoop in on Home Deals', *Tampa Bay Online*, 22 Nov. At: www2.tbo.com/content/2009/nov/22/ 220033/na-snowbirds-swoop. (17 June 2010)

22. Jarvis, Eric. 2002. 'Florida's Forgotten Ethnic Culture: Patterns of Canadian Immigration, Tourism, and Investment since 1920', *Florida Historical Quarterly* 81, 2: 186–97.

23. Jedwab, Jack. 2010. 'Are Good Fences Necessary to Preserve Good Neighbours? Migration Flows across the Canada–US Border and the Desired Proximity of Canadians to the United States', *Canadian Issues/Themes Canadiens* 7, 2: 33–8.

24. Jones, Joseph. 2005. *Contending Statistics: The Numbers for US Vietnam War Resistors in Canada.* Vancouver: Quarter Sheaf.

25. Kasinsky, Renee. 1976. *Refugees from Militarism: Draft-age Americans in Canada.* New Brunswick, NJ: Transactions Books.

26. Kobayashi, Audrey, and Brian Ray. 2005. *Placing American Emigration to Canada in Context.* Washington: Migration Policy Institute. At: www.migrationinformation. org/Feature/display.cfm?ID=279. (12 Dec. 2006)

27. Konrad, Victor, and Heather N. Nicol. 2008. *Beyond Walls: Re-inventing the Canada–United States Borderlands.* Aldershot, Hampshire, UK: Ashgate.

28. Markels, Alex. 2004. 'After Election Day: A Vote to Leave', *New York Times*, 1 Nov., 1–2.

29. Mathews, Robin, and James Arthur Steele. 1969. *The Struggle for Canadian Universities: A Dossier.* Toronto: New Press.

30. Medipac and the Canadian Snowbird Association. 2010. 'Snowbird Extravaganza Advertising Flier'. At: www. snowbirdextravaganza.com/pdf/Florida MarketingKit.pdf. (1 July 2010)

31. Michalowski, Margaret, and Kelly Tran. 2008. 'Canadians Abroad', *Canadian Social Trends* (13 Mar.). Statistics Canada Catalogue no. 11–008.

32. Nadeau, Laurent. 2008. 'Snowbirds: Get Your House in Order before Heading South for Winter'. At: www.ahcom.ca/mailer/ documents/0810_QuebecsnowbirdsOct 2008_4.doc. (1 July 2010)

33. Nicol, Heather. 2005. 'Resiliency or Change? The Contemporary Canada–US Border', *Geopolitics* 10: 767–90.

34. Ramirez, Bruno. 2001. 'Canada in the United States: Perspectives on Migration and Continental History', *Journal of American Ethnic History* 20, 3: 50–70.

35. Sasso, Michael. 2010. 'Canadian Snowbirds Head South, Take Advantage of the Exchange Rate', *Tampa Bay Online*, 15 Feb. At: www.tbo.com/content/2010/feb/15/canadian-snowbirds-may-be-back-buying-power. (10 Mar. 2010)

36. Shillington, Richard. 2000. 'Canada's "Brain Drain" a Trickle not a Flood', *Straight Goods*. At: www.straightgoods.com/Analyze/0018.shtml. (6 Nov. 2009)

37. Simpson, Jeffrey. 2000. *Star-Spangled Canadians: Canadians Living the American Dream*. Toronto: HarperCollins.

38. Smith, Stanley, and Mark House. 2006. 'Snowbirds, Sunbirds, and Stayers: Seasonal Migration of Elderly Adults in Florida', *Journal of Gerontology* 61, 5: S232–9.

39. Statistics Canada. 2007. 'Winter . . . by the Numbers'. At: www42.statcan.ca/smr08/2007/smr08_097_2007-eng.htm. (17 June 2010)

40. ———. 1971–2006. *Canadian Census of Population*. Ottawa: Statistics Canada.

41. Stewart-Patterson, David. 1999. 'The Drain Will Be a Torrent If We Don't Staunch It Now', *Policy Options* (Sept.): 30–3.

42. US Bureau of the Census. 1970–2007. *United States Census of Population*. Washington: US Bureau of the Census.

43. Vedder, R.K., and L.E. Gallaway. 1970. 'Settlement Patterns of Canadian Emigrants to the United States, 1850–1960', *Canadian Journal of Economics* 3, 3: 476–86.

44. Watson, William. 1999. 'If We're Number One, Why Would Anyone Leave?', *Policy Options* (Sept.): 39–43.

45. Widdis, Randy W. 1997. 'Borders, Borderlands, and Canadian Identity: A Canadian Perspective', *International Journal of Canadian Studies* 15, 1: 49–66.

46. Winks, Robin. 1997. *Slaves in Canada*. Montreal and Kingston: McGill-Queen's University Press.

47. Zarifa, David, and David Walters. 2008. 'Revisiting Canada's Brian Drain: Evidence from the 2000 Cohort of Canadian University Graduates', paper presented at the annual meeting of the American Sociological Association, Boston, 31 July. At: allacademic.com/meta/p239505_index.html. (6 Nov. 2009)

A REVIEW AND SOME SIGNIFICANT FINDINGS

James Allen and Carlos Teixeira

Our purpose here is to highlight the most important findings from each of the preceding chapters. The first three sections here follow the three parts of the book. In the last section we consider especially compelling or unexpected findings and suggest their implications for policy or for further research.

PART I: THE INTERNATIONALIZATION OF NORTH AMERICAN CITIES AND SUBURBS

In Part I the authors compare the immigration policies and regulations of the United States and Canada as these have evolved over the last century and more. Chapter 1, by Helga Leitner and Valerie Preston, probes several aspects of contemporary Canadian and US immigration policies, including some features not widely known. One key point is the fact that policies and regulations concerning immigrants are not completely controlled by the federal governments of Canada and the US. The central governments are still the primary makers of laws and policies related to immigrants; however, in Canada, authority for immigrant admission policies and settlement assistance has been decentralized to some extent to lower levels of government, especially to the provinces, as under the Quebec–Canada Accord and provincial nominee programs. Canadian provinces may select some of their own immigrants and temporary migrants based on their own criteria rather than on the national government's point system, although such immigrants may not be eligible for some forms of assistance like job and language training. This decentralization follows the traditional provincial control over social welfare and natural resources that has been in place since Canada was formed in 1867. With greater provincial authority concerning immigrants, provinces are beginning to differ somewhat in the characteristics of immigrants they attract and the funding available to assist them.

Decentralization is much less advanced in the US, but a controversial US government program—287(g)—encourages local and state police to share some immigration enforcement functions. Also, a few states and many municipalities have passed regulations that affect immigrants, their rights, and the services available to them. While some cities and states have attempted to deny rights to undocumented migrants, others have established themselves officially as sanctuary cities.

Both countries have been admitting increasing numbers of newcomers, and in both countries the proportion of new residents who have visas only for temporary stays has increased. Leitner and Preston offer a useful interpretation of these trends, which they see as resulting from the growing influence of a neo-liberal ideology. From this ideological view, immigrants are good for Canada and the US primarily because they bring economic benefits as workers and consumers. The market—rather than state regulations—should be the key determinant of who and how many are admitted. At the same time, under neo-liberalism the role of the nation-state and its borders should be diminished as international markets become more dominant influences. Because government in Canada traditionally has played a larger role in people's lives than has been the case in the US, the diminished government role accompanying neo-liberalism represents a greater shift for Canadians than it does for Americans.

Leitner and Preston rightly point out that the trend towards neo-liberalism clashes with another powerful ideology—neo-conservatism—because in one respect those ideologies are inconsistent. Neo-conservatives believe that the state should have sufficient authority to set and enforce its own laws pertaining to immigration. If a country's borders remain effectively controlled as advocated by neo-conservatives, the global movement of people and goods cannot be fully determined by market forces, as favoured by neo-liberals. The authors also make the point that over the last 20–30 years these two ideologies have worked together to promote larger numbers of undocumented migrants residing in the US, based on the fact that the neo-conservative emphasis on the rule of law and its enforcement actually ends up promoting the growth of an 'unorganized, cheap, and flexible labour force' that appears to benefit the market and those espousing neo-liberalism. The result is that neo-liberal and neo-conservative approaches, one using economic arguments, the other using so-called moral arguments, actually end up working together to regulate the composition of the immigrant population.

Dirk Hoerder and Scott Walker present in Chapter 2 the main features of the two countries' immigration history. Their territories and governments have been key players for over three centuries in an evolving system of migration that at first overwhelmingly linked areas on either side of the Atlantic Ocean but had become global by the 1970s. Immigration has been so basic to each country's character and to its larger political and economic characteristics that one cannot understand Canada and the United States without knowing the basics of their immigration histories.

Hoerder and Walker show how, in general, the two countries have had somewhat similar policies and trends regarding immigration during various historical stages. Leaders and residents of both countries pushed aside indigenous peoples and welcomed immigrants to aid in expansion and development; however, in the late nineteenth century both countries, for essentially racist and nativist reasons, began to restrict the entry of immigrants from certain parts of the world, albeit in different ways. Both countries have been ambivalent with respect to non-white immigrants, preferring those whose race and cultural backgrounds were similar to those of residents already present, but acknowledging the need for the labour of others when immigrants of the preferred origins did not come in sufficient numbers.

There has been a general similarity between the US and Canada in regard to source countries and regions for immigrants at any given historical time, and this similarity has been based on widely shared racial and ethnic preferences and on the timing of more rapid population growth in source countries. Of course, some differences exist in immigrant source countries as a result of the specifics and timing of immigration policies as each country has tried to balance its needs and preferences for immigrants. The greatest difference is the fact that so many immigrants to the US have come from Mexico, which has long been a source of cheap labour. The importance of the US to Mexican immigrants is a function of the much higher wages available in the US relative to Mexico, and the long land border between these two countries has made jobs fairly accessible to Mexicans. Deliberate American policies such as the Bracero Program, which lasted from 1942 to 1964, also have been instrumental in attracting Mexican immigrants northward.

Canada and the United States radically changed their immigration policies in the 1960s, eliminating racial and ethnic preferences regarding source countries. Changes in immigration law at this time led to steady increases in immigration from a growing range of countries—effects not anticipated when the laws were enacted. These changes, together with Canada's enactment in 1971 of a policy of multiculturalism, underlie the contemporary situations of immigrants that are the subject of this book. Because many of the most ambitious and educated people in less developed countries are the ones who immigrate to North America, such migration flows have been correctly called 'brain drains'. While this loss of skilled people is of concern in some of these countries of origin, the money sent home (remittances) by immigrants is very important to the economies of these countries and especially to the relatives to whom the money is sent. For the Philippines, for example, annual remittances from overseas migrants account for over 10 per cent of the country's $160 billion economy. This is why most source countries do not attempt to restrict the movement of their citizens to the US or Canada, or to other receiving countries.

PART II: THE IMPRINT OF IMMIGRATION IN NORTH AMERICAN CITIES AND SUBURBS

The chapters in Part II deal with various aspects of immigrant situations. In Chapter 3, Robert Murdie and Emily Skop tackle some complex issues regarding the spatial pattern of immigrant settlement in the two countries and how such patterns relate to the social integration of immigrants. They begin by explaining the model of immigrant spatial assimilation that was based on the experience of European immigrants in the US in the early twentieth century. An aspect of US culture is the notion that the success of immigrant groups is based partly on reducing any spatial concentration they may have had at one time, whether such concentration resulted from housing discrimination by the larger society or self-segregation by members of the immigrant group. In other words, the spatial assimilation, as well as cultural and economic assimilation, of groups over time is expected, as each group becomes part of the 'melting pot'.

In contrast, Canadians tend to see their country as a 'spatial mosaic' in which immigrant groups form their own settlement concentrations, and these are not assumed or expected to diminish over time. Also, in Canada the existence of such concentrations are not viewed as indicating a lack of successful integration on the part of the group. Considering the greater acceptance of immigrant residential concentrations in Canada, one might expect immigrant-group concentrations to be more common in Canada than the US. This may be the case—at least it appears so among Indo-Canadian (South Asian) and Italian immigrants, as discussed by the authors. But perhaps ideology and national policies are not the key reasons for these patterns. As Murdie and Skop explain with many examples, the actual facts on the ground with respect to contemporary immigrant group concentrations and their significance do not fit very well with simple conceptions or historical models. In any case, there seems to be no consistent evidence that the contrasting ideologies and government policies of the two countries are significantly related to the presence and size of immigrant settlement concentrations.

The typology of urban and suburban settlements (Figure 3.4) provides a useful organizing framework for much of the last part of the chapter, and Murdie and Skop explain the range of settlement types shown in that diagram. In general, suburbs in recent years have become much more important in immigrant settlement; however, one cannot assume that moving to a suburb necessarily represents upward mobility. Suburbs vary greatly in their housing, employment, and income characteristics so that they should no longer be thought of simply as newer, attractive, and more affluent residential areas. Many poor immigrants live in suburbs, often because investment in older neighbourhoods of central cities (gentrification) has made such areas unaffordable for them.

The expectation that ethnic concentrations should be more common in Canada because of that country's multiculturalism is not supported by evidence from Chapter 4, in which Joe Darden and Eric Fong assess the relative levels of residential segregation and socio-economic status of six non-European immigrant groups for 10 large metropolitan areas in each country that represent the leading immigrant destinations. Darden and Fong found that four out of six of the major immigrant groups they tested were somewhat more segregated from whites in the US than in Canada (Table 4.1). Groups differ in how much more segregated they are in one country than the other, making generalization difficult. It is perhaps ironic that Iranians in the US, who are racialized as white, are the most highly segregated group, probably due to their own decisions to live close to each other for mutual support. Immigrant Caribbean blacks in the US are probably even more segregated from whites than Table 4.1 indicates, because Cubans constitute 30 per cent of Caribbean immigrants to the US but nearly all Cuban immigrants are light-skinned and considered white in the US context. The very high level of segregation of Latin Americans in Canada is unexpected and may be partly due to their very small populations in some metropolitan areas.

Some of Darden and Fong's most interesting results concern the relative educational, occupational, income, home ownership, and poverty status of the immigrant groups in Canada and the US (Table 4.2). All immigrant groups except Latin Americans have higher percentages of four-year college graduates in the US than in Canada; however, one should not assume that the average immigrant who came to the US is more educated than the average Canadian immigrant because in Chapter 6, Lucia Lo and Wei Li show that when all immigrants are considered together, regardless of country of origin or decade of arrival, the percentage of adults with bachelor's degrees is slightly higher in Canada (25.5 per cent) than in the US (24.5 per cent) (Table 6.1). A key factor involved in immigrant educational comparisons between the countries is the very large number of poorly educated immigrant Latinos in the US.

Darden and Fong use ratios to compare the status of each immigrant group to the average for the entire Canada-born white or US-born white populations (Table 4.3). It is striking that the median household incomes of all four Asian immigrant groups in the US are much higher than is the median household income of the US-born white population. Although part of this income differential can be explained by the greater average number of workers in Asian immigrant households, this finding reflects more the much higher percentage of college graduates among these immigrants than among US-born whites. It is consistent with the finding of Wang and Wang (Chapter 10) for Asians as a whole. Also, the fact that in Canada the median individual incomes of immigrants from the four Asian groups are much lower than that of Canadian-born whites is a powerful finding, again confirming the analysis in Chapter 10. Because these immigrants in Canada have a much higher percentage of college graduates than do Canadian-born whites, their low median incomes suggest any of several possible explanations, as discussed in other chapters. On the other hand, the fact that Latin American and Caribbean immigrants in Canada also have lower incomes than Canadian-born whites is not surprising because of their lower educational attainment.

Darden and Fong reject a spatial assimilation model in favour of 'differential incorporation', a two-way process that examines both the internal and external factors that influence immigrant experiences. Their overall conclusion is that there are significant gaps between those groups racialized as white and non-white, but in the US the gaps appear more strongly in the housing market while in Canada they occur in the labour market. They cite a range of potential explanations for these variations, including land-use policy, zoning regulations, relative community size, and timing of arrival, but clearly their findings raise more questions than can be answered without much further work in specific communities.

We cannot understand housing markets, particularly those in large urban and suburban areas of Canada and the United States, without addressing the impact of immigration on these markets. In Chapter 5, Thomas Carter and Domenic Vitiello examine the different housing experiences of new immigrants during their initial settlement and the integration process. Immigrants are a driving force of housing demand and neighbourhood change. Since 1995 the foreign-born population has accounted for one-third of household growth in the United States and close to two-thirds in Canada. Within this context, it is important to recognize the diversity of immigrant settlement patterns in both countries. In some regions, particularly in the larger metropolitan areas, immigrant settlement is heavily weighted towards rental housing, while settlement in small and mid-sized cities happens chiefly in the ownership sector. These patterns appear to reflect differences not only in the Canadian and American markets but also in the ways various immigrant groups in both countries see their long-term settlement prospects.

Any study of housing and immigration must take into account two major differences between Canada and the US. The first is the presence of a large, undocumented population in the US that faces additional barriers in accessing adequate housing. The other is the legacy in the United States of discrimination against African Americans. While visible minorities in Canada have experienced various forms of discrimination, nothing in Canada's history of housing and urban development quite compares to the systematic, historical segregation of this group in the US.

The authors contend that housing affordability is the greatest barrier faced by new immigrants in both countries, although significant differences exist between the rental and ownership markets and between income groups. As rent increases have outpaced income growth, housing affordability challenges have been particularly significant for the lower 20 per cent of the income profile. And while skilled immigrants in both countries have benefited from low mortgage rates in the ownership sector, recent high foreclosure levels have meant falling vacancy levels in the rental sector as people move back into the rental market. The rental supply in Canadian cities has been low for many years—from 3 per cent to below 1 per cent in some cases—a phenomenon now also occurring in American gateway cities such as New York, where current surveys show a vacancy rate of 2.9 per cent.

Carter and Vitiello also note that 'the housing experiences of immigrants and refugees are similar in Canada and the United States, varying more by ethnic group and other characteristics than between the two countries.' Thus, in both countries, the average income and wealth of the different immigrant groups and the cost of housing impact their housing experiences. This being said, many groups—particularly members of black or Hispanic immigrant groups—are likely to encounter discrimination as a barrier, although this is more likely to occur in the rental as opposed to the ownership market.

Interestingly, homelessness does not appear to be as common among immigrants and refugees as among non-immigrant groups (i.e., Canadian Aboriginal people or African Americans), due in part to the existence of immigrant social networks; however, the existence of camps of undocumented Mexicans and hidden homelessness among African immigrants in major cities suggest that this problem may be more significant than is documented at present. Immigrants have been a key driver of housing demand in both countries, and represent the major reason for increased housing prices, especially in Canada. Such price increases may explain why home ownership rates among immigrants in both countries are lower than 30 years ago. Details on which immigrant groups in the two countries have higher rates of home ownership are shown in Table 4.2.

Analyzing the economic performance of immigrants is the subject of Chapter 6 by Lucia Lo and Wei Li. Various factors affect how immigrants are doing in each country. Perhaps most basic is their human capital: the notion that immigrant earnings should reflect their relative educational attainment, job experience, and proficiency in the dominant language of the country or region of settlement. Many immigrant professionals are discouraged when they realize that the credentials they earned in their prior country of residence are unrecognized in Canada or the US, thus reducing their earnings. Learning a new language also can be difficult. Nevertheless, many ultimately do obtain credentials and learn English or French, demonstrating that an immigrant's human capital is not static but typically grows with experience in the new country. Apart from human capital, Lo and Li explain how the relative success of immigrants in each country and in specific metropolitan areas can also be influenced by various aspects of the context of their reception in the places where they settle. These include the timing of arrival in relation to general economic expansion or contraction, the level and type of assistance provided by government or private organizations, job openings in various occupations, and the degree of discrimination experienced by immigrants. Another contextual aspect is the greater regulation of businesses and the professions in Canada.

Through a series of carefully constructed comparisons, the authors investigate different dimensions of the human capital and economic situations of immigrants to produce a rich set of results. For example, the proportion of immigrants without even a high school diploma or its equivalent is much higher in the US than in Canada, presumably a result of the family reunification emphasis of US immigration policy as well as the large number of poorly educated, undocumented migrants in the US; however, the poverty rate among immigrants in Canada (19.3 per cent) is twice that of those born in Canada and higher than that among immigrants in the US (15.2 per cent). In large Canadian cities, poverty is clearly associated with immigrant neighbourhoods (Smith and Ley, 2008).

Lo and Li also compare the educational attainment of immigrants who arrived during different decades, finding that the average educational attainment of immigrants to Canada has increased sharply in recent years: 42 per cent of those arriving in the years 2001 through 2006 had at least four-year college degrees. This remarkable recent increase appears to reflect Canada's success in selecting immigrants according to their human capital. Nevertheless, Lo and Li are concerned about the economic situation facing this most recent cohort of Canadian immigrants because their earnings average only 78 per cent of that of the 1991–2000 immigrant cohort.

An especially original part of Lo and Li's study is their analysis of variations among eight metropolitan areas in each country in the degree of immigrant economic success. Immigrants in Canada's largest cities are earning only about four-fifths of what Canadian-born workers earn, but in smaller places (Halifax, Sudbury, and Fredericton) immigrants earn more than Canadian-born residents (Table 6.5). Lo and Li point out that such smaller places have often been attractive to the better-educated immigrants with greater English-language facility, and in such places immigrants represent a smaller proportion of the total population and are more likely to be racialized white. In contrast, less-educated immigrants tend to stay in Canada's largest metropolitan areas where there are more jobs in manufacturing and services. In the US immigrant earnings compared to those of US-born workers show much less variation among metropolitan areas and in no city do immigrants earn more than the US-born; however, immigrants in the two largest metropolitan areas (New York and Los Angeles) are not doing quite as well as in the other six places. These patterns of city differences in immigrant earnings are fascinating and provocative, although it is extremely difficult to assess the relative importance of various factors behind these differences.

The importance of gender in understanding immigrants is developed by Damaris Rose and Brian Ray in Chapter 7. Although aspects of gender are too often an 'invisible part of the cultural tapestry' of immigrant groups, the authors point out significant ways in which immigrant men and women are differentially affected by aspects of life in the two countries. The authors see more similarities than differences between the US and Canada in the situations of immigrant men and women. In both countries women tend to be dependent on some male family member for migration decisions and sponsorship. Women are more often accepted as refugees and asylum seekers, partly because many are trying to escape persecution from human sex trafficking and domestic violence. Many women immigrants from poorer countries are admitted as nurses or domestic workers, but they often find less success in the US or Canada than expected. Women usually bear a greater burden than men in transnational obligations to their families in the country of origin, especially in caregiving for aging parents back home. In both Canada and the US, women earn an average of 20 per cent less than men, although the gap is much less for more educated workers. But for financially struggling immigrant women and families, tightening of laws regarding welfare and support have particularly hurt. On the other hand, among black immigrants, the search for rental housing in middle-class neighbourhoods is likely more difficult for men than for women because popular discourses have created widespread fear of black men among many Canadians and Americans.

There are also some differences between the countries. Despite reductions in government social programs that began in the 1990s, Canadian systems of support for immigrants remain stronger than those in the US, where private organizations have been expected to take up the slack. In Canada gays and lesbians can formally sponsor a spouse or partner as an immigrant, something not possible in the US.

Rose and Ray also point out that where immigrants live can be significant. Men and women who settle in large cities that have traditionally been immigrant destinations can find public transportation to help them move around, but women in smaller cities and suburbs of large cities can be isolated socially and have little access to jobs and services if they do not have a car. Altogether, this chapter alerts the reader to gendered dimensions of immigrant life that might not have been apparent before.

In Chapter 8, Lu Wang, Elizabeth Chacko, and Lindsay Withers look at immigrant access to health care in the two countries and the relative healthiness of immigrants. Their findings are of special interest because the health systems of the US and Canada are so very different: Canadian provinces and territories attempt to provide taxpayer-supported health care for everyone for 'medically necessary' procedures, with coverage often enhanced through a health plan paid for by employers. Nevertheless, a shortage of physicians and other factors mean that Canada's goal of universal health care is not fully realized. In contrast, health care in the US is highly privatized and expensive, with some Americans paying high premiums for health insurance while many others lack insurance. Although government programs offer subsidized care to the elderly and certain other groups, the general result is that many Americans lack access to adequate care. Under these two contrasting health systems, a major question concerns their comparative health outcomes for immigrants.

Table 8.1 provides some answers to this question. It shows that Canadians, both immigrants and non-immigrants, are slightly healthier than Americans. For non-immigrants, the only large country difference is the much higher incidence of high blood pressure in the US. Canadian immigrants appear slightly less healthy than the general Canadian population, but among Chinese the much better health of those born in Canada is evident in Table 8.3. A breakdown by white and black groups (Table 8.1) indicates that white immigrants to Canada are much more likely to have high blood pressure than Canadian-born whites. By far the lowest rates of these diseases are found among Canadian-born blacks, although the small sample size suggests that this finding should be considered with great caution. All the above may reflect the superior quality of health care in Canada compared to both the US and the prior countries of residence of Canadian immigrants. Some of the findings, too, may suggest a more aggressive, test-oriented medical system and personal ethos in the US in regard to human health, illness, and diagnosis.

In the US, in contrast to Canada, immigrants are often healthier than the general population. This is indicated by differences in rates of high blood pressure. Similar indications of better health status among US immigrants than among the general population have been found elsewhere and have been dubbed the 'healthy immigrant effect', which shows that immigrants arrive healthier but that over time their health status converges with that of the larger population. The effect has been much studied among Latinos, as shown in Table 8.3, and has often been called the Hispanic or Latino paradox, but this label is simply a variation on the concept of the healthy immigrant.

These are surprising findings for Americans, many of whom assume a superiority for American culture that may not be warranted. We suspect that the poorer health of the American-born compared to immigrants is mostly due to the poor diet and inactive lifestyle of many Americans, aspects of the culture that many people are trying to change. The highest disease rates in the table characterize US-born blacks, a likely reflection of the especially serious social and economic problems facing many black Americans.

Wang, Chacko, and Withers explain certain problems that immigrants can encounter in accessing health care. In the US system, having a low income often means poorer access to health care, and in both countries care facilities are often difficult to get to. Hospitals are typically located in the larger central cities so that immigrants living in many suburbs—including ethnoburbs—and lacking cars or good public transportation often find access difficult. Among Latinos in Washington, DC, the process of using their social networks to become aware of local health facilities is slow, and many immigrants are not aware of nearby health centres that they could use. In both countries differences of language and culture between many patients and their providers pose a problem in communication. The authors illustrate the situation by pointing out that among Chinese immigrants in Toronto, only two-thirds of those who preferred to speak in Mandarin with their physician were able to find a Mandarin-speaking physician.

Chapter 9 focuses on comparing the ways in which immigrants become involved in the political life of the two countries. Els de Graauw and Caroline Andrew investigate rates of naturalization, voting, representation on local government bodies, and other types of formal and informal political participation as they appear in five of the largest central cities in each country. Differences between the countries in data collected, the role of parties in local politics, the nature of local immigrant organizations, and numbers of undocumented migrants make it difficult to compare the degree of immigrant political incorporation in the two countries.

Becoming a citizen of Canada is easier and less expensive than it is to become a citizen in the US. This is probably one reason why rates of naturalization are substantially higher among immigrants to Canada. Canada also permits dual citizenship whereas the US does not generally permit its immigrants to retain their former citizenships and passports when they become citizens. In the US, immigrants who are citizens and registered to vote actually vote at rates slightly lower than non-immigrants, but in Canada's federal elections immigrants are more likely than the Canadian-born to vote.

In the US, in regard to both the US-born and immigrants, race is more significant politically than birth status, and in large cities increasing numbers of blacks and other minorities have been elected to political office over the last few decades. This is reflected in the fact that racial and identifiably ethnic minorities in the US are much better represented as city councillors than are immigrants in the five cities studied. In contrast, immigrants are better represented on city councils than visible minorities in Canada's largest cities.

An underlying difference between the US and Canada is that civic and electoral participation, including direct advocacy, is an important aspect of American political life that should involve qualified immigrants. In Canada the administrative or bureaucratic dimension of the government is much more important than electoral activities and advocacy, so that the focus of Canadian political activity is on the provision of services to immigrants. Also, in the US the estimated nearly 12 million undocumented migrants, who tend to be concentrated in central cities, would appear to reduce the potential political incorporation of immigrants in those places, although the presence of undocumented people may prompt grassroots political mobilization and larger advocacy roles for community-based organizations in those places.

Although much of the evidence seems to suggest greater political integration of immigrants in Canada, the authors conclude that the substantial cultural and political differences between Canada and the US and the different dimensions by which incorporation can be measured mean that they do not find immigrants are clearly better integrated into the political activities and systems of one country than the other.

PART III: IMMIGRANT GROUPS IN NORTH AMERICAN CITIES AND SUBURBS

The chapters in Part III investigate in more detail the characteristics and situations of immigrants who originated in various regions of the world. In Chapter 10, Shuguang Wang and Qingfang Wang consider immigrants from Asia. Until the 1990s Asians represented about one-quarter of immigrants to both countries, but by 2006 the share of Asians among immigrants in Canada had increased to 41 per cent. The authors point out how the very large numbers of Filipinos, Koreans, and Vietnamese who have come to the US reflect the important linkages established under either colonial conquest or more recent wars.

Asian immigrants are doing much better economically in the US than in Canada. This finding is consistent with Darden and Fong's results for Iranians and Chinese. The average personal income for all immigrants in both Canada and the US is much less than that for non-immigrants. In Canada, the personal income of Asian immigrants averages 23 per cent lower than that of non-immigrants, but in the US the average income of Asian immigrants is 2 per cent above that of all US-born workers (Tables 10.1 and 10.2). Perhaps a more telling indicator is earned income, because it measures only income derived from work and excludes income from other sources, such as investments and public assistance. The average personal earned income of Asian immigrants is 45 per cent higher than that of US-born workers, whereas in Canada it is 14 per cent lower than that of Canadian-born workers (Tables 10.3 and 10.4). Why is there such a difference between Canada and the US in the relative economic experience of Asian immigrants?

To try to answer this question it makes sense to examine, first, any major differences in the human capital of immigrants in the two countries. Although Asian immigrants in both Canada and the US have more than twice the percentage of college graduates as non-immigrants, the higher earned income that might be expected from their higher average level of education is much less in Canada. Tables 10.1 and 10.2 indicate the Asian immigrants in Canada have a lower percentage of college graduates (32 per cent) than in the US (42 per cent), which could be part of the explanation. The average English- or French-language skill of Asian immigrants in the two countries is probably not very different, although this is difficult to assess because of differences between the countries in defining that census variable. A recent Canadian study shows that the income advantages of the US for all highly educated immigrants increased greatly since 1990, but the country differences could not be well explained by measurable characteristics of recent immigrants in the two countries (Bonikowska et al., 2011). It is possible that Canadians are more reluctant to recognize education earned in Asia or that employment discrimination against Asians is greater in Canada than in the US. Because we cannot demonstrate the role of any of the above, we must conclude tentatively that some unknown factors play key roles in enabling Asian immigrants to achieve much higher incomes relative to the wider general population in the US than in Canada.

As expected, the economic performance of different immigrant groups is correlated with their human capital, with immigrants from India, Hong Kong, and the Philippines earning more in both countries than other Asian immigrants. The high income level of immigrants to the US from Hong Kong is a surprise for many Americans because US researchers do not usually distinguish these immigrants from other Chinese. In Canada, Chinese from Hong Kong have much higher socio-economic status than Chinese from the People's Republic of China (mainland China), perhaps partly because so many mainland Chinese obtained their university degrees outside Canada. In general, immigrants from

Asia as well as other parts of the world have had to cope with what they believe is insufficient recognition—by North American employers, professional organizations, and other institutions—of their experience and education prior to immigrating.

In contrast to the poor Japantowns, Chinatowns, and Manilatowns into which many Asians were segregated 60 and more years ago, newer concentrations of Asian nationality groups reflect mostly voluntary choices on the part of immigrants and tend to be located in suburbs rather than the old central parts of cities. 'Ethnoburb' is a useful term for such modern settlements that include both residences and ethnic business activities but are also quite diverse in ethnic composition. Because ethnoburbs are of average or moderately high socio-economic status, they certainly are very different from the old Asian ghettos of the past.

In Chapter 11, Thomas Boswell and Brian Ray consider foreign-born black immigrants, most of whom have come from Africa or the Caribbean region. After explaining the differing histories of blacks and black migration in and to the two countries, the authors compare the contemporary geographical distributions and socio-economic status of foreign-born blacks, US- and Canadian-born blacks, and immigrants in other groups. In both countries, immigrant blacks have settled primarily in large cities. In the US, New York dominates the distribution of both African and Caribbean immigrants, with almost one-third of all the country's foreign-born blacks living in the New York metropolitan area. Within New York, the authors explain four reasons why blacks, especially those from the Caribbean, have tended to settle in predominantly West Indian and Haitian neighbourhoods. In Canada there is even a stronger concentration in a single city, Toronto, where 47 per cent of black immigrants live. Those in Toronto are mostly from English-speaking islands of the Caribbean, whereas most in Montreal are Haitians, who speak French or a French-based Creole. The settlement of immigrant blacks in these and other very large cities contrasts sharply with the much wider distribution of long-established black communities, especially in the US.

In Tables 11.1 and 11.2, two variables seem particularly useful as indicators of socio-economic status: the percentage of adults with bachelor's degrees and the median income of individuals. In the US, foreign-born blacks have a substantially higher percentage of college graduates than US-born blacks and a median income 46 per cent higher than that of US-born blacks. This difference documents the greater average success of black immigrants in the US compared to the descendants of slaves. In comparison to Hispanic, Asian, and white immigrants, immigrant blacks are clearly doing much better than immigrant Latinos but not as well as the other immigrant groups.

Boswell and Ray find that in Canada, also, median incomes are higher for foreign-born blacks than for blacks born in the country. Much of this difference is due to the younger ages and, for that reason, the predictably lower incomes of Canadian-born blacks; however, Boswell and Ray found that even when they restricted their analysis to people of ages 40–54, immigrant blacks still had higher incomes. This finding is surprising because the greater Canadian experience of those born in Canada should translate into higher incomes. In addition, the expected relationship of income to educational attainment did not hold because immigrant blacks were slightly less likely to have a bachelor's degree than Canadian-born blacks. In Canada the percentage of adults with at least a bachelor's degree was slightly lower among black immigrants than among Latin American and Asian immigrants, but the median income of black immigrants was higher than that of these other two groups. Explaining this anomaly would require much more research and is beyond the scope of this chapter.

Luisa Veronis and Heather Smith examine immigrants from Latin America in Chapter 12. A major contrast between the two countries is the fact that Latin Americans comprise 44 per cent of all immigrants in the US but only 6 per cent of all immigrants in Canada. Another striking contrast is that 69 per cent of Latin Americans in the US are of Mexican origin, whereas in Canada 65 per cent are from some South American countries. This difference is significant because immigrants from Mexico tend to be poorer and much less educated than immigrants from South America, due in large part to the much greater costs involved in moving from South America. To illustrate, South Americans in both countries tend to be professionals or managers, in contrast to Mexicans and Central Americans, who most commonly work in construction and low-level service occupations.

Not surprisingly, the demographic impact of Latin Americans is much smaller in Canada's cities, but locally it can be significant. Sixty per cent of Latin Americans in Canada live in Toronto and Montreal. Many of those from South America are professionals who came as refugees from dictatorships during the 1970s, and Salvadorans and Guatemalans also arrived after fleeing violence in their countries. More recently, both skilled immigrants and newer refugees have been settling in mid-sized places like Calgary, Alberta, and London, Ontario; others have dispersed to smaller metropolitan areas. The net effect is that Latin American settlements in Canada are quite mixed in their composition and not differentiated by their socio-economic characteristics.

American cities in Texas and California have long been the leading destinations of Mexican immigrants, and today they tend to have the highest percentages of Latin Americans among their immigrants. New York City continues to attract a diversity of immigrants, particularly from South America. In these and other places many Latin Americans have found more and better jobs and housing in suburbs than in central cities.

In the 1990s there was a major addition to the Latin American settlement geography in the US, when Mexican immigration was partially redirected from traditional destinations to new and smaller places outside the Southwest, perhaps a reflection in part of the migrants' desire to avoid the low wages and high housing costs of the large traditional centres. Mexicans are now a significant population in many small cities and towns, especially in the Midwest and South, which formerly had few immigrants and certainly no Spanish speakers. Localities have had to rethink how to provide appropriate services in schools, hospitals, policing, etc.; and cultural tensions between the new arrivals and local whites and, in the South, blacks have been pronounced. In Canada's rich farm country of southwestern Ontario are similar small settlements of Mexicans, many of whom have permits for temporary farm work, usually lasting about eight months.

From interviews with community leaders in Charlotte, North Carolina, and stories in local print media, Smith determined that after initially receptive and positive attitudes the established community came to see the poorer and less-educated Latin Americans in their midst as 'undeserving' and as 'sinister and threatening aliens' (Box 12.1). Such feelings may be widespread across the US, although they are usually not acknowledged in public. A different situation occurs in Ottawa–Gatineau, Canada's National Capital Region, where newly arrived Latin American refugees and skilled workers do not appear to face the animosity found in Charlotte. It may be that tensions are less where Latin American immigrants have higher status and where they constitute smaller proportions in local populations.

Latin Americans in Canada have been much more likely to naturalize than have those in the US, echoing the findings of de Graauw and Andrew in Chapter 9 for the total immigrant population. This may be partly because of Canada's support of multiculturalism and the availability of dual citizenship, but also because in the US the very large Mexican population has been the most reluctant of immigrant groups to become US citizens. Income levels are lower and poverty rates higher for Latin Americans than for Canadians or Americans in general, a likely reflection of the lower average human capital of Latin Americans.

In Chapter 13 Susan Hardwick and Heather Smith explore recent migrations between Canada and the US. Their interviews with recent immigrants to each country from the other provide valuable insights into the personal motivations behind these migrations. Although these flows are not widely recognized, there has been a substantial migration of Canadians to the US for over a century, as Canadians have sought higher wages and salaries, better jobs, and greater opportunities for career advancement. As of the year 2000 Canada was the eighth leading source country of immigrants to the US. The loss to Canada of some of its more educated and highly skilled individuals—including immigrants to Canada from other countries—has been seen as a problem but one with no easy solution. In addition, many retired Canadians move temporarily to Florida, Arizona, and other Sunbelt states to escape Canadian winters, but return to Canada each spring.

Migration flows from the US to Canada have been smaller but distinctive in motivation because economic opportunities have generally been fewer in Canada. Employment in the many facets of Alberta's oil industry has attracted some Americans, but more Americans have moved northward to other destinations to take advantage of Canada's universal health care and other services, its policies of multiculturalism and support for gays and lesbians, and a greater sense of safety in Canada. The numbers leaving the US have often fluctuated in response to the country's changing political situation, especially as Americans have fled policies repugnant to them. The largest such flow northward was during the Vietnam War, fuelled by young Americans avoiding the military draft by seeking and finding refuge in Canada. Since 2004, there has again been a resurge of emigration from the US to Canada, so that at the present time the number of US citizens heading north exceeds the number of Canadian citizens heading south.

Hardwick and Smith emphasize that most migrants from one country to the other have been fluent in English, with little or no recognizable accent, and have not been visible minorities, making them largely invisible as immigrants in the new country. Because at least 92 per cent of Canadian-born immigrants to the US have been white, they are particularly invisible in the middle- and upper-income suburbs of America where Canadians are typically found. Neighbourhoods in suburban Atlanta with higher than usual percentages of Canadians document the relatively high status of Canadian immigrants, just as the tendency of Americans in Toronto to live in more upscale, mostly white neighbourhoods is illustrated by mapping their residences. The authors suggest that such white-on-white patterns of settlement and the fact of unspoken white privilege in both countries is a key feature that enables immigrants to adjust relatively easily in the other country.

UNEXPECTED FINDINGS AND IMPLICATIONS FOR RESEARCH AND POLICY

Do the findings of this book suggest that ideological, attitudinal, and policy differences between the United States and Canada are important determinants of how immigrants are doing? We think that many Americans and Canadians expect that attitudes and policies regarding immigrants should relate to the relative success of their adaptation in each country. The US implicitly espouses the assimilation of immigrants into the dominant American society, which many conceive of as a melting pot. In addition, Americans believe that achieving assimilation and success is the responsibility of individuals and families rather than the government. On the other hand, Canada's official policy of multi-culturalism discourages the notion of a dominant culture to which others are expected to assimilate. Canadians also differ from Americans in believing that government should play some role in helping immigrants to integrate successfully in their new country. Although some American–Canadian differences in the situations of immigrants might well be partly explained by these contrasting values, authors occasionally obtain results not consistent with these putative national differences. The net effect is that the Canadian policy of multiculturalism and the greater Canadian support of immigrants do not appear of primary importance in explaining differences between the relative success of immigrants in the US and Canada. It appears that cultural, social, and economic processes at work in both countries are more fundamental than prevailing attitudes and government policies with respect to immigrants. Apart from this generalization, let us now highlight what seem to us to be some of the book's more surprising findings and those with implications for future research and government policy.

In matters of health, Wang, Chacko, and Withers found that Canadians are somewhat healthier on average than Americans. The difference is heightened by the fact that the health of immigrants to Canada is not as good as that of those born in Canada, while immigrants in the US tend to have better health than the US-born, even of members of the same ethnic or race group. These findings are consistent with the efforts of Canadian governments to provide universal health care paid for mostly by taxpayers and the fact that health care is inadequate for many Americans. The demonstrably poorer health of US-born Americans compared both to immigrants and to Canadians raises numerous policy questions concerning the health care of Americans. Some of these were addressed in the US during debates that led to passage of the Patient Protection and Affordable Care Act of 2010; however, many factors behind the poorer health of Americans were either not dealt with in this new federal law or not thought by some Americans to be appropriate matters for government policy, leaving any solution to this problem unfinished and still very controversial. The analysis here shows that immigrants are profoundly influenced by the wider public policy context.

With respect to residential segregation, Darden and Fong demonstrate in Chapter 4 that most of the immigrant groups they studied are slightly more segregated in the US than in Canada. It is interesting that this result is inconsistent with the two countries' differing attitudes and policies concerning minorities and immigrants. With a fundamental emphasis on assimilation in the US and the explicit support of multiculturalism in Canada, one would expect groups to be less segregated in America than in Canada. But, except for Latinos and Caribbeans, that was not the case. It appears that the Canadian emphasis on multiculturalism does not result in any more of a *spatial mosaic* in Canada than in the US.

Traditionally, segregation has been assumed by scholars to be the result of discrimination by the majority white group, with results that are detrimental to the minority or immigrant group. Older immigrant and minority concentrations created prior to around 1970 were imposed by the larger society using a variety of discriminatory practices. More recently, however, scholars have realized that newer residential and business concentrations have some advantages for the groups involved and that such concentrations now result more from voluntary choices on the part of immigrants than from discrimination. Thus, like Murdie and Skop in Chapter 3, we do not assume that residential segregation is necessarily bad or that it indicates injustices or problems necessarily needing correction. Rather, the contemporary clustering of immigrant groups in certain sections of cities and the resulting concentration of ethnic resources can be viewed as generally useful to those immigrants who prefer such neighbourhoods. Thus, our book reflects a more up-to-date and appropriate perspective on residential segregation.

The *ethnoburb* is such an essentially voluntary clustering although it, like other ethnic concentrations, is not sharply defined spatially and contains much ethnic diversity. As described by Murdie and Skop, when a residential concentration of an immigrant group is large and suburban and contains substantial business activities of the immigrant group, it may be called an ethnoburb. This concept was coined by Wei Li, based on her research on Chinese immigrants in greater Los Angeles, and has been developed in both articles and a book (Li, 2009). The increasing numbers of immigrants arriving in US and Canadian metropolitan areas suggest that ethnoburbs in both countries will become larger and more numerous. An ethnoburb seems a useful development in helping immigrants adapt to their new lives in metropolitan areas, but the degree to which groups other than Chinese have created ethnoburbs is not known. Although it is diffi-cult to obtain data on immigrant business activities at the scale of ethnoburbs, scholars should attempt to test comparatively the extent to which different nationalities from Europe, Asia, the Middle East, and Latin America create ethno-burbs and whether ethnoburbs produced by different groups have different characteristics.

The chapters focused on the socio-economic status of immigrants compared to US- or Canadian-born populations are especially valuable. Lo and Li in Chapter 6 note the exceptionally high percentage of university graduates in the post-2000 cohort of immigrants to Canada and their poorer than expected earnings. It appears that Canada has been more successful in attracting immigrants with a higher level of human capital than its economy can support, or chooses to support. Other research has documented the extreme frustrations faced by employed and underemployed immigrants admitted on the basis of advanced levels of education and experience (Smith and Ley, 2008). In earlier decades employers gave much less value to the advanced education credentials of immigrants, apart from some British and European degrees, than they did to the higher education credentials of those born in Canada (Reitz, 2003). Lo and Li's findings suggest that this devaluation of immigrant credentials may have become more severe since 2000, although the lowered earnings of this cohort may also have resulted from a downturn in jobs in the economic sectors for which many in this cohort were admitted. And, as Rose and Ray remind us in Chapter 7, it is important to keep gender in mind when examining groups for whom the circumstances of male- and female-led households may be signifi-cantly different.

We do not assume that this recent surplus of highly educated immigrants neces-sarily requires a short-term change in Canada's immigration policies, although it may. The situation may resolve itself in time. Moreover, evidence indicates that immigrants who are highly educated adapt better than others to the vagaries of the Canadian economy and are a key reason for Canadians' strong support of present immigration policies (Reitz, 2010). Then, too, some better-educated immigrants in the largest metropolitan areas may move to medium-sized metro-politan areas, where the authors' research suggests that greater earning oppor-tunities may exist. Some may gain success by opening their own new businesses. Others may return to their countries of origin or try to move to the US.

Nevertheless, Canadian leaders need to investigate this surplus of highly educated immigrants in Canada to determine the reasons behind it and its effects. Also, as Lo and Li conclude, the disappointing economic experiences of immigrants in Canada suggests the value of reviewing Canada's policies and programs designed to provide assistance after initial settlement. Part of this review should concern the more general question of how credentials earned outside Canada are fairly and appropriately evaluated by employers. In such a review it might be useful to compare the Canadian policies with any relevant policies and programs in the US.

Perhaps the most dramatic and provocative contrasts between the situations of immigrants in Canada and the US concern the relative income levels of immigrants from Asia. The comparative analyses in Tables 10.1 and 10.2 of the chapter by Wang and Wang make it clear that immigrants as a whole earn less than the non-immigrant residents in both countries. That is expected due to the variety of difficulties of adjusting to a new country discussed by these authors and others in the book. In the US, however, Asian immigrants have an average earned income that is 45 per cent higher than that of non-immigrants, whereas in Canada Asian immigrants have an average earned income that is 14 per cent below that of non-immigrants. In terms of specific Asian countries of origin, Darden and Fong found similar results in terms of large discrepancies between the US and Canada in Asian immigrant economic success.

As already discussed in this Conclusion, it is not at all clear why Asian immigrants to the US are doing so much better than those in Canada. In addition to the possible explanations mentioned earlier, perhaps Asian immigrants to the US have been selected positively on the basis of some quality other than education and language, or perhaps the much larger American economy has provided better entrepreneurial and other opportunities than those available in Canada for the sorts of skills and ambitions brought by Asian newcomers. Because the variety of possible factors involved makes potential explanations speculative, this question of why the relative incomes of Asian immigrants in the two countries differ so much is an important topic for further research.

Increasingly, the immigrant experience in both countries takes place in suburbs. Because so many immigrants have been settling in inner and outer suburbs of various types, future research needs to probe the economic, land-use, and social diversity of those suburbs and how their characteristics relate to the adjustment and integration of immigrants and their children.

And so the saga of immigration continues. All of the chapters here have provided rich, detailed examinations of the various dimensions of immigrant characteristics and situations, making this book a comprehensive and up-to-date description and analysis of how immigrants are doing in the United States and Canada. How they fare in the years to come will depend on government immigration policies, public attitudes, and, especially, the political and socio-economic climates not only in the US and Canada but in the major source countries of immigration to North America. The topsy-turvy world of globalized markets and peoples, of stagnant and growing economies, and of civil strife in many parts of the world will mean that millions of people will continue to be on the move. Their impact on the United States and Canada, and the impact of these two countries on newcomers to their shores, will be a story of increasing significance.

REFERENCES

1. Bonikowska, A., F. Hou, and G. Picot. 2011. 'Do Highly Educated Immigrants Perform Differently in the Canadian and U.S. Labour Markets', Research Paper. Ottawa: Statistics Canada.

2. Li, W. 2009. *Ethnoburb: The New Ethnic Community in Urban America*. Honolulu: University of Hawai'i Press.

3. Reitz, J.G. 2003. 'Educational Expansion and the Employment Success of Immigrants in the United States and Canada', in J.G. Reitz, ed., *Host Societies and the Reception of Immigrants*. San Diego: Center for Comparative Immigration Studies, University of California, San Diego, 159–80.

4. ———. 2010. 'Selecting Immigrants for the Short Term: Is It Smart in the Long Run?', *Policy Options* 31, 7: 12–16.

5. Smith, H., and D. Ley. 2008. 'Even in Canada? The Multiscalar Construction and Experience of Concentrated Immigrant Poverty in Gateway Cities', *Annals, Association of American Geographers* 98, 3: 686–713.

GLOSSARY

Acadians A group of seventeenth-century French settlers, and their descendants, who were part of the French colonies in the Atlantic region (now Nova Scotia, New Brunswick, and Prince Edward Island in Canada and Maine in the US). The territory was acquired by the British in 1713 and in the expulsions of 1755 and 1758, because they did not swear allegiance to the British Crown, 14,000 Acadians were transported to the American colonies or Britain (though many of the latter group were lost at sea). Many of the expelled Acadians eventually settled in Louisiana and became known as *Cajuns*. Others avoided expulsion by migrating to northern parts of New Brunswick or to other less settled parts of the region.

Acculturation hypothesis A theory that assumes that immigrants will change their cultural practices to become more like the dominant group of their destination country. The theory is criticized for its assumption of a normative, white culture as representing a nation.

Affordable housing Housing costing less than 30 per cent of before-tax household income. For renters, shelter costs include rent and payments for electricity, fuel, water, and other municipal services. For owners, shelter costs include mortgage payments (principal and interest), property taxes, and condominium fees along with payments for electricity, fuel, water, and other municipal services.

Angel Island An island in San Francisco Bay that served as an immigration processing station for approximately one million Asian immigrants to the US between 1910 and 1940. It was designated a National Historic Landmark in 1997.

Assimilation The absorption into a larger group of a smaller group as the members of the smaller group adopt the behaviours, language, and values of the larger group.

Assimilation theory Similar to the *acculturation hypothesis*, assumes that immigrants will gradually become more like the dominant culture of the receiving society. Classical assimilation theory assumes that the immigrants and their descendants will eventually become indistinguishable from others in society. The theory does not take into account the *differential integration* of those who are not racialized as white, nor the complex effects of class and discrimination as a basis for creating difference. It is also criticized for its assumption of a normative culture to which others are expected to assimilate. See also *spatial assimilation*.

Astronaut families Immigrant families, usually from the Asia-Pacific Rim, in which one spouse, usually the husband, makes regular and prolonged trips back to the country of origin to maintain a business.

Baby boomers The generation born after World War II (1946–66). The large number of people born during this period has had very significant effects on housing and labour markets, consumption, demographics, and public services as they work their way through the various stages of the life cycle.

Border control All efforts to manage entry into the territory of a nation-state. Some of these activities occur at the locations marking the physical boundary of the nation-state but, increasingly, border control takes place throughout the nation-state's territory.

Bouchard-Taylor Commission Following a controversial 'code of conduct' passed by the town council of Hérouxville in the province of Quebec, the provincial government established an inquiry into the 'reasonable accommodation' that could be afforded to immigrants and minority groups. Accommodation is guaranteed under the Canadian Charter of Rights and Freedoms, and was supported by the commissioners, but it has become a topic of contention in public opinion, especially in Quebec where the official federal policy of *multiculturalism* has been recast as 'interculturalism'.

Bracero Program (Mexican Farm Labour Program) A program to attract manual labourers from Mexico to the US in response to agricultural labour shortages during World War II that lasted from 1942 to 1964. The program was intended as an alternative to undocumented migration.

Brain drain The emigration of the most educated citizens of a source country (usually in the global South) to more advanced countries (usually in the global North), thus enhancing the human capital of the destination country. This also has been a periodic concern in Canada when highly educated and trained people have moved to the US.

Britishness In Canada, during the first century following Confederation, an emphasis on and valuing of all things British.

Cajun US descendants of the *Acadians*, mainly in Louisiana, where a rich historical culture remains and has been encouraged and marketed by the state tourism industry.

Canada Citizenship Act (1946) Legislation that came into effect in 1947, giving citizenship to those who had been British subjects and/or legal immigrants to Canada until that date, as well as indigenous peoples. Full citizenship rights were given to all except indigenous peoples and Japanese Canadians, for whom rights were limited for an additional period.

Census metropolitan area (CMA) A Canadian census geostatistical definition describing a major urban core with a population of at least 100,000 surrounded by one or more adjacent municipalities.

Chain migration A process whereby emigrants follow earlier migrants from the same family, town, or village, often contributing to large social networks and *ethnic enclaves* in destination cities. It has often been an informal historic process, related to specific labour market niches (such as the construction industry for Italian immigrants to North America). It is a social process that can occur informally, but can be encouraged by official programs such as *family reunification*.

Chicanos Originally used as a short form for Mexicanos, referring to Mexican immigrants in the US. The term has taken on a cultural and political meaning through the Chicano Movement, which began as a civil rights movement, particularly among agricultural workers, during the 1960s and 1970s, and today extends to a range of cultural expressions including music, art, and literature.

Chinese Immigration Act (1923) Legislation that excluded all immigration from China to Canada except for diplomats, students, children of legal immigrants, and certain businessmen. It extended an earlier Order-in-Council that prohibited 'any immigrant of any Asiatic race' except agriculturalists, farm labourers, female domestic servants, and wives and children of legal immigrants. The Order-in-Council was renewed in 1930; the Act was repealed in 1947.

Civic engagement The extent to which an individual or group participates in communal activities through religious institutions, non-profit organizations, labour unions, etc.

Context of reception The social climate in the receiving society towards immigrants.

Control policies   Policies that determine the numbers and characteristics of immigrants allowed to enter a country, and the measures taken to police and adjudicate entry.

Cultural competency   The ability of those in the larger society to understand and relate to members of a minority population, such as a medical practitioner working within and relating to a Chinese or South Asian immigrant community; also, the extent to which an immigrant has adapted to the receiving society and is able to function successfully within it in regard to language, etc.

Day labourers   Men and women who are hired on a daily basis for work at sites that usually are not regulated, so the work is often dangerous, dirty, and difficult and the workers are vulnerable to exploitation by their employers. Many day labourers are undocumented.

Demographic representation   A type of political representation where the characteristics of legislators (including their race, ethnicity, and socio-economic background) proportionally reflect those of the population they represent.

Differential incorporation (integration)   Process whereby the white majority incorporates some non-white, non-European groups into the mainstream society to a greater extent than other groups. The rate at which immigrants achieve integration in the destination society differs because of internal factors such as differences in *human capital* and access to social networks, and external factors that include discrimination and *racism*.

Dillingham Commission   A US Special Congressional Committee (1907–11) formed to investigate immigrants. Its detailed 41 volumes identified 'undesirable' immigrants, especially those from Eastern and Southern Europe, and provided a justification for the dramatic

decrease in immigration to the US during the 1920s.

Ellis Island   An island in New York harbour that housed the largest immigrant processing station in the US between 1892 and 1954. In 1990 the facilities were refurbished and are now part of the Statue of Liberty National Monument.

Emergency Quota Act (1921)   Established country-based restrictions on immigration to the US following the recommendations of the *Dillingham Commission* during a period of American isolationism. It was replaced by the Immigration Act of 1924 (Johnson-Reed Act).

Empire Settlement Act (1922)   An Act of the British Parliament to give assisted passage and training to married couples, single agricultural labourers, domestic workers, and juveniles aged 14–17, resulting in about 130,000 immigrants to Canada over the next decade.

Ethnic and racial minorities   American designation for individuals who are racialized as non-white, including blacks, Asians, and Latinos.

Ethnic density effects   Positive or negative health outcomes resulting from ethnic residential segregation.

Ethnic enclaves   Spatially segregated neighbourhoods that retain some cultural distinction from a larger, surrounding area. Usually the enclave revolves around businesses and social institutions run by members of the community for members of the community. The formation of enclaves may be voluntary or involuntary, depending on the social, economic, and racial status of group members, as well as perceived differences from the surrounding community.

Ethnoburb   Highly visible residential and business concentrations in the suburbs. Unlike traditionally segregated enclaves, ethnoburbs are multiracial, multi-ethnic, multicultural, multilingual, and often multinational communities in which one ethnic minority group has a significant concentration but does not necessarily constitute a majority. These communities are distinguished by high levels of daily interactions among multi-ethnic and multilingual neighbours.

Family reunification   A justification for immigrant admission based on the objective of reuniting families in the destination country. In the US, family reunification is the major public policy governing immigrant admissibility, while Canada admits larger numbers under a *points system.*

First Nations   Aboriginal peoples and groups in Canada who are 'status Indians' and 'treaty Indians' with a land base (reserve) or claimed land base, who aspire to or have achieved a degree of self-government, and who are listed individually in the Indian Registry maintained by Indian and Northern Affairs Canada.

49th parallel   A common though not entirely accurate name for the border between Canada and the US, referring to latitude 49° north of the equator. This boundary line, established and confirmed by the Oregon Boundary Treaty of 1846, extends from west of Lake of the Woods (situated in northwestern Ontario and straddling the borders of Minnesota and Manitoba) to the Strait of Georgia separating the British Columbia mainland from Vancouver Island, the southern portion of which is south of 49° N. Many of Canada's major cities, in Ontario and Quebec, as well as the Maritimes, and about 80 per cent of the Canadian population are located south of the 49th parallel.

Francophones   People whose mother tongue is French. Canada has two official languages, English and French, and francophones make up just under one-quarter of the population, most of whom live in Quebec, where French is the official language.

Gateway cities   Cities that have traditionally been the first point of entry (and settlement) for new immigrants, especially when most immigrants arrived by ship. They include New York, Miami, and San Francisco in the US, and Toronto, Montreal, and Vancouver in Canada. With the advent of air travel and the expansion of international destinations, more and more cities are gateways for immigration.

Gender   The socially constructed roles, activities, and responsibilities considered to be appropriate for, if not definitive of, women and men.

Gentrification   The restoration and revitalization of an older working-class area of the central city by more affluent residents who purchase and upgrade existing housing stock and displace the original working-class population.

Ghettos   Originally, areas to which Jewish people were confined in European cities; in today's context, impoverished, neglected, or otherwise disadvantaged residential areas with concentrations of non-whites. The surrounding community views these areas as undifferentiated slums, a drain on resources, a threat to social peace, and useless places.

Green Card   An informal name for the document that conveys the right to permanent residency and employment for immigrants to the US.

Hart-Celler   Immigration   and Naturalization Act (1965)   US legislation that replaced the National Origins Formula of 1924 with a set of preferences in employment or family reunification categories, with a maximum of 20,000 from any one country. Its passage has resulted in a considerably greater diversity of immigrants from different parts of the world.

Head tax   Colloquial term for a tariff applied against all Chinese immigrants set at $50 in 1885, rising to $500 by

1903. In 1895, the Canadian government imposed a Right of Landing Fee, which is often referred to as the 'new head tax'.

Healthy immigrant effects Immigrants tend to be healthier on average than those born in Canada or the US, but over time their health status tends to converge with that of the general population. Scholars debate the explanation for this effect, considering the role of immigrant selection, cultural practices, lifestyle, diet, and health-care provision.

Hidden homeless Those living in high-risk housing circumstances, such as with friends or relatives, often on a non-permanent basis.

Hispanic or Latino paradox The occurrence of better health indicators for Latinos than non-Latino whites despite the lower socio-economic status of the former group and their lesser access to health care.

Homeland Security A range of security initiatives organized through the newly organized US Department of Homeland Security after 11 September 2001 to protect the country against terrorism. The initiatives increase federal jurisdiction in a number of areas including immigration, naturalization, border security, and transportation.

Homestead Acts From 1862 until the 1970s, legislation that granted approximately 10 per cent of American territory outside the original Thirteen Colonies to individuals who agreed to develop the land. It provided a means of establishment for many European immigrants. Homesteading occurred in conjunction with the removal of indigenous peoples and the establishment of the reservation system.

Human capital The collection of attributes, including education, language, skills, and other traits, that define an individual's economic value. Human capital is a significant factor in immigration policy, and is the basis for Canada's *points system* of immigration selection.

Immigrant integration The extent to which immigrants are able to achieve their needs and fulfill their interests in the new country. Integration is a potentially complex process incorporating a variety of variables such as housing, language, education, employment, citizenship, and civic participation. It can also involve more subjective factors such as identification with the new country and satisfaction with the immigration and settlement process. Official policies often suggest that integration should be a negotiated outcome between new immigrants and the receiving society. In reality, however, it is often a 'one-way' process whereby immigrants are expected to conform to the norms of the receiving society. In contrast to *assimilation theory*, integration assumes people of both recent and long-settled status living together under equitable conditions. Recent public policy defines immigrant integration as a public policy goal.

Immigration Act (1924) See *Johnson-Reed Act*.

Immigration Act (1952) See *McCarran-Walter Act*.

Immigration and Naturalization Act (1965) See *Hart-Celler Act*.

Immigration Reform and Control Act (IRCA, 1986) See *Simpson-Mazzoli Act*.

Immigration policies Policies concerned with the definition, selection, and admission of immigrants and with immigrant settlement and integration.

Incorporation policies Policies setting the terms by which immigrants become full members of the national community and polity. See also *Immigrant integration*.

Indentured labour An exploitive form of contractual servitude whereby immigrants must work off the cost of their passage and other support (food, clothing, shelter) for those who provided or paid for the passage, etc.

Institutional completeness   Concept developed by Canadian sociologist Raymond Breton, referring to situations in which a given ethnic community has developed virtually all of the economic and cultural institutions it needs for the well-being and advancement of its members, thus reducing dependence on access to those of the receiving society.

Invasion and succession   In urban ecology, the process of neighbourhood change whereby newly arrived immigrants occupy and ultimately dominate a neigh-bourhood, replacing the group that formerly lived there. Over time, as these newcomers move upward economically and outward spatially they are replaced by another newly arrived lower-income group and the process begins again. Although normally applied to ethnic minorities the neighbourhood also may experience *gentrification*, which could be viewed as a form of 'reverse' succession with a more affluent group replacing (displacing) a working-class population.

Invisiburb   Recent immigrants arrive and enter an area, then quickly adopt a dispersed pattern of residential location, primarily in suburban settings. Unlike the traditional enclave, businesses and social institutions are also scattered, resulting in a kind of invisibility because there are few significant, conspicuous clusters, unlike traditional *ethnic enclaves*, which are distinctive and highly visible neigh-bourhoods. Despite the absence of spatial propinquity, strong community ties are maintained via telecommunications, visits, and other methods at the metropol-itan, regional, national, and even inter-national scales.

Johnson-Reed Act (1924)   Officially the **Immigration Act of 1924**, including the National Origins Act, and the Asian Exclusion Act, this legislative parcel effectively stopped Asian immigration to the US and severely reduced immigra-tion from Southern and Eastern Europe by limiting numbers to 2 per cent of those who were already in the country in 1890.

Labour force participation rate   The ratio between the number of people in the labour force (those who are able to work, but not including the military, the retired, students, children, and those who choose not to enter the labour force) and the number of unemployed who are seeking entry to the labour force. Historically, the labour force participation rate of women has been lower than that for men, and it varies by immigrant group and by racial-ized status.

Labour market segmentation   The split-ting of the labour market into discrete groups who may work under very different conditions and at different rates of pay. Historically, the labour market has been divided according to gender and ethno-racial status. Many immigrant groups have found themselves segregated into specific labour market segments.

Landed immigrants   A Canadian term to refer to those individuals who have the right to permanent residency. A landed immigrant may apply for full citizenship after three years of residence.

Latino/a   An individual whose national or ancestral origins are in Latin America.

Live-in Caregiver Program   Began in 1981 as the Foreign Domestic Workers Program in Canada. Live-in caregivers are individuals who are qualified to provide care for children, elderly persons, or persons with disabilities in private homes without supervision. They must live in the homes of their employers, and may obtain permanent residence after meeting residency requirements. The program is widely criticized because of suspected abuse.

Localization   The devolution of immigra-tion policies from the national level to lower tiers of government: state and provincial governments, as well as county and municipal governments.

McCarran-Walter Act (1952)   US legislation that expanded the definition of citizenship to include all those born on island protectorates and maintained the quota system for immigrants. The Act has undergone several controversial revisions and attempts at revision.

Metropolitan statistical areas (MSAs) Metropolitan and micropolitan statistical areas (metro and micro areas) are geographic entities defined by the US Office of Management and Budget (OMB) for use by federal statistical agencies in collecting, tabulating, and publishing federal statistics. The term 'core-based statistical area' (CBSA) is a collective term for both metro and micro areas. A metro area contains a core urban area of 50,000 or more population, and a micro area contains an urban core of at least 10,000 (but less than 50,000) population. Each metro or micro area consists of one or more counties and includes the counties containing the core urban area, as well as any adjacent counties that have a high degree of social and economic integration (as measured by commuting to work) with the urban core (www.census.gov/population/www/metroareas/metroarea.html).

Model minority   A minority group assumed to have become successfully assimilated and that therefore presents a model for other groups. This concept has been largely discredited because it depends on a process whereby a group comes to be known as highly successful and hard-working by broader society, when in reality they may be underemployed and overworked. The majority population rewards this 'model' behaviour and attitude, while unfairly using this 'model' as the standard to which all other immigrant groups should aspire.

Multiculturalism   Multiculturalism became official federal policy in Canada in 1971, with the intent to nurture the increasingly pluralistic character of the country while recognizing the importance of equality for all Canadians in economic, social, and political life. This policy assumes that although Canada has two official languages, it has no official culture, and therefore all cultural identities should be treated equally. The Canadian Multiculturalism Act (1988) requires the government to consider the multicultural character of the population in all official policies. Later, the term was used in the US and other parts of the world as a recognition of and commitment to the inclusion of minority groups in their societies. There are exceptions, however, notably France, where emphasis is placed on the cultural assimilation of ethnic minorities. The policy has been criticized from the political right as a deterrent to the development of a single, unified nation and from the left as a means of concealing economic and other inequalities.

Nation-state   The set of governing institutions that have authority over a bounded territory and that derive their legitimacy from the claim that the territory is home to a nation, people who are deemed to share common customs, origins, and history.

Nativism   An attitude that privileges those born in a country over immigrants, but also adheres to a set of normative assumptions about the qualities of citizens. Nativism ranges from mild dislike of 'foreigners' to radical, and sometimes violent, means of maintaining a traditional and racially pure population.

Naturalization   The legal but voluntary process by which an immigrant acquires US or Canadian citizenship after birth.

Neo-conservatism   A moral-political rationality favouring government intervention to address security and morality issues, often expressed as a support for 'family values', based on a concept of the nation defined at an earlier time in history; social conservatism.

Neo-liberalism A political and economic philosophy propounded by some US economists beginning in the early 1970s that seeks to return contemporary liberalism to its classical roots in seventeenth- and eighteenth-century England. It breaks with the idea (enshrined in post-World War II welfare states) that society as a whole, via the state, should assume some responsibility for the risks that people and households are exposed to as a result of their integration into the market system. These risks are redefined as the responsibility of the individual. Individual freedom and economic advancement are seen as being best served by a return to an unfettered market system. Responsibility for the poor is pushed back onto families and communities. What remains of state social policy must respect economic logics of efficiency and market rationality.

Newcomers An alternate term for immigrants used, especially in Canada, when the term 'immigrant' may be construed as derogatory.

NEXUS cards A joint initiative of the US Customs and Border Protection and the Canadian Border Services Agency, which gives frequent, low-risk travellers faster processing time at the international border, using special kiosks in airports, special lanes at land crossings, and the ability to report by telephone at ports.

North American Free Trade Agreement (NAFTA) Signed by the US, Canada, and Mexico, a trade agreement that came into effect in 1994 to encourage open access to economic markets and trade among the three nations. The supplemental North American Agreement on Labour Cooperation (NAALC) facilitates the movement of temporary workers under special NAFTA visas.

Permanent residents Immigrants who have the right to live permanently in the destination country. In Canada this means they have *landed immigrant* status;

in the US it means acquiring a *Green Card*.

Points system The Canadian system of granting *permanent resident* status to those who apply as independent immigrants, based on their human capital, including knowledge of official languages, age, education, occupation, and labour force experience. The system is based on an economic rationale for immigration policy.

Provincial nominee programs Canadian immigration policies under which provinces select applicants for permanent residence according to provincial rather than federal criteria.

Pull factors Factors in a destination country—such as higher standard of living and job opportunities, relatives and friends, more benign climate/natural environment, and political freedom—that attract the immigrant to that country.

Push factors Factors in a source country—such as lack of job opportunities, loss of land, war and civil strife, poor climate/natural disaster, and ruthless political regime—that make a potential immigrant want to emigrate.

Racialization The process by which racial identities are created and reaffirmed through time by the continuous construction of specific group images based on a set of racial assumptions and/or racial stereotypes. These racial categories and the meaning of race are given concrete expression by the specific social relations and historical context in which they are embedded.

Racialized To be categorized and labelled by others and by the society in which one lives as 'Other', as different from the majority societal norm, because of phenotypical characteristics such as skin colour, shape/colour of eyes, colour/texture of hair, etc.

Racial minorities In the US, those who are racialized as different from the white majority.

Racism The practice of discrimination based on processes of *racialization*. Racism may be either personal, based on beliefs that skin colour and other physical characteristics make others inferior or undeserving, or structural, in which historical effects of differentiation are reproduced through institutions and social practices, often with unintended consequences. Racism is based on the false assumption that there are essential, rather than socially constructed, differences among people.

Refugee A person who is recognized as a refugee by the nation-state where he or she has sought asylum. The 1967 UN Protocol defines a refugee as 'A person who owing to a well-founded fear of being persecuted for reasons of race, religion, nationality, membership of a particular social group or political opinion, is outside the country of his nationality and is unable or, owing to such fear, is unwilling to avail himself of the protection of that country; or who, not having a nationality and being outside the country of his former habitual residence as a result of such events, is unable or, owing to such fear, is unwilling to return to it.' Canada is a signatory to the United Nations Convention Relating to the Status of Refugees (1951) and the 1967 Protocol, while the US signed only the Protocol.

Sanctuary Movement In the 1980s, American churches provided refuge to undocumented Central American migrants fleeing civil wars in their home countries. In the present day a new Sanctuary Movement works for the rights of undocumented workers.

Seasonal Agricultural Workers Program A Canadian program that admits temporary agricultural workers from Latin America and the Caribbean under agreements with the sending countries. Those who enter the country under this program are guaranteed work for at least 240 hours within a period of six weeks or less, and can stay in Canada for up to a maximum of eight months.

Secondary migrants Migrants who have relocated within the receiving country from the locale where they initially stopped and settled or lived for a while.

Self-selection The theory of immigrant self-selection suggests that those who choose to emigrate from their original homes are those with strong ambition, initiative, and human capital, and are therefore more likely to achieve success in the receiving country. It is often held in conjunction with the *model minority* theory.

Settlement services Services such as language training, job search workshops, and housing counselling designed to facilitate the speedy and successful integration of newcomers. In Canada such services are funded by the federal government and often administered at the provincial or local level. In the US, except for refugee settlement services, all services are funded by state and local governments or by NGOs, and the onus is on immigrants to find and fund the services they need.

Simpson-Mazzoli Act (1986) Officially, the **Immigration Reform and Control Act (IRCA)** 1986, which required employers to confirm the immigration status of their employees and made it illegal to knowingly hire or recruit undocumented migrants. It also made provisions to legalize the status of some agricultural and seasonal workers and other undocumented migrants and their families already in the US.

Slavery System of bondage and forced labour whereby human beings are owned by other human beings, primarily for economic purposes. The Thirteenth Amendment to the Constitution abolished the institution of slavery in the US in 1865. In Upper Canada, where slavery never had the economic base that it had in the plantation South, legislation made it illegal to buy new slaves in 1793, and court decisions and the abolishment of the slave trade by Britain in 1808 and of slavery in 1834 ended what little slave-holding existed in British North America.

Snowbirds Colloquial term for Canadians who reside in the US, especially in warmer areas such as Florida and Arizona, during the winter months. Many own second homes in the US. To maintain their right to Canadian health care, they must meet minimum provincial residence requirements.

Social reproduction The labour and resources necessary to ensure both the day-to-day survival, nurturing, and development of children, working people, and families and the longer-term survival of cohesive societies. Although the capitalist market system is based on monetary exchange and profit, much of what is needed for social reproduction cannot be turned into a commodity and sold or operated for a profit without longer-term negative effects on workers (e.g., burdens of elder care), the next genera-tion (e.g., schooling), the economy (e.g., urban infrastructure), and the environ-ment (e.g., water resources). Many of the key activities of social reproduction rely on unpaid work, mostly done by women.

Social rights The income security programs in Canada and the US, and social services such as public education to which people are entitled by virtue of their citizenship or residency.

Spatial assimilation The concept that as immigrants improve their economic status and move outward to the suburbs and better housing they will also disperse spatially and achieve residential integra-tion with the receiving society. Recent research in the US, however, suggests that many immigrant groups experience delayed spatial assimilation, partially because they take longer to integrate economically; or choose to concentrate in

suburban areas among those who arrive with financial resources. Other factors, such as the desire to continue living near co-ethnics and maintain institutional completeness, discriminatory practices by the receiving society, and government policies that encourage groups to retain their cultural identity, also may impede spatial assimilation.

Spatial mosaic The spatial patterning of ethnic groups in a city. The term implies that many immigrant groups are spatially concentrated rather than spatially assimilated, thereby creating a diverse and somewhat persistent ethno-cultural mosaic in the city. The reasons are similar to those that discourage spatial assimilation.

Temporary migrants Foreign-born who have the right to reside in Canada or the US legally for a fixed period, after which they are expected to leave.

Temporary Protected Status (TPS) A humanitarian classification given to undocumented migrants in the US who, if deported to the country of origin, would suffer undue hardship.

Through-migration Using one receiving country as a stepping stone for entry to another receiving country.

Transnationalism A variety of economic, cultural, and social practices of (im)migrants that embed them in and commit them simultaneously to both their countries of origin and their new countries of settlement, creating social spaces or 'fields' that cross borders. This is a more fluid construction of the spaces of belonging and citizenship, compared with traditional linear and dichotom-ized conceptualizations of migratory processes, and both reflects and creates economic and social globalization. For refugees and other forced migrants, their relationships with their countries of origin are more highly conflicted, involving what Catherine Nolin, in her

2006 book, *Transnational Ruptures*, has described as 'ruptures' and 'sutures' rather than relatively smooth connections.

287(g) program A federal program in the US designed to devolve responsibility for immigration law enforcement to states and local authorities by authorizing local law enforcement officials to arrest and detain undocumented immigrants.

Unauthorized immigrants *Undocumented migrants* as well as those who have unauthorized documents, i.e., forged documents or documents that are no longer valid.

Underground Railroad A network of abolitionists who assisted escaped American slaves to find their way to Canada during the period before the Civil War.

Undocumented migrants People living in a country without a legal right to residency; also known as illegal migrants and people without status. An estimated 11–12 million undocumented migrants, mainly from Mexico, are in the US.

United Empire Loyalists Those Americans who supported the British during the War of Independence (1775–83) and migrated to the British territory that was to become Canada. The Loyalists played a significant role in the founding of several Canadian cities, including Saint John, New Brunswick, and Kingston, Ontario.

Visible minorities A Canadian term, first used in the Employment Equity Act of 1986, that refers to 'all persons, other than Aboriginal peoples, who are non-Caucasian in race and non-white in colour.' The term is intended to be non-discriminatory, and aspects of Canadian law allow for self-identification of visible minorities for equity purposes, but some, including a United Nations report, view the term as discriminatory because it identifies difference on the basis of skin colour.

War resisters Draft resisters, deserters, and those who oppose war for ethical reasons. During the Vietnam War an estimated 100,000 war resisters and their families left the US for Canada and then Prime Minister Trudeau allowed nearly all of them to receive permanent residency. Most were highly educated and became established in professional occupations. The post-9/11 'War on Terror' sparked another, much smaller round of emigration from the US.

White flight A trend wherein the white majority population flees urban communities for suburban communities as the minority population increases. Various institutional practices, including governmental policy, redlining, mortgage discrimination, and racially restrictive covenants have been credited with accelerating white flight to the suburbs.

INDEX

9/11, 3, 4, 289
49th parallel, 288, 335
287(g) program, 10–11, 12, 14, 15, 313, 342

Abolition Act (Upper Canada), 237
Aboriginal peoples, xxxiv, 23, 99
Abu-Laban, Yasmeen, 141
academics: American, 292
Acadia/Acadians, 23, 295, 332, 333
acculturation, xxxv, 53
acculturation hypothesis, 164, 173, 332
Across Boundaries Ethnoracial Mental Health
    Centre, Toronto, 166
Act Respecting Aliens, 1794, 24
advocacy organizations, 194
Afghanistan: refugees from, xviii
African immigrants, xxxv, 39, 54, 99, 235, 239;
    gender and, 246; spatial concentrations of,
    235–6; see also black immigrants
age: of foreign-born black population, 242;
    housing and, 96–7
Alberta, 10, 116
Alien and Sedition Act of 1798, 24
Alpharetta, GA, 303
American Community Survey, 166, 184, 209,
    223, 241, 243
American Competitiveness and Workforce
    Improvement Act, xxv
American Competitiveness in the Twenty-First
    Century Act, 2000 (US), 73
American immigrants to Canada, 288, 292–3,
    302, 326–7; Canadian pull factors for,
    326; conservative climate as push factor,
    298; urban social geographies of, 301–5;
    'whiteness' of, 305–6, 307, 327
American Revolution, 24
Andean wave of Latin American immigration,
    277, 278
Angel Island, 30, 332
anglophones, xxiv, 40, 300
anti-American sentiment, 289, 292–3
anti-immigrant sentiment, 4, 31, 289
anti-immigration ordinances, 11–12, 14, 15, 16,
    17, 138
architectural design, 102–3
Argentinean immigrants, 259

Arizona, 15, 16, 138
Asia: as source region, xvi
Asian immigrants, xvi, xix, 30, 54, 208–28,
    322–4; bilateral agreements for, 37; in
    Canada, 209, 214, 215; changes in origins
    of, 210; changing patterns by origins,
    209–12; characteristics of contemporary,
    212–18; citizenship rights of, xxi;
    economic performance of, 323; educational
    attainment, 214, 216, 217–18, 226, 227–8;
    employment sectors and, 218; government
    welfare and, 214; heterogeneity of, 209;
    human capital, 212–14, 216, 227, 323;
    income, 214, 217, 218, 220, 316–17,
    323, 330; increase in, 39; labour market
    performance, 218–23, 227, 228; as model
    minorities, 225, 226; political connections
    to destination countries, 211; political
    incorporation, 225; proficiency in official
    languages, 212, 214, 216, 217; reasons for
    migrating, 211; self-employment, 222;
    settlement patterns of, 223–5, 227; socio-
    economic status of, 216–17; suburban
    concentrations of, 61–2; unemployment
    rates, 218, 220, 222; in US, 209, 213, 216
Asian Indian immigrants, 58–60, 72, 74, 76
assimilation, xxiii–xxiv, xxxv, 48, 49, 62–3, 327,
    328, 332; 'segmented', 62
assimilation theory, 113, 332
asylum seekers, 246, 273; women as, 144, 319
Atlanta, 42; Canadian immigrants in, 303, 304,
    307, 327
Attewell, P., et al., 243, 250, 251
Australia, xvi
Avenarius, Christine B., 151

baby boomers, 299, 332
Bangladeshi immigrants, 64, 209, 212, 217, 222,
    227
Basok, Tanya, 283
Beauregard, R.A., 238
Berlin, Ira, 233, 234
black immigrants, 231–51, 324–5; in Canada,
    232, 236–41; characteristics of, 244–5,
    249; civil unrest and, 238–9; demographic
    profiles of, 241–50; education and

employment of, 242–3, 243–6, 325; gender and, 247–8, 249, 250–1; geographic distribution of, 242; heterogeneity of, 239, 240–1; income, 243, 248, 324–5; increase in US, 234–5; internal migration in US, 233–4; as model minorities, 243; in Ontario, 237–8; poverty and, 248; rental housing and, 320; restrictions on, 238; second generation, 250, 251; segregation of, 233–4, 251; socio-economic status of, 242, 246–50, 324; spatial concentrations of, 240–1, 251; in US, 232–6; women heading lone-parent families, 248–50
Blalock, H., 85
Bloemraad, Irene, 264
'boomburgs', 102
borders: control of, 3, 4–5, 332; gender and illegal crossings, 144
Boston, 72, 303
Bouchard-Taylor Commission, 4, 332–3
Bracero Program (Mexican Farm Labour Program), xxi, 36, 272, 314, 333
Bracero wave of Latin American immigration, 278
brain drain, 296, 314, 333
brain 'waste', 116
Brampton, Ontario, 224
Brazilian immigrants, 275
'breadwinner' role, 6, 142, 143, 145, 146, 151
Bridgeport, Pennsylvania, 11
British Columbia, 10, 27, 148
British Empire, 23, 32
British North America Act, 40
Britishness, 31, 333
Broder, Tanya, 12
Brooks, Alberta, 93–4
Brown, P., 159
burkas, 141
Bush, George W., 293, 298
Business Immigrant Program (Canada), 211, 212
businesses, immigrant, 122; see also entrepreneurs

Cajuns, 295, 333
Calgary, xxxi, 42, 56, 72, 128, 180; American-born residents of, 305; Asian immigrants in, 223; demographic characteristics of foreign-born population, 198–9; demographic representation in, 189, 190, 191; Latin Americans in, 281; naturalization rates in, 184, 200; public-private partnerships in, 193
California, 93; Proposition 187, 11, 17
Canada, xiv, xv, xvi; agricultural opportunities in, 292; American settlement in,

288, 291, 292–3; black immigration and settlement in, 232, 236–41; characteristics of foreign-born population in, 198–9; 'charter groups' in, xxiii; economy, 118–19; immigrant affairs offices in, 193; immigration rates to, xvi, xvii; naturalization rates in, 183, 184–5, 322; residential segregation and socio-economic status in, 76–81
Canada Citizenship Act (1946), 333
Canada Health Act, 160–1
Canada–Ontario–Toronto Memorandum of Understanding on Immigration and Settlement, 193
Canada–US border, xiv, 289, 298, 307
Canada–US Smart Border Declaration, 5
Canadian Broadcasting Corporation, 41
Canadian census, 243
Canadian Community Health Survey (CCHS), 162
Canadian Experience Class, xxii, xxv, 115
Canadian immigrants to US, 288, 293–7, 302, 326–7: declining rates of, 298; pull factors in US, 295; urban social geographies of, 301–5; 'whiteness' of, 297, 306–7, 327
Canadian Pacific Railway, 50, 77
Canadian Snowbird Association, 299–300
Cardoso, Lawrence, 34
Caribbean immigrants, xix, xxxv, 30, 54, 72, 234, 235, 316; female, 38; gender of, 246; public opinion of, 238–9; residential segregation and socio-economic status, 75, 80; spatial concentrations of, 235–6; women as domestic workers, 247; women as temporary labour migrants, 247; women in labour market, 246–7
Carter, Jimmy, 294
Carter, T., 94
Carter, T. and J. Osborne, 94, 99
Carter, T., et al., 99
categories, immigration: differential profiles, xxvi–xxvii; gender and, 140
Census, Canadian, 241
census metropolitan areas (CMAs), 72, 181, 223, 260, 281, 333
Central American immigrants, 271
Central American wave of Latin American immigration, 279
central cities. See inner cities
chain migration, 52, 282, 333
Chappell, N. and D. Lai, 164
Charlotte, NC, 129; Latin American immigrants in, 274–5, 326
Chen, Lily Lee, 225

Chi (*Qi*), 170–1

Chicago, 56, 72, 149, 180, 195; Asian immigrants in, 223; Commission on Human Relations, Advisory Council on Immigrant and Refugee Affairs, 192; demographic characteristics of foreign-born population, 196–7; demographic representation in, 189, 190; immigrant voting in, 188; immigration protests in, 194; Latin Americans in, 259; naturalization rates in, 184

Chicago School 'race relations cycle', xxiii

Chicano/a, xxxv, 39, 333

Children's Health Insurance Program (CHIP), 162

Chilean immigrants, 259

China: as source region, xvi

Chinatown (Vancouver), 50–1

Chinese Exclusion Act (US), xxiv, xxv

Chinese Exclusion Act (Canada), xxv

Chinese immigrants, xvi, xx, 24–5, 29, 54, 72, 106, 209, 211–12, 321; in Canada, 159; discrimination against, 50–1; ending of exclusion of, 37; exclusion of, xx; head tax, xx, xxiv, 32, 335–6; health experiences in Canada, 168–73; mainland, 209, 218, 220, 227, 323; refugees, xviii; residential segregation and socio-economic status, 75, 77–80, 81; in Toronto, 170–1, 321

Chinese Immigration Act, 1923 (Canada), 77, 333

Chinese traditional medicines, 170–1

cities: black immigrants in, 235, 250; Canadian, 180–1; characteristics of selected US and Canadian, 128–9, 130–1; immigrant population in, 130; medium-sized, 131; political incorporation and, 179; US, 181

citizenship, xvii–xviii, 183; in Canada, 38; *see also* naturalization

Citizenship Act (Canada), xxv

city councils, 189, 190–1

City for All Women Initiative (CAWI), 194

civic engagement, 180, 193–5, 322, 333

civic organizations, 201–2

civil rights movement, 40, 296

Cleveland, Grover, 34

co-ethnic networks, 147–8

Cold War, 37

Colombian immigrants, 257, 258, 259, 271, 275–6, 281–4; in Ottawa-Gatineau, 282–3; refugee requests from, xviii

Committee of Public Information (US), 35

communication technologies, 58

community development corporations (CDCs), 104

community health centres, 162, 168

community-based immigrant organizations, 103–4, 193; US and Canadian, 182–3

COMPAC, 193

'condominiumization', 95

context of reception, 114, 333

continuous passage policy, xxi, 32–3

control policies, 2, 334

'coolie system', 26

core-based statistical areas (CBSAs), 338

Coup wave of Latin American immigration, 279

court cases, xxiv–xxv

creative economy, 55

credentials, foreign, xxvii, 86, 119, 123, 133, 161, 318, 329

Cuban immigrants, 316

cultural competency, 160, 166, 173, 334

cultural mosaic, xxxiv

Dallas–Fort Worth, 72, 259

day labourers, 11, 271, 334

decentralization, 312–13

decolonization movement, 39

Democratic Party, 182

demographic representation, 180, 189–91, 334

Department of Citizenship and Immigration (Canada), 238

Department of Manpower and Immigration (Canada), 239

deportation measures, 36

Desrosiers-Lauzon, Godefroy, 300

destinations, new, 276–7

Detroit, 303

devolution, 7, 9, 10, 313

Díaz, Porfirio, 34

differential incorporation, 70–1, 86, 317, 334

Dillingham Commission, 34, 36, 334

discrimination, xxvii, 37–8, 82; black immigrants and, 240; Chinese and, 50–1; housing and, 97, 98–9, 107, 317, 318; institutional, 83–4, 85; legislation against, 12, 14; racial, 84; spatial concentration and, 61

disenfranchisement, 116, 139

Diversity Immigrants Program (US), 212

domestic violence, 144

domestic workers, 320; Caribbean women as, 247; live-in paid, 143, 150–1; *see also* Live-In Caregiver Program

Doméstica (Hondagneu-Sotelo), 63

Dooley, John, 294

downloading/downscaling, 7

dual citizenship (Canada), 183, 200, 201, 264, 326

earnings. *See* income

EB-5 visa (US), xxii
economic class immigrants, 3, 115, 140–1, 211
economic crisis of 2008, 41, 95–6, 161
economic enclaves, xxx
economic experiences: theories on, 113–14;
    urban variations in, 130–1
economic growth, xxix–xxx, xxxiii–xxxiv, 3, 28
economic well-being, 112–13, 114, 132, 318, 319
Economy Food Plan (US), 127
Ecuadorian immigrants, 275
Edmonton, 56, 72; Asian immigrants in, 223
educational attainment, 54, 73, 131, 316, 319;
    in Canada, 115; immigration status and,
    117; labour market outcomes and, 127–30
Einfrank, Aaron, 238–9
ejido system (Mexico), 34, 35
El Paso, 259
elderly immigrants, 97
elections: local, 188–9; national, 185–8
Elections Canada, 185
Ellis Island, 30, 334
Emancipation Proclamation, 233
Emergency Quota Act, 1921 (US), 36, 334
emergency shelters, 97
Empire Settlement Act (1922), 334
employers, 7–8, 12, 130; anti-immigration
    ordinances and, 12; health insurance and,
    161; housing support and, 105
employment: barriers to, xxx; bifurcation
    of, 124–5; day labourers, 11, 271, 334;
    discrimination and, 82, 85; gender and, 139;
    industrial sectors, 122–4; in manufacturing,
    55
Employment Equity Act of 1986, 342
enhanced driver's licences, 289
entrepreneurs, xxi, xxx, 122, 211
entry effect, 113, 127
entry myths (US), 30
entry papers, 23; *see also* passports
Equal Access to Services Ordinance (San
    Francisco), 192
ethnic and racial minorities, 189, 334
ethnic density effects, 159
ethnic enclaves, xxviii, xxix, xxx–xxxi, 32, 52–3,
    60, 61, 65, 74, 149, 334; Asian immigrants
    and, 227; suburban, 56; *see also* ethnoburbs
ethnoburbs, xxix, 61, 63, 65, 102, 160, 223–5,
    324, 328–9, 334–5; Asian-dominated, 223–4
Etobicoke, Ontario, 224
European immigrants, xix, xxiv, xxvi, 23, 24–5,
    27; decline in, 39, 54–5; exclusion of, 33;
    post-World War II, 52–4, 65
exclusion, 32–4; era, xxiv–xxv
exit polls surveys, 188

Fair Housing Act (US), 235
families, 32; astronaut, 151, 332; lone-parent,
    140, 248–50
family-class immigrants, xxvi, 115, 141
family reunification, 38, 54, 115, 131, 140, 141,
    246, 333, 335; gender and, 142; same-sex
    couples and, 142; in US, 3
Fannie Mae (Federal National Mortgage
    Association), 100
farming, 27, 92–4, 123, 292
Farmingville, 11
federal benefits, 147; *see also* social services
federal government: Canadian, 7; US, 10–11
Federal Housing Administration, 57
Federal Welfare Reform Act (US), 11
federalism, 3, 15
Filipino immigrants, xxxv, 25, 33, 72, 209, 211,
    214, 217, 218, 220–2; self-employment, 222;
    residential segregation and socio-economic
    status, 74–5, 80
First Nations, xxix, xxxiv, 335
First Peoples, 27
Florida: snowbirds in, 299–300
Florida Historical Quarterly, 300
Foner, Nancy, 246
foreign worker program (Canada), 123
France, xvi
francophones, xxiii, 40, 194, 200, 300, 335
Fredericton, 128
French Canadians, 23–4
French immigrants, xx
French Revolution, 24
Frey, William, 56
fugitive slave laws, 292

Gamio, Manuel, 28
gateway cities, 42, 56, 72, 148–51, 335;
    established or traditional, 56, 148, 149;
    Latin American immigrants in, 261;
    medium-sized, 56; new or non-traditional,
    148, 150; pre-emerging and emerging, 274,
    276; twenty-first-century, 56
gender, 138–52, 319–20, 335; access to
    services, 146, 147; black immigrants and,
    246–50, 250–1; border crossings and, 144;
    employment and, 125, 139; in gateway cities
    and, 149–51; health outcomes and, 165;
    housing and, 96–7; income and 139; labour
    market inequalities and, 248; language
    training and, 146; migration decision-
    making and, 141–5, 151; poverty and, 127,
    147; refugees and, 144; regionalization and,
    150–1; undocumented migrants and, 143–4
gender roles, 143

'gendering of survival', 143
'Gentlemen's Agreement,' US and Japan, xxi, xxiv, 33
gentrification, 57–8, 95, 149, 316, 335
Germany, xvi, xvii, 8
ghettos, xxix, 49–50, 61, 65, 240, 335
Ghosh, S., 64
globalization, xvi, xxi–xxii, 228, 281
gold rushes, 30
Goldring, Luin, 280
Goodman, J., 95
Google, 116
Grand Rapids, 129
Grant, Madison, 26
Great Depression, 36, 295
Great Society program, 38
Greek immigrants, 54
Green Card, 212, 335, 339
gross domestic product (GDP), 118
Guatemalan immigrants, 257

Haitian immigrants, 235, 240, 247, 324; refugees, xviii
Halifax, 30, 128, 130, 131; black immigrants in, 239–40
Hamilton, 56, 72, 223
Handlin, Oscar, 32
Hart-Celler Act (Immigration and Naturalization Act, 1965), xxi, xxv, 272–3, 296, 335
Haymarket Square, bomb at, 31
Hazelton, Pennsylvania, 11, 102
head tax, xx, xxiv, 32, 33, 35, 335–6
health, determinants of, 159; literature on, 159–60
health care: access to, 159, 173, 320, 321; barriers to, 160; in Canada, 158, 160–1, 166, 298, 320; for immigrant children, 12; literature on, 159–60; medically necessary services, 161; uninsured services, 161; in US, 158, 161–2, 166, 320; use of, 159, 165–7, 171; waiting period for services, 161
health insurance, population without, 162, 166
health outcomes, 320–1, 328; ethnic variations in, 162–5; geographic regions and, 164–5
health-care providers: cultural competency and, 160; see also physicians
healthy immigrant effects, 159–60, 164, 173, 321, 336
Hérouxville, Quebec, 9, 16
heterolocalism, xxix, 58
Hiebert, D.S. and P. Mendez, 94
Hiebert, D.S., et al., 99
Hiemstra, Nancy, 6
Higham, John, 31

hijab, 141
Hispanic or Latino paradox, 164, 321, 336
Hispanic/Latino ethnic group, xxxv, 257
Hispanics: health outcomes and, 165
home ownership, 53, 96, 100–1, 107, 317; changes in, 94–5; counselling, 106
Homeland Security, 307, 336
homelessness, 99–100, 318; hidden, 100, 318, 336
Homestead Acts, 27, 336
Hondagneu-Sotelo, P., 63
Hong Kong immigrants, 64, 106, 209, 212, 214, 216, 217, 218, 220, 222, 323
House of Commons: foreign-born in, 189
Household Servant Scheme (Canada), 247
housing, xxix, 64, 91–108, 317; affordability of, 97–8, 105, 107, 317; affordable, 91, 95, 149, 150, 332; black immigrants and, 235; changes in, 102–3; class of entry and, 96–7; community-based non-profit organizations and, 103–4; core need, 97; counselling, 106; demand, 92–4, 317, 318; discrimination and, 82, 84, 85, 97, 98–9, 107, 317, 318; diversity of, 92, 107; experiences of newcomers, 96–100; government policy and programs, 105–6, 107–8; household size, 101–2; income allocated to, 100–1; increased prices, 318; institutional discrimination and, 83–4; in migrant-sending communities, 92; mortgages and, 84, 100–1, 317–18; overcrowding and, 101–2, 150; rental, 57, 93, 94, 95, 150, 317–18; social, 84, 98, 99, 105; social profile and, 96–7; social support networks and, 105; suburban, 55–7; support from employers, 105; trailer, 94; transitional, 105–6; transnational strategies, 106
Houston, 72, 180, 195; Asian immigrants in, 223; demographic characteristics of foreign-born population, 196–7; demographic representation in, 189–91, 190; Latin Americans in, 259; Mayor's Office of Immigrant and Refugee Affairs, 192; naturalization rates in, 184
Hudson's Bay Company, 27
human agency, 42
human capital, xix, 54, 115, 131, 139, 144, 246, 336; differential levels, 115–18; transferability of, 114
humanitarian immigrants, 115, 140
Huntington, Samuel, 42
Hurricane Mitch, 276
H-1B visa (US), xxii, xxv, xxvi–xxvii, 73, 115, 116, 212, 296

illegal aliens, 5; *see also* undocumented migrants
Illegal Immigration Reform and Immigrant
Responsibility Act (IIRIRA), 5, 11, 161–2
Illegal Immigration Relief Act Ordinance, 11–12,
102
illegality, 6
immigrant affairs offices, 192
immigrant integration, xxxv, 49, 64, 66, 336;
ideals, xxii–xxiv, 48–9, 181–2
immigrant-receiving countries, xxvi–xvii
immigrants, 23; differential profiles, xxiv–xxix;
'preferred', 26; surplus of highly educated,
329–30
immigration: cost of, 29; history of, 23–42;
reasons for, xviii–xix, 28
Immigration Act (Canada): 1967 amendments,
xxi, xxv, 114
Immigration Act (1924). *See* Johnson-Reed Act
(Immigration Act, 1924)
Immigration Act (1952). *See* McCarran-Walter
Act (Immigration Act, 1952)
Immigration Act (Canada): *1869*, 24; *1952*,
xxv; *1953*, 37; *1965*, 38; *1976*, xxv; *1978*,
xxi
Immigration Act, 1990 (US), xxii, xxv, 116,
212
Immigration and Nationality Amendments Act,
1965 (US), 38, 72, 73, 114, 234
Immigration and Nationalization Act, 1952
(US), 73
Immigration and Naturalization Act (1965).
See Hart-Celler Act (Immigration and
Naturalization Act, 1965)
Immigration and Refugee Protection Act, 2002
(Canada), xxii, xxv
immigration policies, xx, xxiv–xxv, 2–17, 6,
201, 312, 336; Canadian, xxxii, 39, 73,
131; differential incorporation and, 70–1;
discrimination and, xx–xxi; economic
experiences and, 114–15; localization of,
5, 7–10, 10-15, 17, 337–8; removal of race-
based restrictions, 239, 247; restrictive, 26,
314; in US, 73, 131
immigration process, xix
immigration protests (2006, US), 194
Immigration Reform and Control Act (IRCA). *See*
Simpson-Mazzoli Act (Immigration Reform
and Control Act)
income, 126–7, 131, 316–17, 318; allocated to
housing, 100–1; gender and, 139
incorporation policies, xxxv, 2, 336
indentured labour, 232, 336–7
index of dissimilarity, 72–3, 81, 87
India, xvi

Indian immigrants, 209, 212, 214, 217, 218, 222
Industrial Workers of the World, 36
industrialization, 27, 28, 29–30
inner cities, 52, 57, 70, 102, 251; enclaves and,
60
institutional completeness, 114, 147, 337
integration. *See* immigrant integration
'interculturalism,' 333
interest rates: housing and, 100
International Organization for Migration,
xviii–xix
Internet, 58
invasion and succession, 52, 337
investor stream, xxi, xxii
invisiburbs, xxix, 58, 337
Iranians, 72, 316; residential segregation and
socio-economic status, 75, 76, 77
Iraq: refugees from, xviii
IRCA wave of Latin American immigration, 279
Ireland, xvi
Israel, xvi
Italian immigrants, 25, 54, 61; spatial
concentration for a limited time and, 60
Italy: as source region, xvi

Jamaican immigrants, 165
Japanese immigrants, xvi, 25
Jewish population, 52, 61
jobs. *See* employment
Johnson, Albert, 36
Johnson, Lyndon, 38
Johnson-Reed Act (Immigration Act, 1924), 36,
337
Joint Centre for Housing Studies, 100
Jones, Terry-Ann, 243
Jones-Correa, Michael, 201
jus sanguinis, xvii
jus soli, xvii–xviii

Kasinitz, P., et al., 250
Kelowna, BC, 94
Khan, A.A. and S. Bhardwaj, 169
kin-based networks, 150
kirpan, 16
Kitchener, 56, 72
knowledge-based economy, xviii
Korean immigrants, 25, 33, 209, 211, 218, 220;
self-employment, 222
Ku Klux Klan, 36

labour, division of, 123, 125; *see also*
employment
labour force, xxix, xxx, 32; participation rates,
119–21, 337

labour market, 85, 119–25; discrimination in, 86; ethnic niching, 123–4; outcomes, 125–30; post-industrial restructuring and, 55; segmentation, 122–4, 337

Lai, D.W., 164

laissez-faire policy, 118–19

Lam, L., 164

landed immigrants, 293, 337, 339

landlords, 12, 98, 99, 102

land-use policies, 84, 86

language training, 4, 146, 150

languages, proficiency in official, 40, 73, 115–18, 166, 173, 318

Lankin, Frances, 133

Latin American immigrants, xvi, xix, xxxv, 39, 54, 235, 256–84, 325–6; in Canada, 81, 259–61, 263, 277–84; community health-center use in Washington, DC, 172–3; community organizing, 281; day labourers, 271; defining, 256–7; demographic characteristics of, 265–70; educational attainment, 264, 265, 268, 270; employment and occupational status, 265, 266, 268–9, 270–2; health experiences in US, 166, 168–73; immigration status, 261–4, 265, 267; income, 266, 269–70, 271, 317; integration and, 281; labour force activity, 268; metropolitan/urban scale geographic distribution, 259–61; national scale geographic distribution, 259; in non-traditional destinations, 273–5, 281; poverty and, 266–7, 271; prenatal care, 172; pull and push factors, 272; regions and countries of origin, 257–9; residential segregation and socio-economic status, 76; size of populations, 257; undocumented, 264; unemployment rates, 270; in US, 159, 259, 260, 262, 272–7; in Washington, DC, 321; waves of, 277–80

Latino/a, xxxv, 72, 337; *see also* Latin American immigrants

Lazarus, Emma, 24, 28

Lead wave of Latin American immigration, 277, 278

Leamington, Ontario, 261

Lethbridge, Alberta, 261

Lewis, Bernard, 42

LGBT (people who identify as lesbian, gay, bisexual, transgendered), 142

Li, Wei, 61–2, 63, 224, 329

Lim, J.H., 226

literacy, 34, 35

Little Portugal (Toronto), 64–5

Live-in Caregiver Program, xxii, 80, 143, 337

living conditions, of immigrants, 30, 31, 32, 36, 49

localization, 5, 337–8; in Canada, 15, 7–10; impact of, 17; in US, 10–15

London, Ontario, 72, 281, 282

lone-parent families, 140; black immigrant women and, 248–50

Longitudinal Survey of Immigrants to Canada (LSIC), 97

Los Angeles, xv, 56, 61, 72, 129, 130, 180, 195; Asian immigrants in, 223; Canadian-born residents of, 303; demographic characteristics of foreign-born population, 196–7; demographic representation in, 189, 190; immigrant voting in, 188; immigration protests in, 194; Latin Americans in, 259; Mayor's Office of Immigrant Affairs, 192; naturalization rates in, 184; vacancy rate in, 95

Louisiana Territory, 295

Lower East Side Tenement Museum, 49–50

low-income cut-offs (LICO), 127

LSIC survey, 101

Luce-Celler Bill (US), xxv

Luxembourg, xvi

Manitoba, 8, 27, 94

Marcuse, P., 61

Markham, Ontario, 224

Massey, Douglas and Chiara Capoferro, 273

Massey, D.S., 53

Maytree Foundation, 195

McCarran-Walter Act (Immigration Act, 1952), 37, 338

Medicaid, 147, 161, 162

Medicare, 161

melting pot, xxiii, xxxiv, 315, 327; *see also* assimilation

men: breadwinner role of, 6, 142, 143, 145, 151; health outcomes and, 165

mental health, 164, 166

Métis, 27

metropolitan statistical areas (MSAs), 259, 303, 338

Mexican immigrants, xxxiii, 29, 34, 54, 257, 258, 259, 261, 283, 314, 325; deportation of, 36; naturalization rates, 264; refugees, xviii; temporary farm labourers, xxi, 41, 272; undocumented, 41; women, 144

Mexico, 34, 35

Miami, xv, 56, 72, 129, 130, 259

Microsoft Inc., 116

Middle Eastern immigrants, 54, 209, 212

Minimum Wage Ordinance (San Francisco), 192

minority relations, theory of, 85
Minuteman Project, 4
Mississauga, 224
model minority, 63, 225, 226, 243, 338, 340
money: immigrants arriving with, 28; sent home, 314–15
Montreal, xv, 30, 42, 56, 72, 128, 130, 149, 180; American-born residents of, 305; Asian immigrants in, 223; black immigrants in, 239, 240, 240–1, 324; demographic characteristics of foreign-born population, 198–9; demographic representation in, 189, 190, 191; Italian immigrants in, 60, 61; Latin Americans in, 260; naturalization rates in, 184, 200; suburbanization in, 54
mortgages, 84; flexible, 100, 101; payments, 100–1; rates, 317–18
multiculturalism, xxiii–xxiv, xxxii, 40, 48–9, 65, 264, 314, 316, 326, 327, 328, 333, 338
Multiculturalism Act, xxiii
municipalities: boundaries, 180–1; housing discrimination and, 98; settlement services and, 9
Murdie, R., 94, 99

National Center for Health Statistics (NCHS) (US), 162, 166
National Health Care Interviews (US), 162
National Health Care Surveys (US), 162
National Housing Discrimination Study, 98–9
National Origins Act, 1924 (US), 73
National Population Health Survey (NPHS) (Canada), 162
nation-building, xx, xxxii, 27
nation-states, 2, 338
Native Americans, xxxiv
Native peoples, 23, 24; *see also* Aboriginal peoples
nativism, 27, 31, 37, 338
naturalization, 179, 183–5, 322, 338; Latin American immigrants and, 261–4; rates, 195–200, 200–1
Nazi forced labour camps: displaced persons from, 37
neighbourhood houses, 193
neighbourhoods: changes in, 102–3, 317
'neighbourhoods of hope', 62, 64
Nelson, BC, 293, 294–5
neo-conservatism, 2–3, 5–7, 15, 313, 338–9
neo-liberalism, 2, 3, 5–7, 15, 16, 313, 339; gender and, 146
New France: immigration to, 23
New York City, 56, 72, 129, 149, 180, 195; Asian immigrants in, 223; black immigrants in, 235; Canadian-born residents of, 303; Caribbean immigrants in Harlem, 234; Colombian immigrants in, 276; demographic characteristics of foreign-born population, 196–7; demographic representation in, 189, 190; foreign-born blacks in, 237; immigrant population in, xv; immigrant voting in, 188; Italian immigrants in, 60; Latin Americans in, 259; Mayor's Office of Immigrant Affairs, 192; naturalization rates in, 184; vacancy rate in, 95; voting in local elections, 188
New Zealand: international migrants in total population, xvi
newcomers, 23, 339
NEXUS cards, 289, 339
niqab, 141
Nolin, Catherine, 280
North American Free Trade Agreement (NAFTA), 296, 339
North York, Ontario, 224
Nova Scotia, 23; black Loyalist refugees in, 236–7
nursing, 143, 320

Obama, Barack, 162, 185
occupational concentration, 124–5, 126; gender and, 125; time of arrival and, 124–5; urban variations in, 130–1
OECD countries, xvi–xvii
Oh, B.C., 211
Ontario, xxxi, 283; involvement in settlement of immigrants, 10; Latin Americans in, 281; Mexican immigrants in, 283; provincial nominee program in, 8
Open wave of Latin American immigration, 278
open-door period, xxv
'Orientals', 25, 26
'Others', 31
Ottawa, xxxi, 180; Asian immigrants in, 223; demographic representation in, 189, 190, 191; immigrant integration in, 194; naturalization rates in, 184, 200; public-private partnerships in, 193
Ottawa Police Department, 193
Ottawa-Gatineau, 56, 72, 282–3; black immigrants in, 239, 240; demographic characteristics of foreign-born population, 198–9; Latin American immigrants in, 326
overcrowding, 101–2, 150

Page Act (US), xxiv, 33
Pakistani immigrants, 209, 212, 217, 222, 227
Park, Robert E., 69

Passing of the Great Race, Or the Racial Basis of European History (Grant), 26
passports, 23, 289
Patient Protection and Affordable Care Act, 2010 (US), 328
Peace of Paris of 1783, 24
Peach, C., 49, 60, 62
permanent residents, xxvi, 3, 140, 339
Personal Responsibility and Work Opportunity Reconciliation Act (PRWORA), 4, 11, 147, 161–2
Peruvian immigrants, 259, 275
Peterson, W., 226
Philadelphia, 103–4
Philadelphia Chinatown Development Corporation, 104
Philippines, xvi, 214, 315
physicians, 168–9; cultural competency, 166, 173; linguistic barriers, 166; shortage of, 161; in Toronto, 170–1
'places of stigma', 62, 64
pluralism, xxiii
points system, xxi, xxii, xxvi, 38, 73, 115, 211, 339
political incorporation, 179–202, 321–2; challenges of US–Canada comparison, 180–3; civic organizations and, 201–2; data on, 181–2; indicators of, 179–80, 183–95; role of policy and contextual factors in, 200–1; spatial concentration and, xxx–xxxi; suburbs and, 181
political parties, local, 182, 193, 201–2
population health surveys, 162–4
Portes, A. and M. Zhou, 62
Portuguese immigrants, 52, 54
poverty, xxix, 127, 319; black immigrants and, 248; gender and, 127, 147, 248; Latin American immigrants and, 271
Prehn, A.W., et al., 160
Price, M., et al., 58
primary sector, 122, 125, 218
'prohibited classes', 33
pro-immigration ordinances, 12–14, 14, 15
Proposition 187 (California), 11, 17
provincial jurisdiction, 7, 15, 312–13
provincial nominee program, 7–8, 312–13, 339
Pruegger, Valerie and Alina Tanasescu, 97
public benefits: access to, 11; undocumented immigrants and, 162
public housing. *See* social housing
public opinion, 112; of Caribbean immigrants, 238–9; surveys, xxxii, xxxiv
public services: access to, 12–14, 138

public transit, 320; gender and access to, 149–50; in new gateway cities, 150
public–private partnerships, 193
pull factors, xix, 272, 339
Punjabi immigrants, 25
purchasing power, 94, 95
Puritans, xx
push factors, xix, 38–9, 272, 339

Quadeer, M. and S. Kumar, 225
quaternary sector, 122
Quebec, 7, 17; Latin Americans in, 281; services to immigrants, 283
Quebec–Canada Accord, 10, 312
quota system, 38, 115, 272–3

race, xxxv; inequality and, 82, 83, 86; removal of restrictions on, 54, 247, 289, 314
racial minorities, xvi, xxxiv, 72, 340; *see also* ethnic and racial minorities; visible minorities
racial profiling, 16
racial segregation: black immigrants and, 240, 251
racialization, xix, xx, 32–4, 34, 63, 64, 339; gender and, 141
racialized minorities, xxix, xxxi, xxxvi, 64–5, 339
racism, 34, 36, 73, 152, 234, 340; gender and, 141; living conditions and, 49–50
railway construction, xvi, xx, 24, 26, 50, 209
'Red scares', 31, 36
Refugee Act (US), xxi
refugees, xviii, 6, 40, 140, 246, 340; to Canada, xxi; claiming status as, 4–5; Convention, 144, 150; gender and, 144, 319; in new gateway cities, 150; in rental housing, 94; to US, xxi, 273
regionalism, 40
regionalization, 150–1
Reid, Ira, 233
religious organizations: in Canada, 202
Reno, 129
rental and real estate agents: discriminatory practices of, 98–9
rental housing, 94, 95, 96; affordability of, 97–8; black immigrants and, 320; changes in, 95; discrimination and, 98; household size and, 101–2; supply, 95, 318; *see also* housing
Repatriation Act (US), xxv
Republican Party, 6, 182
residential segregation, 81, 83–6, 316, 328;

difficulty of conducting comparative studies, 71; health and, 160; inequality and, 83–6; socio-economic status and, 69–70, 78, 79, 81–2; voluntary, 328–9

Richmond Hill, Ontario, 224

Riverside–San Bernadino, 259

Rogers, Reuel R., 201

Royal Commission on Bilingualism and Biculturalism, 40

Russian Federation, xvi, xviii

Salvadoran immigrants, xviii, 257, 258, 276

same-sex couples, 142

San Diego, 72, 223

San Francisco, 72, 180, 195; Asian immigrants in, 223; Canadian-born residents of, 303; demographic characteristics of foreign-born population, 196–7; demographic representation in, 189, 190; Equal Access to Services Ordinance, 192; immigrant integration in, 192; immigrant voting in, 188; Minimum Wage Ordinance, 192; naturalization rates in, 184, 195–200; Office of Civic Engagement and Immigrant Affairs, 192; sanctuary ordinances, 12–13; vacancy rate, 95

San Gabriel Valley, California, 62, 223, 225

San Jose, 223

Sanctuary Movement, 12, 340

sanctuary ordinances, 12–13, 16

Saudi Arabia, xvi

Saxenian, A.L., 228

SB1070 law (Arizona), xxxiii

Scarborough, Ontario, 224

Scott, Katherine, Kevin Selbee, and Paul Reed, 194–5

Seasonal Agricultural Workers Program, xxii, 283, 340

Seattle, 129, 223

secondary migrants, 274, 340

secondary sector, 122, 123

securitization, 3

selective period, xxv

self-employment, xxx, 122, 130, 131; of Asian immigrants, 222; health insurance and, 161

self-selection, 114, 340

settlement houses, 193, 201

settlement services, 9, 132–4, 146, 202, 340; funding of, xxxi, xxiii, 9, 10; gender and, 146; provincial and municipal involvement in, 8–10; in US, 9; see also social services

settlements: differential immigrant profiles, xxvii–xxix; typology of urban and suburban ethnic, 58–62

Sherbrooke, Quebec, 281

shopping centres, ethnically oriented, 102

Sifton, Clifford, 27, 28

Simpson, Jeffrey, 296

Simpson-Mazzoli Act (Immigration Reform and Control Act, 1986), 40–1, 272, 340

Sinclaire, B., 99

Singapore, xvi

Singer, Audrey, 276

Singh, P. and T.V. Thomas, 212

skilled class of immigrants, xxvi

skilled labour recruitment program (US), 123

Skop, E., 58

Skop, E. and C.E. Altman, 60

slavery, 232, 233, 292, 340–1; in Canada, 236–8; in US, 232–3

Slavery Abolition Act (Canada), 236

small businesses, xxx

Smith, H. and D. Ley, 62

Smith, Heather A. and Owen Furuseth, 274

snowbirds, 298, 299–300, 303, 341

social housing, 84, 98, 99, 105

social networks, 66, 105, 147–8, 149

social reproduction, 142, 143, 341

social rights, 4, 341

social services, 4, 145–8, 320; cutbacks to, 147–8; devolution of funding for, 148; gender and, 147; gender and access to, 152

social welfare policies, 146

socio-economic status (SES), 81–2, 316, 329; black immigrants and, 240; health and, 159, 165; occupational status and, 82; political incorporation and, 195–200; residential segregation and, 69–70; time of arrival and, 85

sojourners, 31

Somalia, xviii

source regions, xxviii; differential immigrant profiles, xxiv–xxvi; non-traditional, xvi

South American immigrants, 257, 258, 264, 271

South Asian immigrants, 72, 76

South Korean immigrants, 211

Spain, xviii

spatial assimilation, 49, 52–4, 58, 65, 69–70, 315, 317, 341

spatial concentration, xxx–xxxi, 315; consequences of, 62–5; housing and, 101; for a limited time, 59, 60, 63, 65; long-term or permanent, 59, 60–2, 65; low, 58; no, 58–60, 63, 65

spatial mosaic, 49, 315, 328, 341

spatial outcomes, 59, 65

sponsorship, 142

spousal abuse, 138
Sri Lankan immigrants, 4, 209, 217, 222, 227
State Children's Health Insurance Program
    (SCHIP), 161
state governments: immigration laws, 15
Statistics Canada, xxxi, 224; 2001 Census Profile
    File, 71; 2002 Ethnic Diversity Survey, 185;
    Canadian Census Individual Public Use
    Microdata File, 71
Statue of Liberty, 30
Steinbach, Manitoba, 93
substantive representation, 180, 191–3
suburbanization, 54
suburbs, xxviii, xxxi, 54–8, 65, 70, 74, 102, 303,
    315–16, 330; Asian immigrants in, 223;
    housing diversity in, 55–7; isolation in,
    63–4; political incorporation in, 181
Sudbury, 128, 130
Suffolk County, NY, 12
Supplementary Security Income (SSI), 147
Suro, Roberto, 274
Switzerland, xvi

Taiwanese immigrants, 151
Takoma Park, Maryland, 13–14, 188
Taoism, 170
Taylor, Frederick Winslow, 30
Taylorization, 28
Temporary Assistance for Needy Families (TANF),
    147
temporary foreign worker program, 3, 8, 94
temporary migrants, 3–4, 115, 131, 144, 281,
    341
Temporary Protected Status (TPS), 276, 341
terra nullius, 23
tertiary sector, 122
Third Safe Country Agreement, 4–5
Thirteenth Amendment to the US Constitution,
    233
through-migration, 295, 341
Tiananmen Square, 211
time of arrival: economic well-being and, 132;
    occupational concentration and, 124–5
Toronto, xv, xxxi, 42, 56, 72, 128, 130, 131,
    149, 159, 180; American immigrants in,
    327; American-born residents of, 305,
    306, 307; Asian immigrants in, 223, 224,
    225; Bangladeshi immigrants in, 64; black
    immigrants in, 239, 241, 324; demographic
    characteristics of foreign-born population,
    198–9; demographic representation in, 189,
    190, 191; Diversity Management and Civic
    Engagement Unit, 193; health experiences
    of Chinese in, 168–73, 321; homelessness

in, 99; housing in, xxix, 94; Indo-Canadians
    in, 60; Italian immigrants in, 60, 61; Latin
    Americans in, 260; Little Portugal, 64–5;
    naturalization rates, 184; pro-immigration
    legislation, 13; settlement services in, 9;
    suburban newcomers in, 54, 61; visible
    minorities in, 224
Toronto District School Board, 13
Toronto Police, 13
Tossutti, Livianna, 195
Trade NAFTA (TN) status and TN visas, 296
transition period, of immigration, xxv
transnationalism, xix, 143, 151, 341–2; health
    care and, 173; Latin American immigration
    and, 280–1
Trudeau, Pierre, 40, 293, 342
Truman, Harry, 37
Tydings-McDuffie Act (US), xxv

unauthorized immigrants, 22, 94, 342; see also
    undocumented migrants
Underground Railroad, 237, 292, 305, 342
undocumented migrants, xxxv, 3–4, 106, 182,
    342; access to federal benefits, 147, 162; in
    Canada, 264; efforts to stem, 277; gender
    and, 143–4; Hispanic, 40; homelessness
    and, 318; housing and, 94, 98, 103, 107;
    local law enforcement and, 11; in US; Latin
    American, 264, 272, 273; local measures and
    ordinances against, 11–12; in San Francisco,
    192; in US, xxxiii, 4, 55, 143, 264, 317
unemployment rates, 119, 125–6, 222
United Empire Loyalists, 24, 292, 305, 342
United Kingdom, xvi
United Nations Convention Relating to the
    Status of Refugees, xviii, 340
United States: assimilation in, xxiii–xxiv;
    attitudes towards immigrants, xxxii–xxxiii;
    black population in, 232, 236; calculating
    voter turnout, 186; Canadian migration
    and settlement in, 288, 290, 293–7;
    characteristics of foreign-born population in,
    196–7; civic engagement in, 194; devolution
    in, 10–11; economy, 118–19; health-care
    system, 158, 161–2, 166, 320; immigrant
    affairs offices in, 192; immigration rates
    to, xvi, xvii; migrant population, xiv, xv,
    xvi; national identity, xxxiii; naturalization
    rates in, 183, 184, 322; out-migration due
    to ultra-conservative climate, 293; refugee
    requests, xviii; residential segregation and
    socio-economic status in, 74–6
urban areas, xiv–xv, xxvii; see also cities
Urban Studies (Li), 224

US Bureau of the Census: *Profile of Foreign-Born Population in the United States 2000*, 71; *Survey of Business Owners*, 222; *Summary File 4*, 71
US Congress: foreign-born in, 189
US–Canada border, xiv, 289, 298, 307
US–Canadian migration, post-9/11, 297–301
US–Mexico border, 4, 41, 272, 277

vacancy rates, 95, 96
Valley Park, Missouri, 11
Vancouver, xxxi, 42, 56, 128, 130, 72, 149, 180; American-born residents of, 305; Asian immigrants in, 223; black immigrants in, 240; Chinatown, 50–1; demographic characteristics of foreign-born population, 198–9; demographic representation in, 189, 190, 191; homelessness in, 99; Hong Kong immigrants in, 64; housing in, xxix; immigrant population in, xv; Indo-Canadians in, 60; naturalization rates in, 184; neighbourhood houses, 193
'vertical mosaic', 40
Vietnam, 216
Vietnam War, 293, 294, 296, 327, 342
Vietnamese immigrant, 209s, 211, 216, 220, 227
visible minorities, xvi, xxxiv, 72, 224, 342; housing and, 97, 98; in social housing, 99; *see also* racial minorities
voting, 179–80, 185–9, 322; in 2000 Canadian federal election, 187; data on immigrant, 185; in local elections, 188–9; in national elections, 185–8; non-citizen, 188–9; turnout at elections, 186–7; in US, 185, 186

Waco, **129**
Walker, Kyle and Helga Leitner, 12, 14
Walton-Roberts, Margaret, 148
Wang, L., 168
Wang, L. and D. Roisman, 168
Wang, L., et al., 168
Wang, S. and J. Zhong, 224
'War on Terror', 342

war resisters, 293, 294–5, 305, 327, 342
Ware, Caroline, 32
Washington, DC, 58–60, 72, 159; Asian immigrants in, 223; health experiences of Latinos in, 168–73, 321; Salvadorans in, 276
welfare reform (US), 147
welfare states: key premises of, 145–6
West Asians, 72; residential segregation and socio-economic status in Canada, 77
West Indies, immigrants from, 235
Western Hemisphere Travel Initiative, 289
white flight, 57, 342
white supremacy, 83, 86
whites, xxxv
Wilding, Carol, 133
Windsor, 56
Winnipeg, 30, 56, 72, 128, 131, 223
Winnipeg General Strike, 36
Withers, L., 172
women, 28; access to services, 138, 147–8; Afro-Caribbean immigrant, 246–7; banned from wearing veils, 9, 16, 138; black immigrant, 250–1; dependence on male family members, 142; employment and, 139; health outcomes and, 165; labour market participation, 119, 123; migrating alone, 142–3; newcomer outreach for, 145; as nurses or domestic workers, 320; as refugees and asylum seekers, 319; social networks and, 147–8; transnational support, 143; as undocumented migrants, 143–4 social reproduction and, 142, 143, 341
Wong, Janelle S., 201
World War I, 35
World War II, 36, 54; European settlement after, 52–4, 65
Wright, R.A., M. Ellis and V. Parks, 63

yin-yang, 170

Zelinsky, W. and B.A. Lee, 58
zoning regulations, 84, 86